Communications and Control Engineering

Springer
London
Berlin
Heidelberg
New York
Hong Kong
Milan
Paris
Tokyo

Published titles include:

Stability and Stabilization of Infinite Dimensional Systems with Applications
Zheng-Hua Luo, Bao-Zhu Guo and Omer Morgul

Nonsmooth Mechanics (Second edition)
Bernard Brogliato

Nonlinear Control Systems II
Alberto Isidori

L2-Gain and Passivity Techniques in nonlinear Control
Arjan van der Schaft

Control of Linear Systems with Regulation and Input Constraints
Ali Saberi, Anton A. Stoorvogel and Peddapullaiah Sannuti

Robust and H∞ Control
Ben M. Chen

Computer Controlled Systems
Efim N. Rosenwasser and Bernhard P. Lampe

Dissipative Systems Analysis and Control
Rogelio Lozano, Bernard Brogliato, Olav Egeland and Bernhard Maschke

Control of Complex and Uncertain Systems
Stanislav V. Emelyanov and Sergey K. Korovin

Robust Control Design Using H^∞ Methods
Ian R. Petersen, Valery A. Ugrinovski and Andrey V. Savkin

Model Reduction for Control System Design
Goro Obinata and Brian D.O. Anderson

Control Theory for Linear Systems
Harry L. Trentelman, Anton Stoorvogel and Malo Hautus

Functional Adaptive Control
Simon G. Fabri and Visakan Kadirkamanathan

Positive 1D and 2D Systems
Tadeusz Kaczorek

Identification and Control Using Volterra Models
F.J. Doyle III, R.K. Pearson and B.A. Ogunnaike

Non-linear Control for Underactuated Mechanical Systems
Isabelle Fantoni and Rogelio Lozano

Robust Control (Second edition)
Jürgen Ackermann

Flow Control by Feedback
Ole Morten Aamo and Miroslav Krstic

Learning and Generalization (Second edition)
Mathukumalli Vidyasagar

Zdzislaw Bubnicki

Analysis and Decision Making in Uncertain Systems

With 108 Figures

Springer

Professor Zdzislaw Bubnicki, PhD
Institute of Control and Systems Engineering, Wroclaw University of Technology,
Wyb. Wyspianskiego 27, 50-370 Wroclaw, Poland.

Series Editors
E.D. Sontag • M. Thoma • A. Isidori • J.H. van Schuppen

British Library Cataloguing in Publication Data
Bubnicki, Zdzislaw
Analysis and decision making in uncertain systems
(Communications and control engineering)
1.System analysis 2.Uncertainty (Information theory)
3.Decision making – Mathematical models 4. Decision support systems
I.Title
629.8

Library of Congress Cataloging-in-Publication Data
A catalog record for this book is available from the Library of Congress

Communications and Control Engineering Series ISSN 0178-5354
ISBN 978-1-84996-909-3
Springer-Verlag is a part of Springer Science+Business Media
springeronline.com

69/3830-543210 Printed on acid-free paper

Preface

Problems, methods and algorithms of decision making based on an uncertain knowledge now create a large and intensively developing area in the field of knowledge-based decision support systems. The main aim of this book is to present a unified, systematic description of analysis and decision problems in a wide class of uncertain systems described by traditional mathematical models and by relational knowledge representations. A part of the book is devoted to new original ideas introduced and developed by the author: the concept of uncertain variables and the idea of a learning process consisting in knowledge validation and updating. In a certain sense this work may be considered as an extension of the author's monograph *Uncertain Logics, Variables and Systems* (Springer-Verlag, 2002). In this book it has been shown how the different descriptions of uncertainty based on random, uncertain and fuzzy variables may be treated uniformly and applied as tools for general analysis and decision problems, and for specific uncertain systems and problems (dynamical control systems, operation systems, knowledge-based pattern recognition under uncertainty, task allocation in a set of multiprocessors with uncertain execution times, and decision making in an assembly system as an example of an uncertain manufacturing system). The topics and the organization of the text are presented in Chapter 1 (Sects 1.1 and 1.4).

The material presented in the book is self-contained. I hope that the book can be useful for graduate students, researchers and all readers working in the field of control and information science, especially those interested in the problems of uncertain decision support systems and uncertain control systems.

I wish to acknowledge with gratitude the encouragement and help I received from Professor Manfred Thoma, editor of this series. His inspiration and interest have been invaluable in the preparation of the book.

I wish also to express my gratitude to my co-workers at the Institute of Control and Systems Engineering of Wroclaw University of Technology, who assisted in the preparation of the manuscript. My special thanks go to Dr. L. Siwek for the valuable remarks and discussions, and for his work concerning the formatting of the text. Thanks are also due to Dr. D. Orski who assisted in the final phase of preparation of the manuscript.

This work was supported in part by the Polish Committee for Scientific Research under the grant nos 4 T11C 001 22 and 7 T11A 039 20.

Z. Bubnicki

Contents

1 Introduction to Uncertain Systems

1.1 Uncertainty and Uncertain Systems

Uncertainty is one of the main features of complex and intelligent decision making systems. Various approaches, methods and techniques in this field have been developed for several decades, starting with such concepts and tools as adaptation, stochastic optimization and statistical decision theory (see e.g. [2, 3, 68, 79, 80]). The first period of this development was devoted to systems described by traditional mathematical models with unknown parameters. In the past two decades new ideas (such as learning, soft computing, linguistic descriptions and many others) have been developed as a part of modern foundations of knowledge-based Decision Support Systems (DSS) in which the decisions are based on *uncertain knowledge*. Methods and algorithms of decision making under uncertainty are especially important for design of computer control and management systems based on incomplete or imperfect knowledge of a decision plant. Consequently, problems of analysis and decision making in uncertain systems are related to the following fields:

1. General systems theory and engineering.
2. Control and management systems.
3. Information technology (knowledge-based expert systems).

There exists a great variety of definitions and formal models of uncertainties and uncertain systems. The most popular non-probabilistic approaches are based on fuzzy sets theory and related formalisms such as evidence and possibility theory, rough sets theory and fuzzy measures, including a probability measure as a special case (e.g. [4, 7, 9, 64, 65, 67, 69, 71, 74, 75, 78, 81, 83, 84, 96–100, 103, 104]). The different formulations of decision making problems and various proposals for reasoning under uncertainty are adequate for the different formal models of uncertainty. On the other hand, new forms of uncertain knowledge representations require new concepts and methods of information processing: from computing with numbers to granular computing [5, 72] and computing with words [101].

Special approaches have been presented for multiobjective programming and scheduling under uncertainty [91, 92], for uncertain object-oriented databases [63], and for uncertainty in expert systems [89]. A lot of works have been concerned with specific problems of uncertain control systems, including problems of stability and stabilization of uncertain systems and an idea of robust control (e.g. [31, 61, 62, 77, 87, 88]).

In recent years a concept of so-called *uncertain variables* and their applications to analysis and decision problems for a wide class of uncertain systems has been developed [25, 30, 35, 40, 42, 43, 44, 46, 50, 53, 54, 55]. The main aim of this book is to present a unified, comprehensive and compact description of analysis and decision problems in a class of uncertain systems described by traditional mathematical models and by relational knowledge representations. An attempt at a uniform theory of uncertain systems including problems and methods based on different mathematical formalisms may be useful for further research in this large area and for practical applications to the design of knowledge-based decision support systems. The book may be characterized by the following features:

1. The problems and methods are concerned with systems described by traditional mathematical models (with number variables) and by knowledge representations which are treated as an extension of classical functional models. The considerations are then directly related to respective problems and methods in traditional system and control theory.
2. The problems under consideration are formulated for systems with unknown parameters in the known form of the description (*parametric problems*) and for the direct non-deterministic input–output description (*non-parametric problems*). In the first case the unknown parameters are assumed to be values of random or uncertain variables. In the second case the values of input and output variables are assumed to be values of random, uncertain or fuzzy variables.
3. The book presents three new concepts introduced and developed by the author for a wide class of uncertain systems:
   a. Logic-algebraic method for systems with a logical knowledge representation [9 – 14].
   b. Learning process in systems with a relational knowledge representation, consisting in *step by step* knowledge validation and updating (e.g. [18, 22, 25]).
   c. Uncertain variables based on uncertain logics.
4. Special emphasis is placed on uncertain variables as a convenient tool for handling the uncertain systems under consideration. The main part of the book is devoted to the basic theory of uncertain variables and their application in different cases of uncertain systems. One of the main purposes of the book is to present recent developments in this area, a comparison with random and fuzzy variables and the generalization in the form of so-called *soft variables*.
5. Special problems such as pattern recognition and control of a complex of operations under uncertainty are included. Examples concerning the control of manufacturing systems, assembly processes and task distributions in computer systems indicate the possibilities of practical applications of uncertain variables and other approaches to decision making in uncertain systems.

The analysis and decision problems are formulated for input–output plants and two kinds of uncertainty:

1. The plant is non-deterministic, i.e. the output is not determined by the input.

2. The plant is deterministic, but its description (the input–output relationship) is not exactly known.

The different forms of the uncertainty may be used in the description of one plant. For example, the non-deterministic plant may be described by a relation such that the output is not determined by the input (i.e. is not a function of the input). This relation may be considered as a *basic description* of the uncertainty. If the relation contains unknown parameters, their description, e.g. in the form of probability distributions, may be defined as an *additional description* of the uncertainty or the *second-order uncertainty*.

In the wide sense of the word an uncertain system is understood in the book as a system containing any kind and any form of uncertainty in its description. In a narrow sense, an uncertain system is understood as a system with the description based on uncertain variables. In this sense, such names as "random, uncertain and fuzzy knowledge" or "random, uncertain and fuzzy controllers" will be used. Additional remarks will be introduced, if necessary, to avoid misunderstandings. Quite often the name "control" is used in the text instead of decision making for a particular plant. Consequently, the names "control plant, control system, control algorithm, controller" are used instead of "decision plant, decision system, decision algorithm, decision maker", respectively.

1.2 Uncertain Variables

In the traditional case, for a static (memoryless) system described by a function $y = \Phi(u, x)$ where u, y, x are input, output and parameter vectors, respectively, the decision problem may be formulated as follows: to find the decision u^* such that $y = y^*$ (the desirable output value). The decision u^* may be obtained for the known function Φ and the value x. Let us now assume that x is unknown. In the probabilistic approach x is assumed to be a value of a random variable $\tilde{x}$ described by the probability distribution. In the approach based on uncertain variables the unknown parameter x is a value of an uncertain variable $\bar{x}$ for which an expert gives the certainty distribution $h(x) = v(\bar{x} \cong x)$ where v denotes a certainty index of the soft property: "$\bar{x}$ is approximately equal to x" or "x is the approximate value of $\bar{x}$". The certainty distribution evaluates the expert's opinion on approximate values of the uncertain variable. The uncertain variables, related to random variables and fuzzy numbers, are described by the set of values X and their certainty distributions which correspond to probability distributions for the random variables and to membership functions for the fuzzy numbers. To define the uncertain variable, it is necessary to give $h(x)$ and to determine the certainty indexes of the following soft properties:

1. "$\bar{x} \tilde{\in} D_x$" for $D_x \subset X$, which means "the approximate value of $\bar{x}$ belongs to D_x" or "$\bar{x}$ belongs approximately to D_x".

2. "$\bar{x} \,\tilde{\notin}\, D_x$" = "$\neg(\bar{x} \,\tilde{\in}\, D_x)$", which means "$\bar{x}$ does not belong approximately to D_x".

To determine the certainty indexes for the properties: $\neg(\bar{x} \,\tilde{\in}\, D_x)$, $(\bar{x} \,\tilde{\in}\, D_1) \vee (\bar{x} \,\tilde{\in}\, D_2)$ and $(\bar{x} \,\tilde{\in}\, D_1) \wedge (\bar{x} \,\tilde{\in}\, D_2)$ where $D_1, D_2 \subseteq X$, it is necessary to introduce an *uncertain logic*, which deals with the soft predicates of the type "$\bar{x} \,\tilde{\in}\, D_x$". In Chapter 4 four versions of the uncertain logic have been defined and used for the formulation of the respective versions of the uncertain variable.

For the proper interpretation (semantics) of these formalisms it is convenient to consider $\bar{x} = g(\omega)$ as a value assigned to an element $\omega \in \Omega$ (a universal set). For fixed ω its value $\bar{x}$ is determined and $\bar{x} \in D_x$ is a crisp property. The property $\bar{x} \,\tilde{\in}\, D_x = x \in D_x$ = "the approximate value of $\bar{x}$ belongs to D_x" is a soft property because $\bar{x}$ is unknown and the evaluation of "$\bar{x} \,\tilde{\in}\, D_x$" is based on the evaluation of $\bar{x} \cong x$ for the different $x \in X$ given by an expert. In the first version of the uncertain variable, $v(\bar{x} \,\tilde{\in}\, D_x) \neq v(\bar{x} \,\tilde{\notin}\, \overline{D}_x)$ where $\overline{D}_x = X - D_x$ is the complement of D_x. In the version called the *C*-uncertain variable, $v_c(\bar{x} \,\tilde{\notin}\, D_x) = v_c(\bar{x} \,\tilde{\in}\, \overline{D}_x)$ where v_c is the certainty index in this version

$$v_c(\bar{x} \,\tilde{\in}\, D_x) = \frac{1}{2}[v(\bar{x} \,\tilde{\in}\, D_x) + v(\bar{x} \,\tilde{\notin}\, \overline{D}_x)].$$

The uncertain variable in the first version may be considered as a special case of the possibilistic number with a specific interpretation of $h(x)$ described above. In our approach we use soft properties of the type "P is approximately satisfied" where P is a crisp property, in particular $P =$ "$\bar{x} \in D_x$". It allows us to accept the difference between $\bar{x} \,\tilde{\in}\, D_x$ and $\bar{x} \,\tilde{\notin}\, \overline{D}_x$ in the first version. More details concerning the relations to random variables and fuzzy numbers are given in Chapter 6. Now let us pay attention to the following aspects which will be more clear after the presentation of the formalisms and semantics in Chapter 4:

1. To compare the meanings and practical utilities of different formalisms, it is necessary to take into account their semantics. It is specially important in our approach. The definitions of the uncertain logics and consequently the uncertain variables contain not only the formal description but also their interpretation. In particular, the uncertain logics may be considered as special cases of multi-valued predicate logic with a specific semantics of the predicates. It is worth noting that from the formal point of view the probabilistic measure is a special case of the fuzzy measure and the probability distribution is a special case of the membership function in the formal definition of the fuzzy number when the meaning of the membership function is not described.
2. Even if the uncertain variable in the first version may be formally considered as a very special case of the fuzzy number, for simplicity and unification it is better to introduce it independently (as has been done in the book) and not as a special case

of the much more complicated formalism with different semantics and applications.
3. *Uncertainty* is understood here in the narrow sense of the word, and concerns an incomplete or imperfect knowledge of something which is necessary to solve the problem. In our considerations, it is the knowledge of the parameters in the mathematical description of the system or the knowledge of a form of the input–output relationships, and is related to a fixed expert who gives the description of the uncertainty.
4. In the majority of interpretations the value of the membership function means a *degree of truth* of a soft property determining the fuzzy set. In our approach, "$\bar{x} \in D_x$" and "$x \in D_x$" are crisp properties, the soft property "$\bar{x} \,\tilde{\in}\, D_x$" is introduced because the value of $\bar{x}$ is unknown and $h(x)$ is a *degree of certainty* (or $1-h(x)$ is a degree of uncertainty).

1.3 Basic Deterministic Problems

The problems of analysis and decision making under uncertainty described in the book correspond to the respective problems for deterministic (functional) plants with the known mathematical models. Let us consider a static plant described by a function $y = \Phi(u)$ where $u \in U = R^p$ is the input vector, $y \in Y = R^l$ is the output vector, U and Y are p-dimensional and l-dimensional real number vector spaces, respectively. The function Φ may be presented as a set of functions

$$y^{(i)} = \Phi_i(u^{(1)}, u^{(2)}, ..., u^{(p)}); \quad i = 1, 2, ..., l$$

where $y^{(i)}$ is the i-th component of y and $u^{(j)}$ is the j-th component of u.

Analysis problem: Given the function Φ and the value $u = u^*$, find the corresponding output $y^* = \Phi(u^*)$.

Decision problem: For the given function Φ and the value y^* required by a user, find the decision u^* such that $y = y^*$.

The solution of the problem is reduced to solving the equation $y^* = \Phi(u)$ with respect to u. In general, we may obtain a set of decisions

$$D_u = \{u \in U: \quad \Phi(u) = y^*\}.$$

In particular $D_u = \varnothing$ (an empty set), which means that the solution does not exist. For the plant described by a function $y = \Phi(u, z)$ where z is a vector of external disturbances, the set of solutions $D_u(z)$ depends on z. In the case of a unique solution we obtain $u^* \stackrel{\Delta}{=} \Psi(z)$, i.e. the deterministic decision (control) algorithm in an open-loop decision system when z is measured (Fig. 1.1). For the plant described by the function $y = \Phi(u)$, on the assumption that the equation $\Phi(u) = y^*$ has a unique solution, the decision u^* may be determined by the following recursive algorithm:

$$u_{n+1} = u_n - K[y^* - \Phi(u_n)]; \quad n = 0, 1, \ldots \tag{1.1}$$

where u_n denotes the n-th approximation of u^* and K is a matrix of coefficients. Under some conditions concerning Φ and K, the sequence u_n converges to u^* for any u_0. The algorithm (1.1) may be executed in a closed-loop decision system (Fig. 1.2) where the output $y_n = \Phi(u_n)$ is measured. It is worth noting that to assure the convergence, it is not necessary to know exactly the function Φ. Then feedback is a way to achieve the proper decision u^* for the uncertain plant, i.e. it is one of the possible approaches to decision making under uncertainty.

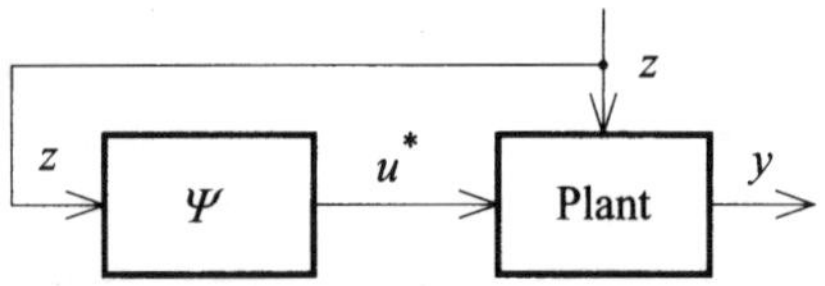

Figure 1.1. Open-loop decision system

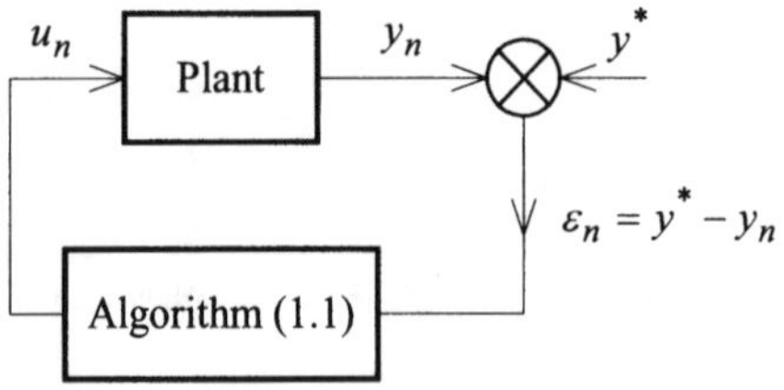

Figure 1.2. Closed-loop decision system

If there are additional constraints and/or the solution of the equation $\Phi(u) = y^*$ does not exist, the decision problem may be formulated as an optimization problem consisting in finding u^* minimizing a quality index $\varphi(y, y^*)$, e.g.

$$\varphi(y, y^*) = (y - y^*)^{\mathrm{T}}(y - y^*)$$

where vectors are presented as one-column matrices and T denotes transposition of a matrix.

The formulations of basic analysis and decision problems may be extended to deterministic dynamical plants. Let us consider a plant described by the equation

$$s_{n+1} = f(s_n, u_n); \quad n = 0, 1, \ldots$$

where s_n is a state vector.

Analysis problem: For the given function *f*, initial state s_0 and the sequence $u_0, u_1, \ldots, u_{N-1}$ one should find the sequence $s_1, s_2, \ldots, s_N$.

One of the possible formulations of a decision problem is the following: for the given function *f*, s_0 and $s_N = s^*$ required by a user, one should determine the sequence of decisions $u_0, u_1, \ldots, u_{N-1}$ such that $s_N = s^*$. The solution exists for sufficiently large N if the plant is controllable. The optimization problem corresponding to the minimization of $\varphi(y, y^*)$ for a static plant may be formulated as follows.

Optimal decision problem: For the given function *f*, state s_0 and a quality index $\varphi(s, s^*)$, one should determine the sequence $u_0, u_1, \ldots, u_{N-1}$ minimizing the global performance index

$$Q_N = \sum_{n=1}^{N} \varphi(s_n, s^*) = \sum_{n=0}^{N-1} \varphi[f(s_n, u_n), s^*].$$

1.4 Structure of the Book

The book consists of two informal parts. The first part containing Chapters 2–7 presents basic analysis and decision problems for static plants. The second part containing Chapters 8–14 concerns dynamical systems and special problems connected with learning and complex systems, pattern recognition and operation systems. The parts are organized as follows.

Chapter 2 presents basic analysis and decision problems for static plants described by relations. A general concept of so-called *determinization*, consisting in replacing an uncertain description by its deterministic representation, is introduced. Two kinds of relational knowledge representation are considered: the knowledge of the plant and the knowledge of the decision making.

Chapter 3 deals with the application of random variables to the description of the uncertainty. In the first part of the chapter, analysis and decision problems are considered for the functional and relational plant with random parameters. The second part is devoted to the respective problems with a non-parametric description of the uncertainty. In this case the knowledge of the plant has a form of conditional probability distribution. In both cases it is shown how the probabilistic knowledge of the decision making (i.e. the random decision algorithm in an open-loop decision system) may be obtained from the probabilistic knowledge of the plant, and how to obtain the deterministic decision algorithm as a result of determinization.

Chapters 4 and 5 are devoted to uncertain variables and their applications to uncertain systems. The basic definitions and properties of the uncertain logics and variables are given in Chapter 4. We consider four versions of the uncertain variables with different definitions of the certainty distributions and operations. The application of the uncertain variables to the formulation and solving of the analysis and decision problems for the functional and relational static plant is the topic of Chapter 5. The chapter is completed with considerations for the non-parametric case in which the knowledge of the plant has the form of conditional certainty distributions. The uncertain decision algorithm is obtained from the uncertain knowledge of the plant.

In the first part of Chapter 6 the applications of fuzzy numbers (fuzzy variables) to non-parametric analysis and decision problems for the static plant are presented. In the second part of the chapter the comparison of uncertain variables with random and fuzzy variables and analogies between the non-parametric problem statements and solutions for the descriptions based on random, uncertain and fuzzy variables are discussed. These analogies lead to a generalization in the form of so-called *soft variables* and their application in analysis and decision problems for the static plant.

Chapter 7 is concerned with relational static plants described by a logical knowledge representation, which may be treated as a special form of the relational knowledge representation that consists of relations in the form of logical formulas concerning input, output and additional variables. Consequently, to formulate and solve the analysis and decision problems, one may apply the so-called *logic-algebraic method.* The modification of this method may be applied to a plant with random and uncertain parameters.

The purpose of Chapter 8 is to show how the approaches and methods presented in the first part of the book for static plants (in particular, the considerations based on the relational knowledge representation and uncertain variables) may be applied to dynamical plants. The application of the presented approach to knowledge-based control of an assembly process is described.

Chapter 9 has a special character, and is devoted to the general idea of parametric optimization and its application to uncertain, random and fuzzy controllers in closed-loop decision systems with dynamical plants. The chapter is completed with remarks concerning so-called descriptive and prescriptive approaches, and the quality of the decisions based on different forms of the knowledge given by an expert.

The idea and algorithms of learning based on *step by step* knowledge validation and updating are presented in Chapter 11. Two cases are considered. In the first case the validation and updating concerns the knowledge of the plant, and in the second case – the knowledge of the decision making. The idea of learning is illustrated by an example of the application to the assembly system considered in Chapter 8.

Chapters 12, 13 and 14 deal with specific problems and systems: the decision problems for plants with three-level uncertainty, complex relational systems (with an application to a complex manufacturing system), control of a complex of operations (with an application to task allocation in a group of parallel processors), and knowledge-based pattern recognition under uncertainty.

2 Relational Systems

This chapter is concerned with analysis and decision making problems for a static input–output plant described by a relation which is not reduced to the function Φ considered in Sect. 1.3. Consequently, for the given relation, the output is not determined by the input. The analysis problem consists in finding the *output property* (or the set of possible outputs) for the given *input property* (or the given set of inputs), and the decision problem consists in finding the *input property* (or the set of possible inputs) for the given *output property* (or the set of acceptable outputs, required by a user). For the functional plant presented in Sect. 1.3, the input and output properties have the form "$u = u^*$" and "$y = y^*$", respectively. For the relational plant the respective properties have the form "$u \in D_u$" and "$y \in D_y$" where D_u and D_y are subsets of U and Y, respectively.

2.1 Relational Knowledge Representation

Let us consider a static plant with input vector $u \in U$ and output vector $y \in Y$, where U and Y are real number vector spaces. The plant is described by a relation

$$u \, \rho \, y \stackrel{\Delta}{=} R(u, y) \subset U \times Y \tag{2.1}$$

which may be called a *relational knowledge representation* of the plant. It is an extension of the traditional functional model $y = \Phi(u)$ considered in Sect. 1.3. The relation $R(u, y)$ denotes a set of all possible pairs (u, y) in the Cartesian product $U \times Y$, which may appear in the plant. In other words, the plant is described by a property (a predicate) concerning (u, y), and $R(u, y)$ denotes the set of all pairs (u, y) for which this property is satisfied, i.e.

$$R(u, y) = \{(u, y) \in U \times Y : w[\varphi(u, y)] = 1\} \stackrel{\Delta}{=} \{(u, y) \in U \times Y : \varphi(u, y)\}$$

where $w[\varphi(u, y)] \in [0, 1]$ for the fixed values (u, y) denotes a logic value of the property $\varphi(u, y)$. When the relation $R(u, y)$ is not a function, the description (2.1) given by an expert may have two practical interpretations:

1. The plant is deterministic, i.e. at every moment n

$$y_n = \Phi(u_n),$$

but the expert has no full knowledge of the plant and for the given u he/she can determine only the set of possible outputs:

$$D_y(u) \subset Y : \{y \in Y : (u, y) \in R(u, y)\}.$$

For example, in a one-dimensional case $y = cu$, the expert knows that $c_1 \le c \le c_2$; $c_1, c_2 > 0$. Then as a description of the plant he/she gives a relation presented in the following form:

$$\left.\begin{array}{lll} c_1 u \le y \le c_2 u & \text{for} & u \ge 0 \\ c_2 u \le y \le c_1 u & \text{for} & u \le 0 \end{array}\right\}. \tag{2.2}$$

The situation is illustrated in Fig. 2.1, in which the set of points (u_n, y_n) is marked.

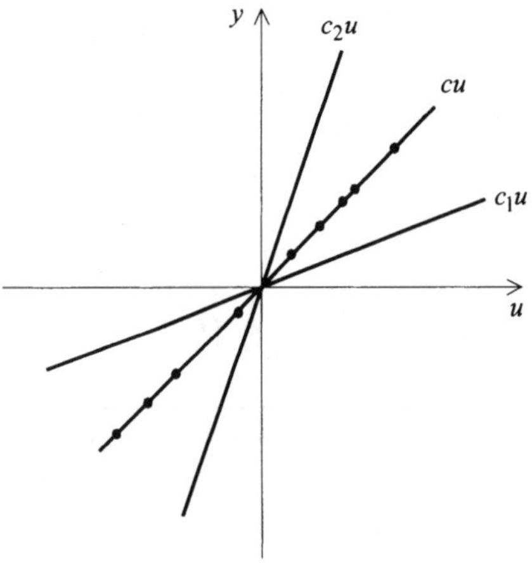

Figure 2.1. Illustration of a relation – the first case

2. The plant is not deterministic, which means that at different n we may observe different values y_n for the same values u_n. Then $R(u, y)$ is a set of all possible points (u_n, y_n), marked for the example (2.2) in Fig. 2.2, and $D_y(u)$ is a set of all possible values which may be observed at the output for the fixed value u.
In the first case the relation (which is not a function) is a result of the expert's uncertainty and in the second case – a result of uncertainty in the plant. For simplicity, in both cases we shall talk about an *uncertain plant*, and the plant described by a relational knowledge representation will be shortly called a *relational plant.*
In more complicated cases the relational knowledge representation given by an expert may have the form of a set of relations:

$$R_i(u, \overline{w}, y) \subset U \times W \times Y, \quad i = 1, 2, ..., k \tag{2.3}$$

where $\overline{w} \in W$ is a vector of additional auxiliary variables used in the description of the knowledge. The set of relations (2.3) may be called a *basic knowledge representation*. It may be reduced to a *resulting knowledge representation* $R(u, y)$:

$$R(u,y) = \{(u,y) \in U \times Y : \bigvee_{\overline{w} \in W} (u, \overline{w}, y) \subset \overline{R}(u, w, y)\}$$

where

$$\overline{R}(u, \overline{w}, y) = \bigcap_{i=1}^{k} R_i(u, \overline{w}, y) .$$

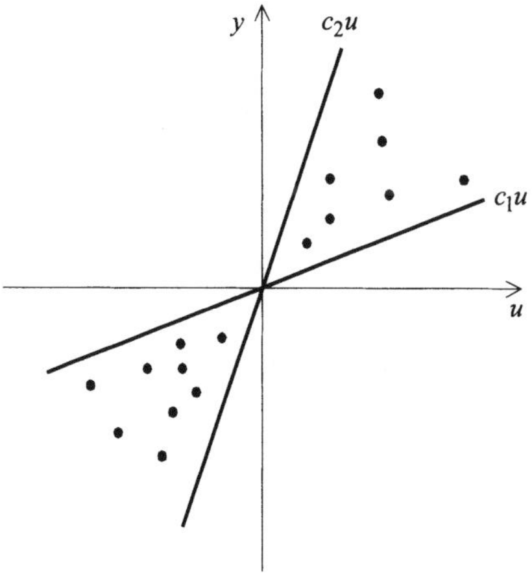

Figure 2.2. Illustration of a relation – the second case

The relations $R_i(u, w, y)$ may have the form of a set of inequalities and/or equalities concerning the components of the vectors u, w, y.

In Chapter 7 we shall consider a special form of the relational knowledge representation, in which the relations (2.3) are expressed by logical formulas concerning $(u, \overline{w}, y)$, and in Chapter 8 the extension of the relational knowledge representation to a dynamical plant will be presented.

The relational knowledge representation has a specific form in a discrete case when U and Y are finite sets of vectors. Assume that U is a finite discrete set

$$U = \{\overline{u}_1, \overline{u}_2, ..., \overline{u}_\alpha\} .$$

Then the relation $R(u, y)$ is reduced to the family of sets

$$D_y(\overline{u}_j) = \{y \in Y : (\overline{u}_j, y) \in R(u, y)\}, \quad j = 1,2,...,\alpha ,$$

i.e. the sets of possible outputs for all inputs $\overline{u}_j$. If $Y = \{\overline{y}_1, \overline{y}_2, ..., \overline{y}_\beta\}$ then $R(u, y)$ is a set of pairs $(\overline{u}_j, \overline{y}_i)$ selected from $U \times Y$ and $D_y(\overline{u}_j)$ is a corresponding finite set of the points $\bar{y}_i$ (a subset of Y).

For the plant with external disturbances, the relational knowledge representation has the form of a relation

$$R(u, y, z) \subset U \times Y \times Z$$

where $z \in Z$ is a vector of the disturbances.

2.2 Analysis and Decision Making for Relational Plants

The formulations of the analysis and decision making problems for a relational plant analogous to those for a functional plant described by a function $y = \Phi(u)$ are adequate for the knowledge of the plant [24].

Analysis problem: For the given $R(u, y)$ and $D_u \subset U$ find the smallest set $D_y \subset Y$ such that the implication

$$u \in D_u \rightarrow y \in D_y \tag{2.4}$$

is satisfied.

The information that $u \in D_u$ may be considered as a result of observation. For the given D_u one should determine the best estimation of y in the form of the set of possible outputs D_y. It is easy to note that

$$D_y = \{y \in Y : \bigvee_{u \in D_u} (u, y) \in R(u, y)\}. \tag{2.5}$$

This is then a set of all such values of y for which there exists $u \in D_u$ such that (u, y) belongs to R. In particular, if the value u is known, i.e. $D_u = \{u\}$ (a singleton), then

$$D_y(u) = \{y \in Y : (u, y) \in R(u, y)\} \tag{2.6}$$

where $D_y(u)$ is a set of all possible y for the given value u. The analysis problem is illustrated in Fig. 2.3 where the shaded area illustrates the relation $R(u, y)$ and the interval D_y denotes the solution for the given interval D_u.

Example 2.1.
Let us consider the plant with two inputs $u^{(1)}$ and $u^{(2)}$, described by the inequality

$$c_1 u^{(1)} + d_1 u^{(2)} \leq y \leq c_2 u^{(1)} + d_2 u^{(2)},$$

and the set D_u is determined by

$$au^{(1)} + bu^{(2)} \le \alpha \tag{2.7}$$

$$u^{(1)} \ge u^{(1)}_{\min}, \quad u^{(2)} \ge u^{(2)}_{\min}. \tag{2.8}$$

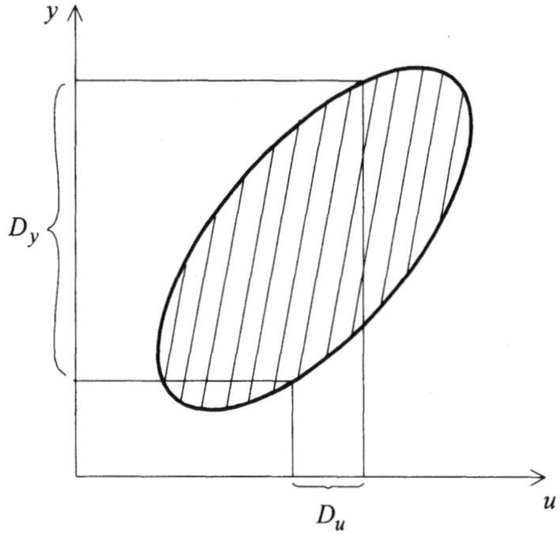

Figure 2.3. Illustration of analysis problem

For example, y may denote the amount of a product in a production process, $u^{(1)}$ and $u^{(2)}$ – amounts of two kinds of raw material, and the value $au^{(1)} + bu^{(2)}$ – the cost of the raw material. Assume that c_1, c_2, d_1, d_2, a, b, α are positive numbers and $c_1 < c_2$, $d_1 < d_2$. It is easy to see that the set (2.5) is described by the inequality

$$c_1 u^{(1)}_{\min} + d_1 u^{(2)}_{\min} \le y \le y_{\max} \tag{2.9}$$

where

$$y_{\max} = \max_{u^{(1)}, u^{(2)}} (c_2 u^{(1)} + d_2 u^{(2)}) \tag{2.10}$$

subject to constraints (2.7) and (2.8).

The maximization in (2.10) leads to the following results:
If

$$\frac{c_2}{d_2} \le \frac{a}{b}$$

then

$$y_{\max} = c_2 u^{(1)}_{\min} + \frac{d_2}{b}(\alpha - a u^{(1)}_{\min}). \tag{2.11}$$

If

$$\frac{c_2}{d_2} \geq \frac{a}{b}$$

then

$$y_{\max} = \frac{c_2}{a}(\alpha - bu_{\min}^{(2)}) + d_2 u_{\min}^{(2)} . \tag{2.12}$$

For the numerical data $c_1 = 1$, $c_2 = 2$, $d_1 = 2$, $d_2 = 4$, $a = 1$, $b = 4$, $\alpha = 3$, $u_{\min}^{(1)} = 1$, $u_{\min}^{(2)} = 0.5$

$$\frac{c_2}{d_2} = \frac{1}{2}, \quad \frac{a}{b} = \frac{1}{4} < \frac{c_2}{d_2} .$$

From (2.12) we obtain $y_{\max} = 4$ and according to (2.9) $y_{\min} = c_1 u_{\min}^{(1)} + d_1 u_{\min}^{(2)} = 2$. The set D_y is then determined by inequality $2 \leq y \leq 4$. □

Now let us consider a decision making problem for the relational plant described by $R(u, y)$ which is not a function. In this case the requirement $y = y^*$ (where y^* is a given value) cannot be satisfied and should be replaced by a weaker requirement $y \in D_y$ for a given set D_y. As a result we may obtain not one particular decision u^*, but a set of possible decisions D_u.

Decision problem: For the given $R(u, y)$ and $D_y \subset Y$ find the largest set $D_u \subset U$ such that the implication (2.4) is satisfied.

The set D_y is given by a user, the property $y \in D_y$ denotes the user's requirement and D_u denotes the set of all possible decisions for which the requirement concerning the output y is satisfied. It is easy to note that

$$D_u = \{u \in U : \quad D_y(u) \subseteq D_y\} \tag{2.13}$$

where $D_y(u)$ is the set of all possible y for the fixed value u, determined by (2.6), or

$$D_u = \{u \in U : (u, y) \in R(u, y) \rightarrow y \in D_y\} .$$

The solution may not exist, i.e. $D_u = \varnothing$ (empty set). Such a case is illustrated in Fig.2.4: for the given interval D_y, a set $D_u \neq \varnothing$ satisfying the implication (2.4) does not exist. This means that the requirement is too strong, i.e. the interval D_y is too small. The requirement may be satisfied for a larger interval D_y (see Fig. 2.3). In the example illustrated in Fig. 2.2, if $D_y = [y_{\min}, y_{\max}]$ and $c_1, c_2 > 0$ then

$$D_u = [\frac{y_{\min}}{c_1}, \frac{y_{\max}}{c_2}]$$

and the solution exists on the condition

$$\frac{y_{\min}}{c_1} \le \frac{y_{\max}}{c_2}.$$

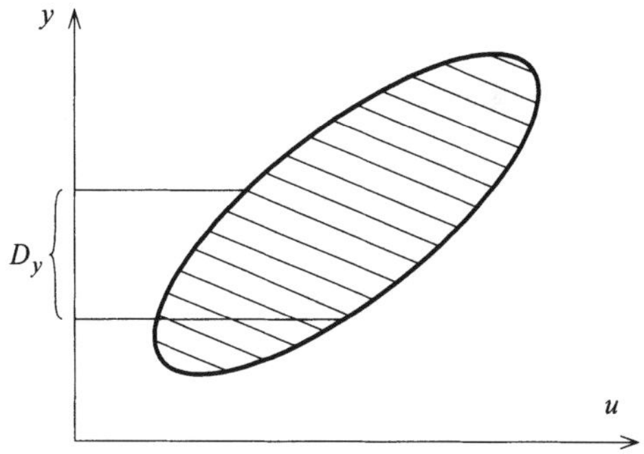

Figure 2.4. Illustration of the case where the solution does not exist

The analysis and decision problems for the relational plant are the extensions of the respective problems for the functional plant, presented in Sect. 1.3. The properties "$u \in D_u$" and "$y \in D_y$" may be called *input and output properties*, respectively. For the functional plant we considered the input and output properties in the form: "$u = u^*$" and "$y = y^*$" where u^*, y^* denote fixed values. For the relational plant the analysis problem consists in finding the best output property (the smallest set D_y) for the given input property, and the decision problem consists in finding the best input property (the largest set D_u) for the given output property required. The procedure for determining the effective solution D_u or D_y based on the general formulas (2.5) or (2.13) depends on the form of $R(u,y)$ and may be very complicated. If $R(u,y)$ and the given property (i.e. the given set D_u or D_y) are described by a set of equalities and/or inequalities concerning the components of the vector u and y, then the procedure is reduced to "solving" this set of equalities and/or inequalities.

Example 2.2.

Consider a plant with a single output, described by a relation

$$G_1(u) \le y \le G_2(u) \tag{2.14}$$

where G_1 and G_2 are the functions

$$G_1 : U \to \mathrm{R}^+, \quad G_2 : U \to \mathrm{R}^+; \quad \mathrm{R}^+ = [0, \infty),$$

and

$$\bigwedge_{u \in U} [G_1(u) \le G_2(u)].$$

For example, y is the amount of a product as in Example 2.1, and the components of the vector u are features of the raw materials. For a user's requirement

$$y_{\min} \le y \le y_{\max}, \tag{2.15}$$

i.e. $D_y = [y_{\min}, y_{\max}]$, we obtain

$$D_u = \{u \in U : [G_1(u) \ge y_{\min}] \wedge [G_2(u) \le y_{\max}]\}.$$

In particular, if the relation (2.14) has the form

$$c_1 u^{\mathrm{T}} u \le y \le c_2 u^{\mathrm{T}} u, \quad c_1 > 0, \ c_2 > c_1$$

where $u \in R^k$ and

$$u^{\mathrm{T}} u = (u^{(1)})^2 + (u^{(2)})^2 + \ldots + (u^{(k)})^2$$

then D_u is described by the inequality

$$\frac{y_{\min}}{c_1} \le u^{\mathrm{T}} u \le \frac{y_{\max}}{c_2}$$

and the decision u satisfying the requirement (2.15) exists if

$$\frac{y_{\max}}{c_2} \ge \frac{y_{\min}}{c_1}.$$

□

2.3 Relational Plant with External Disturbances

The considerations may by extended to a plant with external disturbances, described by a relation $R(u, y, z) \subset U \times Y \times Z$ where $z \in Z$ is a vector of the disturbances which may be observed. The property $z \in D_z$ for the given $D_z \subset Z$ may be considered as a result of observations. Our plant has two inputs (u, z) and the analysis problem is formulated in the same way as for the relation $R(u, y)$, with $(u, z) \in D_u \times D_z$ in place of $u \in D_u$.

Analysis problem: For the given $R(u,y,z)$, D_z and D_u find the smallest set $D_y \subset Y$ such that the implication

$$(u \in D_u) \wedge (z \in D_z) \rightarrow y \in D_y \tag{2.16}$$

is satisfied.

The result analogous to (2.5) is

$$D_y = \{y \in Y : \bigvee_{u \in D_u} \bigvee_{z \in D_z} (u, y, z) \in R(u, y, z)\}.$$

The decision making is an inverse problem consisting in the determination of the set of all decisions *u* such that for every decision from this set and for every $z \in D_z$ the required property $y \in D_y$ is satisfied.

Decision problem: For the given $R(u,y,z)$, D_y (the requirement) and D_z (the result of observations), find the largest set D_u such that the implication (2.16) is satisfied. The general form of the solution is as follows:

$$D_u = \{u \in U : \bigwedge_{z \in D_z} [D_y(u,z) \subseteq D_y]\} \tag{2.17}$$

where

$$D_y(u,z) = \{y \in Y : (u,y,z) \in R(u,y,z)\}. \tag{2.18}$$

It is then the set of all such decisions u that for every $z \in D_z$ the set of possible outputs y belongs to D_y. For the fixed z (the result of measurement) the set D_u is determined by (2.17) with the relation

$$R(u,y,z) \stackrel{\Delta}{=} R(u, y; z) \subset U \times Y.$$

In this notation z is the parameter in the relation $R(u, y; z)$. Then

$$D_u(z) = \{u \in U : D_y(u,z) \subseteq D_y\} \stackrel{\Delta}{=} \overline{R}(z,u) \tag{2.19}$$

where $D_y(u,z)$ is defined by (2.18). The formula (2.19) defines a relation between z and u denoted by $\overline{R}(z,u)$. The relation $\overline{R}(z,u)$ may be called a knowledge representation for the decision making (a description of the knowledge of the decision making) or a *relational decision algorithm*. The block scheme of the open-loop decision system (Fig. 2.5) is analogous to that of Fig. 1.1 for a functional plant. The *knowledge of the decision making*

$$< \overline{R}(z,u) > \stackrel{\Delta}{=} \mathrm{KD}$$

has been obtained for the given *knowledge of the plant*

$$< R(u,y,z) > \stackrel{\Delta}{=} \mathrm{KP}$$

and the given requirement $y \in D_y$.

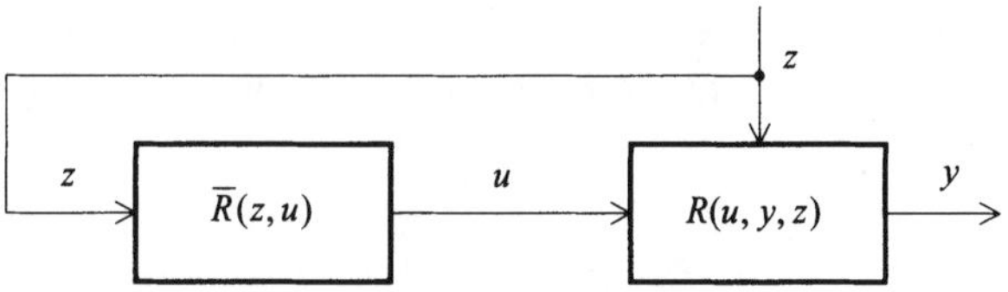

Figure 2.5. Open-loop decision system

Example 2.3.
Consider a plant with one output, described by a relation

$$G_1(u,z) \le y \le G_2(u,z) \tag{2.20}$$

where G_1 and G_2 are the functions

$$G_1 : U \times Z \to \mathrm{R}^+, \quad G_2 : U \times Z \to \mathrm{R}^+; \quad \mathrm{R}^+ = [0,\infty),$$

and

$$\bigwedge_{u \in U} \bigwedge_{z \in Z} [G_1(u,z) \le G_2(u,z)].$$

For example, y is the amount of a product (see Example 2.2), the components of the vector u are the features of the raw material which may be chosen by a decision maker, and the components of the vector z are the features of the raw material which may be observed. For a user's requirement

$$y_{\min} \le y \le y_{\max} , \tag{2.21}$$

i.e. $D_y = [y_{\min}, y_{\max}]$, we obtain

$$D_u : \{u \in U : [\overline{G}_1(u) \ge y_{\min}] \wedge [\overline{G}_2(u) \le y_{\max}]\}$$

where

$$\overline{G_1}(u) = \min_{z \in D_z} G_1(u,z), \qquad \overline{G_2}(u) = \max_{z \in D_z} G_2(u,z).$$

Assume that $z^{\mathrm{T}} = [z^{(1)} \ z^{(2)}]$, the relation (2.20) has the form

$$c_1 z^{(1)} u^{\mathrm{T}} u \le y \le c_2 z^{(2)} u^{\mathrm{T}} u,$$

$$c_1 > 0, \ c_2 > c_1, \quad z_1 > 0, \ z_2 > z_1$$

and D_z is described by the inequality

$$r_{\min}^2 \le (z^{(1)})^2 + (z^{(2)})^2 \le r_{\max}^2.$$

Then

$$\overline{G_1}(u) = c_1 z_{\min}^{(1)} u^{\mathrm{T}} u, \quad \overline{G_2}(u) = c_2 z_{\max}^{(1)} u^{\mathrm{T}} u$$

where

$$z_{\min}^{(1)} = \frac{r_{\min}}{\sqrt{2}}, \quad z_{\max}^{(2)} = \frac{r_{\max}}{\sqrt{2}}.$$

Consequently, the set D_u is described by the inequality

$$\frac{\sqrt{2}\, y_{\min}}{c_1 r_{\min}} \le u^{\mathrm{T}} u \le \frac{\sqrt{2}\, y_{\max}}{c_2 r_{\max}}$$

and the decision u satisfying the requirement (2.21) exists if

$$\frac{y_{\max}}{c_2 r_{\max}} \ge \frac{y_{\min}}{c_1 r_{\min}}. \qquad \square$$

Example 2.4.
A plant with $u, y, z \in R^2$ (two-dimensional vectors) is described by the inequalities

$$z^{(1)} u^{(1)} \le y^{(1)} \le 2 z^{(1)} u^{(1)}$$

$$z^{(2)} u^{(2)} \le y^{(2)} \le 2 z^{(2)} u^{(2)},$$

$z^{(1)}$, $z^{(2)}$, $u^{(1)}$, $u^{(2)} > 0$. The requirement concerning the output is the following

$$\alpha \le (y^{(1)})^2 + (y^{(2)})^2 \le \beta$$

for the given $\alpha, \ \beta > 0$. From the description of the plant we have

$$(z^{(1)})^2\,(u^{(1)})^2 \le (y^{(1)})^2 \le 4(z^{(1)})^2\,(u^{(1)})^2$$

$$(z^{(2)})^2\,(u^{(2)})^2 \le (y^{(2)})^2 \le 4(z^{(2)})^2\,(u^{(2)})^2 .$$

If $z^{(1)} \in [z_{\min}^{(1)}, z_{\max}^{(1)}]$ and $z^{(2)} \in [z_{\min}^{(2)}, z_{\max}^{(2)}]$, then the set D_u is determined by the inequalities

$$4[(z_{\max}^{(1)})^2\,(u^{(1)})^2 + (z_{\max}^{(2)})^2\,(u^{(2)})^2] \le \beta$$

$$(z_{\min}^{(1)})^2\,(u^{(1)})^2 + (z_{\min}^{(2)})^2\,(u^{(2)})^2 \ge \alpha .$$

2.4 Determinization

The deterministic decision algorithm based on the knowledge KD may be obtained as a result of *determinization* (see Sect. 1.4) of the relational decision algorithm $\overline{R}(z,u)$ by using the mean value

$$\widetilde{u}(z) = \int\limits_{D_u(z)} u\,du \cdot [\int\limits_{D_u(z)} du\,]^{-1} \triangleq \widetilde{\Psi}(z).$$

In such a way the relational decision algorithm $\overline{R}(z,u)$ is replaced by the deterministic decision algorithm $\widetilde{\Psi}(z)$.

For the given desirable value y^* we can consider two cases: in the first case the deterministic decision algorithm $\Psi(z)$ is obtained via determinization of the knowledge of the plant KP, and in the second case the deterministic decision algorithm $\Psi_d(z)$ is based on the determinization of the knowledge of the decision making KD obtained from KP for the given y^*. In the first case we determine the mean value

$$\widetilde{y}(z) = \int\limits_{D_y(u,z)} y\,dy \cdot [\int\limits_{D_y(u,z)} dy\,]^{-1} \triangleq \Phi(u,z) \tag{2.22}$$

where $D_y(u,z)$ is described by formula (2.18). Then, by solving the equation

$$\Phi(u,z) = y^* \tag{2.23}$$

with respect to u, we obtain the deterministic decision algorithm $u = \Psi(z)$, on the assumption that Equation (2.23) has a unique solution.

In the second case we use

$$R(u, y^*, z) \triangleq R_d(z,u), \tag{2.24}$$

i.e. the set of all pairs (u,z) for which it is possible that $y = y^*$. The relation $R_d(z,u) \subset Z \times U$ may be considered as the knowledge of the decision making KD, i.e. the relational decision algorithm obtained for the given KP and the value y^*. The determinization of the relational decision algorithm R_d gives the deterministic decision algorithm

$$u_d(z) = \int\limits_{D_{ud}(z)} u\,du \cdot [\int\limits_{D_{ud}(z)} du\,]^{-1} \triangleq \Psi_d(z) \tag{2.25}$$

where

$$D_{ud}(z) = \{u \in U : \ (u,z) \in R_d(z,u)\}.$$

Two cases of the determination of the deterministic decision algorithm are illustrated in Figs. 2.6 and 2.7. The results of these two approaches may be different, i.e. in general $\Psi(z) \neq \Psi_d(z)$ (see Example 2.5).

Example 2.5.

Consider a plant with $u, z, y \in R^1$ (one-dimensional variables), described by the inequality

$$cu + z \le y \le 2cu + z, \quad c > 0. \tag{2.26}$$

For $D_y = [y_{\min}, y_{\max}]$ and the given z, the set (2.19) is determined by the inequality

$$\frac{y_{\min} - z}{c} \le u \le \frac{y_{\max} - z}{2c}.$$

The determinization of the knowledge KP according to (2.22) gives

$$\tilde{y} = \frac{3}{2}cu + z = \Phi(u,z).$$

From the equation $\Phi(u,z) = y^*$ we obtain the decision algorithm

$$u = \Psi(z) = \frac{2(y^* - c)}{3z}.$$

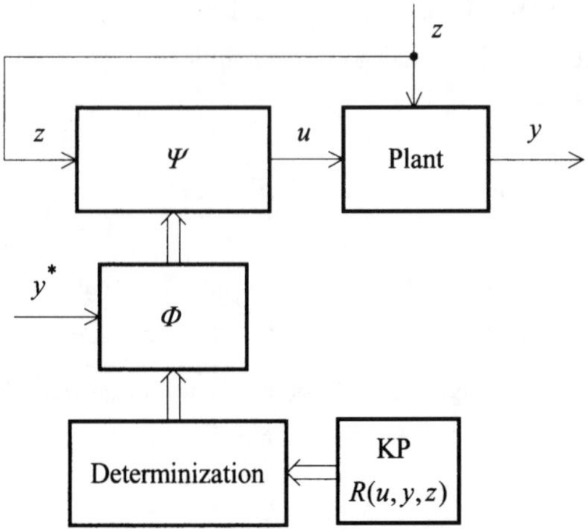

Figure 2.6. Decision system with determinization – the first case

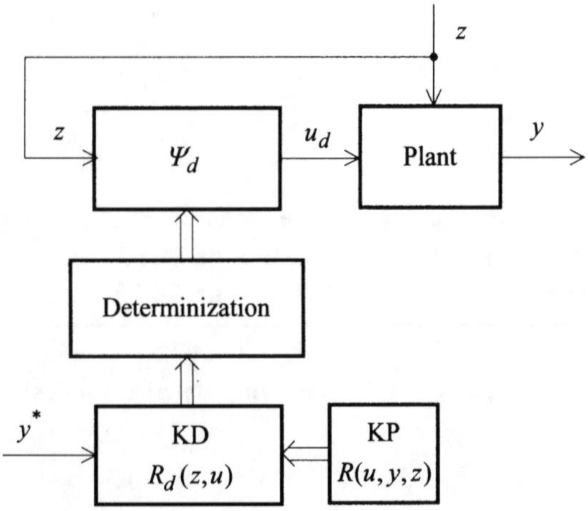

Figure 2.7. Decision system with determinization – the second case

Substituting y^* into (2.26) we obtain the relational decision algorithm $R_d(z,u)$ in the form

$$\frac{y^* - z}{2c} \le u \le \frac{y^* - z}{c}$$

and after the determinization

$$u_d = \Psi_d(z) = \frac{3(y^* - z)}{4c} \ne \Psi(z). \qquad \square$$

2.5 Discrete Case

Assume that

$$U = \{\overline{u}_1, \overline{u}_2, ..., \overline{u}_\alpha\}, \quad Y = \{\overline{y}_1, \overline{y}_2, ..., \overline{y}_\beta\}.$$

Now the relation $R(u, y)$ is a set of pairs $(\overline{u}_j, \overline{y}_i)$ selected from $U \times Y$, and may be described by the zero-one matrix

$$\chi_{ji} = \begin{cases} 1 & \text{if} \quad (\overline{u}_j, \overline{y}_i) \in R \\ 0 & \text{if} \quad (\overline{u}_j, \overline{y}_i) \notin R, \quad j = 1, ..., \alpha, \quad i = 1, ..., \beta. \end{cases}$$

The sets $D_u \subset U$ and $D_y \subset Y$ may be determined by the sets of the respective indexes

$$J \subset \{1, 2, ..., \alpha\} \triangleq S_u, \quad I \subset \{1, 2, ..., \beta\} \triangleq S_y,$$

i.e.

$$\overline{u}_j \in D_u \leftrightarrow j \in J, \quad \overline{y}_j \in D_y \leftrightarrow i \in I.$$

Analysis problem: For the given matrix $[\chi_{ij}]$ and the set J find the smallest set I such that

$$j \in J \rightarrow i \in I. \tag{2.27}$$

According to (2.5)

$$I = \{i \in S_y : \bigvee_{j \in S_u} (\chi_{ji} = 1)\} \tag{2.28}$$

Decision problem: For the given matrix $[\chi_{ij}]$ and the set I required by a user, find the largest set J such that the implication (2.27) is satisfied.

According to (2.13)

$$J = \{j \in S_u : S_y(j) \subseteq I\}$$

where

$$S_y(j) = \{i \in S_y : \chi_{ji} = 1\}, \tag{2.29}$$

or

$$J = \{j \in S_u : \chi_{ji} = 1 \rightarrow i \in I\}. \tag{2.30}$$

It is worth noting that the sets (2.28) and (2.29) may be easily generated by a computer containing the matrix $[\chi_{ij}]$ as a knowledge base. For the plant with external disturbances

$$z \in Z = \{\bar{z}_1, \bar{z}_2, ..., \bar{z}_\gamma\},$$

the relation $R(u, y, z)$ may be described by the three-dimensional zero-one matrix

$$\chi_{jik} = \begin{cases} 1 & \text{if} \quad (\bar{u}_j, \bar{y}_i, \bar{z}_k) \in R \\ 0 & \quad \text{otherwise,} \end{cases}$$

$j = 1, ..., \alpha$, $i = 1, ..., \beta$, $k = 1, ..., \gamma$. The set D_z may be determined by the set of respective indexes $K \subset \{1, 2, ..., \gamma\}$, i.e. $\bar{z}_k \in D_z \leftrightarrow k \in K$.

The decision problem consists in finding the largest set J such that the implication

$$(j \in J) \wedge (k \in K) \rightarrow i \in I$$

is satisfied.

The solution is analogous to (2.17) and (2.18):

$$J = \{j \in S_u : \bigwedge_{k \in K} S_y(j,k) \subseteq I\}$$

where

$$S_y(j,k) = \{i \in S_y : \chi_{jik} = 1\}.$$

The form corresponding to (2.30) is as follows:

$$J = \{j \in S_u : \bigwedge_{k \in K} (\chi_{jik} = 1 \rightarrow i \in I)\}.$$

Remark 2.1. Note that in the discrete case it may be possible to satisfy the requirement $y = y^* \in Y$, i.e. $i = i^*$ for R which is not a function. The solution has the form

$$J = \{j \in S_u : S_y(j,k) = i^*\}.$$ □

Example 2.6.
Let $\alpha = 5$, $\beta = 6$,

$$\chi = \begin{bmatrix} 1 & 0 & 0 & 0 & 0 & 1 \\ 0 & 0 & 0 & 1 & 1 & 0 \\ 0 & 0 & 1 & 1 & 0 & 1 \\ 0 & 0 & 0 & 1 & 0 & 0 \\ 0 & 1 & 1 & 1 & 0 & 1 \end{bmatrix}$$

and the requirement is determined by $I = \{3,4,5\}$, which means that $D_y = \{\bar{y}_3, \bar{y}_4, \bar{y}_5\}$. According to (2.29)

$$S_y(1) = \{1,6\}, \quad S_y(2) = \{4,5\}, \quad S_y(3) = \{3,4,6\},$$

$$S_y(4) = \{4\}, \quad S_y(5) = \{2,3,4,6\}.$$

Then $J = \{2,4\}$, which means that the requirement is satisfied for the decisions $\bar{u}_2$ and $\bar{u}_4$. It is easy to see that for

$$\chi = \begin{bmatrix} 1 & 0 & 0 & 0 & 0 & 1 \\ 0 & 1 & 0 & 0 & 1 & 0 \\ 0 & 0 & 1 & 1 & 0 & 1 \\ 1 & 0 & 0 & 1 & 0 & 0 \\ 0 & 1 & 1 & 1 & 0 & 1 \end{bmatrix}$$

the solution does not exist. □

3 Application of Random Variables

This chapter presents an application of random variables in the analysis and decision problems for a static plant. In the parametric case, the unknown parameters in the function or in the relation describing the plant are assumed to be values of random variables with the given probability distributions. In the non-parametric case, the plant is described by the given conditional probability distribution. The foundations of random variables and probabilistic theory are presented in many books in this classical area (see e.g. [66, 73, 94]). In Sect. 1.1, a very short description of random variables is given to introduce the notation and to bring together formalisms concerning random, uncertain and fuzzy variables in a unified framework.

3.1 Random Variables and Probabilistic Forms of Knowledge Representations

A random variable $\tilde{x}$ is defined by a set of variables $X \subseteq R^k$ (multidimensional vector space) and a probability distribution. In the discrete case when $X = \{\bar{x}_1, \bar{x}_2, ..., \bar{x}_m\}$, the probability distribution is reduced to $P(\tilde{x} = \bar{x}_j)$, i.e. the probabilities that $\tilde{x} = \bar{x}_j$ for $j = 1, 2, ..., m$. In general, the probability distribution may be described by the *distribution function*

$$F(x) = P(\tilde{x} \le x)$$

where $\tilde{x} \le x$ denotes a set of inequalities for the respective components of the vectors, i.e.

$$\text{“}\tilde{x} \le x\text{”} = (\tilde{x}^{(1)} \le x^{(1)}) \wedge (\tilde{x}^{(2)} \le x^{(2)}) \wedge ... \wedge (\tilde{x}^{(k)} \le x^{(k)})$$

and $\tilde{x}^{(i)}$, $x^{(i)}$ are the i-th components of the vectors $\tilde{x}$, x, respectively. If X is a continuous set and the function $F(x)$ is differentiable with respect to the components of x then $\tilde{x}$ is called a continuous random variable and the function

$$f(x) = \frac{\partial F(x)}{\partial x^{(1)} \partial x^{(2)} ... \partial x^{(k)}}$$

is called a *probability density*. For a continuous set $D_x \subseteq X$

$$P(\widetilde{x} \in D_x) = \int_{D_x} f(x)\, dx.$$

For practical applications an *empirical* (or *statistical*) interpretation of probabilities and probability densities is very important. To describe it let us introduce a set Ω with elements ω and a function $g: \Omega \to X$, i.e. $x = g(\omega)$ where x is a vector of features characterizing the element ω. Let $(x_1, x_2, ..., x_n)$ denote a sequence of features corresponding to a sequence of elements $(\omega_1, \omega_2, ..., \omega_n)$ chosen randomly from the set Ω. A *frequency* of the event "$x = \overline{x}_j$" for the discrete case is defined as a ratio between the number of cases $x_i = \overline{x}_j$ in the sequence $(x_1, x_2, ..., x_n)$ and n. In a similar way we define a frequency of the event "$x \in D_x$" for the continuous case as a ratio between the number of cases $x_j \in D_x$ in the sequence and n. If n is sufficiently large then the first frequency is approximately equal to $P(\widetilde{x} = \overline{x}_i)$ and the second frequency is approximately equal to $P(\widetilde{x} \in D_x)$. Precisely speaking, the frequencies converge in a probabilistic sense (in this case converge with probability 1) to the respective probabilities for $n \to \infty$.

Let us consider a static plant with the input vector $u \in U = R^s$ (s-dimensional vector space) and the output vector $y \in Y = R^l$ (l-dimensional vector space), described by a relation $R(u, y)$. For the non-deterministic plant considered in Chapter 2 the relation $R(u, y)$ is not a function. Now, we shall introduce an additional description of the uncertainty in a probabilistic form. Let us assume that the points (u_n, y_n) in the successive moments are chosen randomly from $R(u, y)$. Precisely speaking, we assume that (u_n, y_n) are values of random variables $(\widetilde{u}, \widetilde{y})$. If $(\widetilde{u}, \widetilde{y})$ are continuous random variables, then they may be described by the probability distribution in the form of the *joint probability density* $f(u, y)$, which is equal to zero for $(u, y) \notin R$. An extension of this description leads to the probability density determined for $U \times Y$. Now, the knowledge representation of the plant (or shortly – knowledge of the plant KP) has the form of $f(u, y)$, i.e. $\text{KP} = \langle f(u, y) \rangle$. The joint density determines the conditional probability densities:

$$f_y(y \mid u) = \frac{f(u, y)}{\int_Y f(u, y)\, dy} = \frac{f(u, y)}{f_u(u)}, \tag{3.1}$$

$$f_u(u \mid y) = \frac{f(u, y)}{\int_U f(u, y)\, du} = \frac{f(u, y)}{f_y(y)} \tag{3.2}$$

where $f_u(u)$ and $f_y(y)$ are called *marginal probability densities*. The probability density $f_y(y \mid u)$ characterizes a set of possible outputs for the fixed input u and $f_u(u \mid y)$ characterizes a set of different inputs for which the output is equal to y. Consequently, we can consider three different forms of KP: $f(u,y)$, $f_y(y \mid u)$ and $f_u(u \mid y)$. Taking into account that in typical practical situations the relation between the input and output is a "cause–effect" relation, we can say that $\text{KP}=<f_y(y \mid u)>$ is a typical form of the knowledge representation describing the relation between the input u and the output y. It is easy to note from (3.1) that $\text{KP}=<f(u,y)>$ describes not only the plant itself, but also a set of inputs characterized by $f_u(u)$. It is also worth noting that in the description $\text{KP}=<f_y(y \mid u)>$ it is not necessary to assume that u is a value of a random variable (i.e. the probability distribution $f_u(u)$ may not exist) and in the description $\text{KP}=<f_u(u \mid y)>$ it is not necessary to assume that y is the value of a random variable.

In the discrete case we have finite sets of possible input and output vectors:

$$u \in \{\overline{u}_1, \overline{u}_2, ..., \overline{u}_\alpha\}, \qquad y \in \{\overline{y}_1, \overline{y}_2, ..., \overline{y}_\beta\}.$$

The knowledge representation K(p) analogous to $f(u,y)$ now has a form of the matrix

$$\overline{P} = [\overline{p}_{ij}]_{\substack{i=1,2,...,\beta \\ j=1,2,...,\alpha}}$$

where $\overline{p}_{ij}$ denotes the joint probability that $\widetilde{u} = \overline{u}_j$ and $\widetilde{y} = \overline{y}_i$:

$$\overline{p}_{ij} = P[(\widetilde{u} = \overline{u}_j) \wedge (\widetilde{y} = \overline{y}_i)].$$

The descriptions KP analogous to $f_y(y \mid u)$ and $f_u(u \mid y)$ have the form of the matrices of conditional probabilities:

$$P_y = [p_{ij}]_{\substack{i=1,2,...,\beta, \\ j=1,2,...,\alpha}} \qquad P_u = [q_{ij}]_{\substack{i=1,2,...,\beta, \\ j=1,2,...,\alpha}}$$

respectively, where

$$p_{ij} = P(\widetilde{y} = \overline{y}_i \mid \widetilde{u} = \overline{u}_j), \qquad q_{ij} = P(\widetilde{u} = \overline{u}_j \mid \widetilde{y} = \overline{y}_i).$$

The relationships analogous to (3.1) and (3.2) are as follows:

$$p_{ij} = \frac{\overline{p}_{ij}}{p_j}, \qquad q_{ji} = \frac{\overline{p}_{ij}}{q_i} \tag{3.3}$$

where

$$p_j = P(\widetilde{u} = \overline{u}_j), \qquad q_i = P(\widetilde{y} = \overline{y}_i). \tag{3.4}$$

The forms of KP presented above may be called *non-parametric* descriptions of the uncertain plant. In a *parametric* case the plant is described by a known function $y = \Phi(u, x)$ or a known relation $R(u, y; x) \subset U \times Y$ where $x \in X \subseteq R^k$ denotes an unknown vector parameter which is assumed to be a value of a random variable $\widetilde{x}$ described by the probability density $f_x(x)$ (or by the respective form of the probability distribution in a discrete case). In this case KP = $<\Phi(u, x), f_x(x)>$ for a functional plant and

$$\text{KP} = <R(u, y; x), f_x(x)>$$

for a relational plant. Consequently, in the parametric case we shall consider functional or relational plants with *random parameters*, i.e. with parametric uncertainty described by probability distributions.

The probabilistic description may be extended for the plant with the vector of external disturbances $z \in Z$. We can consider three different forms of KP in the non-parametric case: $f(u, y, z)$, $f_y(y \mid u, z)$ and $f_{uz}(u, z \mid y)$. In the first and the third description, z is assumed to be a value of a random variable $\widetilde{z}$. In the parametric case we have the function $y = \Phi(u, z, x)$ for the functional plant and the relation $R(u, y, z; x) \subset U \times Y \times Z$ for the relational plant.

3.2 Functional Plants with Random Parameters. Continuous Case

Let us start with analysis and decision problems for the parametric case and the functional plant described by the function $y = \Phi(u, x)$, which is assumed to be a continuous function of x.

Analysis problem: For the given KP $=< \Phi, f_x >$ and u find the probability density $f_y(y \mid u) \triangleq f_y(y; u)$ of the random variable $\widetilde{y} = \Phi(u, \widetilde{x})$.

Let us assume that Φ as a function of x is one-to-one mapping and denote by $\overline{\Phi}$ an inverse function, i.e.

$$x = \Phi^{-1}(u, y) \triangleq \overline{\Phi}(u, y)$$

or for the components of x

$$x^{(i)} = \overline{\Phi}_i(u, y), \qquad i = 1, 2, ..., k.$$

It is well known in general probability theory (see e.g. [73]) that

$$f_y(y;\, u) = f_x[\overline{\Phi}(u,y)]\,|\det J_y[\overline{\Phi}(u,y)]| \tag{3.5}$$

where $\det J_y[\overline{\Phi}(u,y)]$ denotes a determinant of the Jacobian matrix

$$\begin{bmatrix} \dfrac{\partial \overline{\Phi}_1(u,y)}{\partial y^{(1)}} & \dfrac{\partial \overline{\Phi}_1(u,y)}{\partial y^{(2)}} & \cdots & \dfrac{\partial \overline{\Phi}_1(u,y)}{\partial y^{(l)}} \\ \dfrac{\partial \overline{\Phi}_2(u,y)}{\partial y^{(1)}} & \dfrac{\partial \overline{\Phi}_2(u,y)}{\partial y^{(2)}} & \cdots & \dfrac{\partial \overline{\Phi}_2(u,y)}{\partial y^{(l)}} \\ \cdots & \cdots & \cdots & \cdots \\ \dfrac{\partial \overline{\Phi}_k(u,y)}{\partial y^{(1)}} & \dfrac{\partial \overline{\Phi}_k(u,y)}{\partial y^{(2)}} & \cdots & \dfrac{\partial \overline{\Phi}_k(u,y)}{\partial y^{(l)}} \end{bmatrix}$$

and $y^{(j)}$ for $j=1,2,...,l$ are the components of y.

Having $f_y(y;u)$ determined according to (3.5), we can determine the mean value (or expected value)

$$\mathrm{E}_y(\tilde{y};u) = \int_Y y\, f_y(y;u)\, dy \tag{3.6}$$

which may be the final result of the analysis problem-solving.

For the plant with external disturbances z described by a function $y = \Phi(u,z,x)$ the analysis problem is as follows: for the given KP = $<\Phi, f_x>$, u and z find the probability density $f_y(y\,|\,u,z) \triangleq f_y(y;u,z)$. Now, we have

$$x = \Phi^{-1}(u,y,z) \triangleq \overline{\Phi}(u,y,z)$$

and the result is the same as (3.5) with $\overline{\Phi}(u,y,z)$ in place of $\overline{\Phi}(u,y)$.

Let us formulate three versions of the decision problem for the plant with external disturbances, on the assumption that z is measured, KP = $<\Phi, f_x>$ and y^* (a desirable output) are given.

Decision problem:

Version I. To find the decision $u \triangleq u_a$ maximizing $f_y(y^*;\, u,z) \triangleq \Phi_a(u,z)$.

Version II. To find the decision $u \triangleq u_b$ such that $\mathrm{E}_y(\tilde{y};u,z) = y^*$ where

$$\mathrm{E}_y(\tilde{y};u,z) = \int_Y y\, f_y(y;u,z)\, dy \triangleq \Phi_b(u,z)\,. \tag{3.7}$$

Version III. To find the decision $u \triangleq u_c$ minimizing $\mathrm{E}_s(\widetilde{s};u,z) \triangleq \Phi_c(u,z)$ where $s = \varphi(y, y^*)$ is a quality index, e.g. $s = (y - y^*)^{\mathrm{T}}(y - y^*)$ where y and y^* are column vectors.
In version I

$$u_a = \arg \max_{u \in U} \Phi_a(u,z) \triangleq \Psi_a(z) \tag{3.8}$$

where $\Phi_a(u,z) = f_y(y^*;u,z)$ and $f_y(y;y,z)$ should be determined in the same way as in the analysis problem. The result u_a is a function of z if u_a is a unique value maximizing Φ_a for the given z.
In version II one should solve the equation

$$\Phi_b(u,z) = y^* \tag{3.9}$$

where the function Φ_b is determined by (3.7). If Equation (3.9) has a unique solution with respect to u for a given z, then as a result one obtains $u_b = \Psi_b(z)$.

In version III for the determination of $\mathrm{E}_s(\widetilde{s};u,z)$ one should find the probability density $f_s(s;u,z)$ for $\widetilde{s}$ (if it exists), using the function

$$\widetilde{s} = \varphi[\Phi(u,z,\widetilde{x}),\ y^*]$$

and the probability density $f_x(x)$. Then

$$u_c = \arg \min_{u \in U} \int_{-\infty}^{\infty} s\, f_s(s;u,z)\, ds = \arg \min_{u \in U} \Phi_c(u,z) \triangleq \Psi_c(z).$$

The problem in version III may be called a *probabilistic optimization*. It is worth noting that the determination of the expected values $\mathrm{E}_y(\widetilde{y};u,z)$ in version II and $\mathrm{E}_s(\widetilde{s};u,z)$ in version III may be based directly on the function Φ and the probability density $f_x(x)$, i.e. the determination of $f_y(y;u,z)$ and $f_s(s;u,z)$ is not necessary:

$$\mathrm{E}_y(\widetilde{y};u,z) = \int_X \Phi(u,z,x) f_x(x)\, dx,$$

$$\mathrm{E}_s(\widetilde{s};u,z) = \int_X \varphi[\Phi(u,z,x),\ y^*] f_x(x)\, dx.$$

The functions Ψ_a, Ψ_b, Ψ_c are three versions of the decision algorithm $u = \Psi(z)$ in an open-loop decision system (Fig 3.1), corresponding to three formulations of the decision problem.

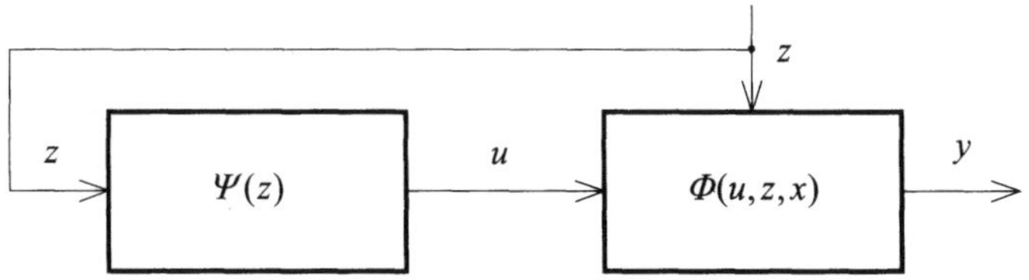

Figure 3.1. Open-loop decision system

The functions Φ_a, Φ_b, Φ_c are the results of three different ways of *determinization* of the random plant, and the functions Ψ_a, Ψ_b, Ψ_c are the respective deterministic decision algorithms based on the knowledge of the plant (KP):

$$\text{KP} = <\Phi, f_x>. \tag{3.10}$$

Assume that the equation

$$\Phi(u,z,x) = y^*$$

has a unique solution with respect to u:

$$u \triangleq \Phi_d(z,x). \tag{3.11}$$

The relationship (3.11) together with the probability density $f_x(x)$ may be considered as a *knowledge of the decision making* (KD):

$$\text{KD} = <\Phi_d, f_x> \tag{3.12}$$

obtained by using KP and y^*. The function (3.11) together with f_x may also be called a *random decision algorithm* in the open-loop decision system. The determinization of this algorithm leads to two versions of the deterministic decision algorithm Ψ_d, corresponding to versions I and II of the decision problem:

Version I.

$$u_{ad} = \arg\max_{u \in U} f_u(u;z) \triangleq \Psi_{ad}(z) \tag{3.13}$$

where the probability density $f_u(u;z)$ should be determined (if it exists) from the function (3.11) and the probability density $f_x(x)$.

Version II.

$$u_{bd} = \mathrm{E}_u(\tilde{u};z) = \int_U u\, f_u(u;z)\,du = \int_X \Phi_d(z,x)\, f_x(x)\,dx \triangleq \Psi_{bd}(z). \tag{3.14}$$

The third version corresponding to version III of the decision problem does not exist because the desirable value u^* for the output of the decision algorithm is not determined. The deterministic decision algorithms Ψ_{ad} and Ψ_{bd} are based directly on the knowledge of the decision making. Two concepts of the determination of deterministic decision algorithms corresponding to general ideas described in Chapter 2 are illustrated in Figs. 3.2 and 3.3. In the first case (Fig. 3.2) the decision algorithms $\Psi_a(z)$, $\Psi_b(z)$ and $\Psi_c(z)$ are obtained via the determinization of the knowledge of the plant KP. In the second case (Fig. 3.3) the decision algorithms are the results of the determinization of KD obtained from KP for the given y^*. The results of these two approaches may be different.

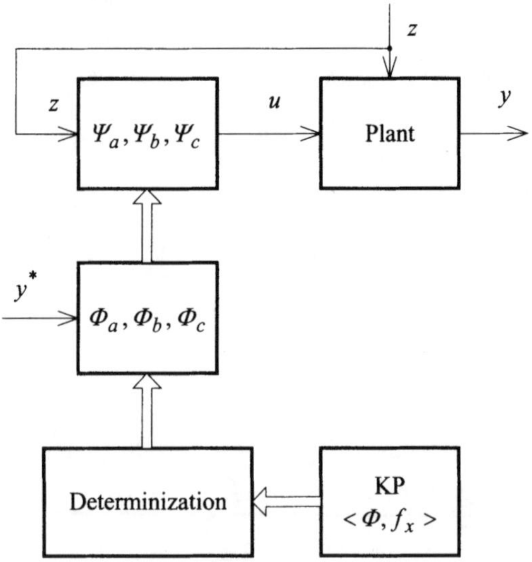

Figure 3.2. Decision system with determinization – the first case

Example 3.1.
Consider a plant with $u, y, z \in R^1$ (one-dimensional case), described by

$$y = \Phi(u,z,x) = xu + z.$$

Let us determine the decision algorithm after the determinization of the plant. Using (3.5) for the one-dimensional case we obtain

$$f_y(y^*;u,z) = f_x(\frac{y^* - z}{u})\frac{1}{|u|} = \Phi_a(u,z).$$

Then in version I, according to (3.8)

$$u_a = \arg\max_{u \in D_u} f_x(\frac{y^* - z}{u})\frac{1}{|u|}.$$

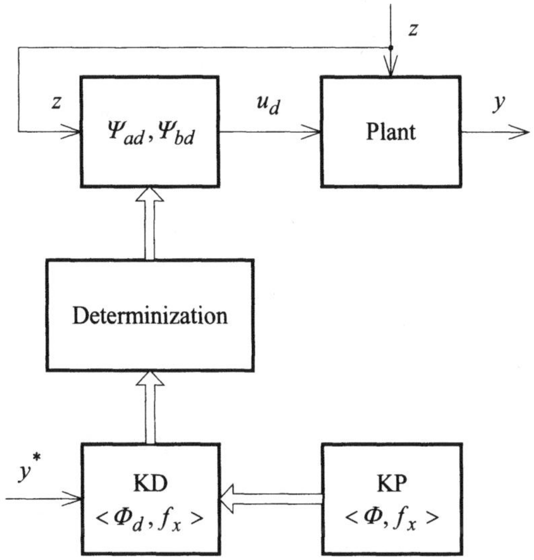

Figure 3.3. Decision system with determinization – the second case

In version II, according to (3.9), (3.7)

$$\mathrm{E}_y(\widetilde{y};u,z) = \Phi_b(u,z) = u\,E_x(\widetilde{x}) + z = y^*$$

and

$$u_b = \Psi_b(z) = \frac{y^* - z}{E(\widetilde{x})}.$$

From the equation $xu + z = y^*$ we obtain the random decision algorithm (3.11)

$$u = \Phi_d(z,x) = \frac{y^* - z}{x}.$$

By applying the determinization of this algorithm one obtains two versions of the deterministic decision algorithm $\Psi_d(z)$. In version I, according to (3.13)

$$u_{ad} = \Psi_{ad}(z) = \arg\max_{u \in D_u} f_x\left(\frac{y^* - z}{u}\right)\frac{|y^* - z|}{u^2} \neq \Psi_a(z).$$

In version II

$$u_{bd} = \Psi_{bd}(z) = \mathrm{E}_u(\widetilde{u};z) = (y^* - z)\,\mathrm{E}_x\left(\frac{1}{\widetilde{x}}\right) \neq \Psi_b(z).$$

This very simple example shows that the deterministic decision algorithms Ψ_a, Ψ_b obtained via the determinization of the random plant may differ from the corresponding deterministic decision algorithms Ψ_{ad}, Ψ_{bd} obtained as a result of the determinization of the random decision algorithm. □

Example 3.2.

Consider a plant with s inputs and one output described by

$$y = x^{\mathrm{T}} u$$

where x and u are column vectors. Let us determine the decision u_c minimizing

$$\mathrm{E}[(\widetilde{y} - y^*)^2] = \mathrm{E}[(\widetilde{x}^{\mathrm{T}} u - y^*)^2] \stackrel{\Delta}{=} Q(u).$$

It is easy to see that

$$Q(u) = u^{\mathrm{T}} \mathrm{E}(\widetilde{x} \cdot \widetilde{x}^{\mathrm{T}}) u - 2y^* u^{\mathrm{T}} \mathrm{E}(\widetilde{x}) + (y^*)^2$$

and

$$\operatorname*{grad}_u Q(u) = 2\mathrm{E}(\widetilde{x} \cdot \widetilde{x}^{\mathrm{T}}) u - 2y^* \mathrm{E}(\widetilde{x})$$

where *grad* denotes the gradient of the function $Q(u)$. From the equation

$$\operatorname*{grad}_u Q(u) = \overline{0}$$

where $\overline{0}$ denotes a vector with zero components, we obtain

$$u_c = [\mathrm{E}(\widetilde{x} \cdot \widetilde{x}^{\mathrm{T}})]^{-1} \mathrm{E}(\widetilde{x}) y^* = [\mathrm{E}(\widetilde{x}) \mathrm{E}(\widetilde{x}^{\mathrm{T}}) + M_{xx}]^{-1} \mathrm{E}(\widetilde{x}) y^*$$

where M_{xx} denotes the covariance matrix of the vector $\widetilde{x}$, i.e.

$$M_{xx} = \begin{bmatrix} \sigma_1^2 & \mu_{12} & \mu_{13} & \cdots & \mu_{1s} \\ \mu_{21} & \sigma_2^2 & \mu_{23} & \cdots & \mu_{2s} \\ \cdots & \cdots & \cdots & \cdots & \cdots \\ \mu_{s1} & \mu_{s2} & \mu_{s3} & \cdots & \sigma_s^2 \end{bmatrix}$$

where

$$\sigma_i^2 = \mathrm{E}(\widetilde{x}^{(i)} - \underline{x}^{(i)})^2, \qquad \mu_{ij} = \mu_{ji} = \mathrm{E}[(\widetilde{x}^{(i)} - \underline{x}^{(i)})(\widetilde{x}^{(j)} - \underline{x}^{(j)})],$$

$\underline{x}^{(i)} = \mathrm{E}(\widetilde{x}^{(i)})$ and $x^{(i)}$ denotes the i-th component of the vector x.

In this case, the probability density $f_x(x)$ is not necessary to solve the probabilistic optimization problem. It is sufficient to know $\mathrm{E}(\widetilde{x})$ and M_{xx}. □

3.3 Functional Plants with Random Parameters. Discrete Case

In the discrete case

$$x \in X = \{\bar{x}_1, \bar{x}_2, ..., \bar{x}_m\}$$

and the probability distribution is determined by probabilities

$$P(\tilde{x} = \bar{x}_j) \triangleq p_{xj}, \quad j = 1, 2, ..., m.$$

Consequently, $\tilde{y}$ is a discrete random variable and if for the fixed u, z the function $\Phi(u,z,x)$ as a function of x is one-to-one mapping then

$$y \in Y(u,z) = \{\bar{y}_1(u,z), \bar{y}_2(u,z), ..., \bar{y}_m(u,z)\}$$

where

$$\bar{y}_j(u,z) = \Phi(u,z,\bar{x}_j).$$

In this case

$$P[\tilde{y} = \bar{y}_j(u,z)] \triangleq p_{yj} = P(\tilde{x} = \bar{x}_j) = p_{xj}.$$

In general, let us consider a set

$$S_i = \{j \colon \Phi(u,z,\bar{x}_j) = \hat{y}_i(u,z)\}, \quad i = 1, 2, ..., v$$

where $\hat{y}_i(u,z) \in Y(u,z)$, if $i \neq l$ then $\hat{y}_i(u,z) \neq \hat{y}_l(u,z)$, and v is a number of different values in the set $Y(u,z)$. Then

$$P[\tilde{y} = \hat{y}_i(u,z)] \triangleq \hat{p}_{yi} = \sum_{j \in S_i} p_{xj}. \tag{3.15}$$

The analysis problem consists in the determination of the probability distribution for $\tilde{y}$ in the form (3.15) and the expected value

$$\mathrm{E}_y(\tilde{y}; u, z) = \sum_{j=1}^{m} \bar{y}_j(u,z) p_{yj} = \sum_{i=1}^{v} \hat{y}_i(u,z) \hat{p}_{yi}. \tag{3.16}$$

Decision problem:

Version I. Assume that for each $x \in \{\bar{x}_1, \bar{x}_2, ..., \bar{x}_m\}$ and for every $z \in Z$, there exists u such that

$$\Phi(u,z,x) = y^*.$$

Then, the decision problem, analogous to that in version I for the continuous case, consists in finding the decision $u \triangleq u_a = \Psi_a(z)$ such that

$$\hat{y}_q(u_a, z) = y^*$$

and $\hat{u}_q(u_a, z)$ is the most probable value of $\hat{u}_i(u, z)$, i.e.

$$q = \arg\max_i \hat{p}_{yi}.$$

In the case when Φ is one-to-one mapping with respect to x, $u = u_a = \Psi_a(z)$ is a solution of the equation

$$\Phi(u, z, \bar{x}_\gamma) = y^*$$

where

$$\gamma = \arg\max_j p_{xj}.$$

Version II. The decision problem consists in finding the decision $u = u_b$ such that

$$\mathrm{E}_y(\hat{y}; u, z) = y^* \tag{3.17}$$

where $\mathrm{E}_y(y^*; u, z)$ is determined by (3.16) or directly by using p_{xj}:

$$\mathrm{E}_y(\tilde{y}; u, z) = \sum_{j=1}^{m} \Phi(u, z, \bar{x}_j) p_{xj}.$$

If Equation (3.17) has a unique solution with respect to u for a given z then as a result one obtains the deterministic decision algorithm $u_b = \Psi_b(z)$.

Version III. For the given form of a quality index $s = \varphi(y, y^*)$ considered in the previous section one should find the decision $u \triangleq u_c$ minimizing the expected value

$$\mathrm{E}_s(\tilde{s}; u, z) = \sum_{j=1}^{m} \varphi[\Phi(u, z, \bar{x}_j), y^*] p_{xj}.$$

Then

$$u_c = \arg\min_{u \in U} \mathrm{E}_s(\tilde{s}; u, z) \triangleq \Psi_c(z). \tag{3.18}$$

The result u_c is a function of z if u_c is a unique value minimizing E_s for the given z.

The functions Ψ_a, Ψ_b, Ψ_c are the deterministic decision algorithms obtained via the determinization of the knowledge of the random plant

$$\text{KP} = <\Phi; p_{x1}, p_{x2}, ..., p_{xm}>. \tag{3.19}$$

The function Φ_d (3.11) and the probability distribution of $\tilde{x}$ form the random decision algorithm or the knowledge of the decision making in our case:

$$\text{KD} = <\Phi_d; p_{x1}, p_{x2}, ..., p_{xm}>. \tag{3.20}$$

The determinization of KD leads to two versions of the deterministic decision algorithms analogous to $\Psi_{ad}(z)$ and $\Psi_{bd}(z)$ in the previous section:

Version I consists in finding the most probable value $u = u_{ad}$ for the given z. Denote by $\hat{u}_1(z), \hat{u}_2(z), ..., \hat{u}_\mu(z)$ all possible values u corresponding to the values $\overline{x}_1, \overline{x}_2, ..., \overline{x}_m$ and introduce the set

$$S_{ui} = \{j: \Phi_d(z, \overline{x}_j) = \hat{u}_i(z)\}, \quad i = 1,2,..., \mu.$$

Then

$$P[\tilde{u} = \hat{u}_i(z)] \stackrel{\Delta}{=} \hat{p}_{ui} = \sum_{j \in S_{ui}} p_{xj}$$

and

$$u_{ad} = \hat{u}_q(z) \stackrel{\Delta}{=} \Psi_{ad}(z)$$

where

$$q = \arg\max_i \hat{p}_{ui}.$$

Version II.

$$u_{bd} = \text{E}_u(\tilde{u}; z) = \sum_{j=1}^{m} \Phi_d(z, \overline{x}_j) p_{xj}.$$

Now the deterministic decision algorithms Ψ_{ad} and Ψ_{bd} are based directly on the knowledge of the decision making KD obtained from KP for the given y^*, and are the results of two different determinizations of KD.

3.4 Empirical Interpretations

Proper empirical interpretations of the probabilistic data and the results of the decision problems under consideration are very important from a practical point of view. The assumption that the unknown parameter x is a value of a random variable $\widetilde{x}$ means that the value x in our plant has been chosen randomly from the set of values described by the probability distribution f_x or p_{xj}. In other words, we may say that the plant under consideration has been chosen randomly from a set of plants with different values x, described by the underlying statistics.

The result $u = u_a$ in version I is a decision maximizing the probability that the output y is equal to the given value y^* in the discrete case, or maximizing the value of the probability density of the output for $y = y^*$ in the continuous cases. It means that u_a is the decision maximizing the frequency of the event " $y = y^*$ " for a large number of random choosing in the discrete case, or maximizing the frequency of the event " $y \in D_y$ " where D_y is a small neighbourhood of y^* in the continuous case.

For the decision $u = u_b$ in version II, the mean value of the outputs for a large number of random choices is approximately equal to the desirable value y^*. In version III (probabilistic optimization), by applying the decision $u = u_c$ we obtain the minimal mean value of the different values of the quality index s for the different values of x and a large number of random choices.

The frequencies or the mean values in the interpretations described above may concern the different values of y for particular plants in a set of plants, with different values of x chosen randomly. Consequently, the results of the decisions are concerned with the set of plants and it is not possible to evaluate the results of the decisions u_a, u_b, u_c for one particular plant under consideration. Precisely speaking, it is not possible to evaluate the results a priori, i.e. before applying these decisions and observing the outputs. In particular, the decision u_c, optimal in a probabilistic sense for the set of plants, may be very poor for one plant for which the value of the quality index s may be large and unacceptable. We can apply another interpretation of u_a, u_b, u_c when x is a time-varying parameter in the particular plant under consideration, and z is constant, i.e. the plant is described by the equation

$$y_n = \Phi(u_n, z, x_n), \quad n = 1, 2, ...$$

where n denotes the discrete time (indexes of successive moments). Now, we assume that x_n is a value of a random variable $\widetilde{x}_n$, and the random variables $\widetilde{x}_n$ for different n have the same probability distribution f_x or p_{xj}. It means that the values x_n in the successive moments are chosen randomly from the same set described by f_x or p_{xj}. The frequencies and mean values concern different moments in the plant under consideration and consequently, the results of the decisions u_a, u_b, u_c are concerned with the moments in this plant, e.g. the decision u_c assures the minimal mean value of the quality index for a long time interval (a large number of moments). The mean value for a set of plants in the former interpretation is now replaced by the mean value for a time interval. It is not possible to evaluate *a priori* the decisions u_a, u_b, u_c for one particular moment.

The situation is more complicated for the plant with time-varying disturbances z_n, described by

$$y_n = \Phi(u_n, z_n, x_n). \tag{3.21}$$

Assume that z_n is a value of a random variable $\tilde{z}_n$ and the random variables $\tilde{z}_n$ for the different n have the same probability distribution $f_z(z)$. This means that the values z_n in the successive moments are chosen randomly from the same set described by f_z. Assume also that $\tilde{x}_n$ and $\tilde{z}_n$ are independent random variables for every n, i.e. $f(x,z) = f_x(x) f_z(z)$. Then in versions I, II, III we have the conditional probability density for $\tilde{y}$, and the conditional expected values for $\tilde{y}$ and $\tilde{s}$, respectively:

$$f_y(y^*; u, z) = f_y(y^* \mid z; u),$$

$$\mathrm{E}_y(\tilde{y}; u, z) = \mathrm{E}_y(\tilde{y} \mid z; u),$$

$$\mathrm{E}_s(\tilde{s}; u, z) = \mathrm{E}_s(\tilde{s} \mid z; u).$$

Using $f_z(z)$ we may introduce unconditional formulations:

$$f_y(y^*; u) = \int_Z f_y(y^* \mid z; u) f_z(z) dz, \tag{3.22}$$

$$\mathrm{E}_y(\tilde{y}; u) = \int_Z \mathrm{E}_y(\tilde{y} \mid z; u) f_z(z) dz, \tag{3.23}$$

$$\mathrm{E}_s(\tilde{s}; u) = \int_Z \mathrm{E}_s(\tilde{s} \mid z; u) f_z(z) dz. \tag{3.24}$$

It is easy to note that finding the algorithm $u = \Psi(z)$ maximizing f_y defined by (3.22) or minimizing E_s defined by (3.24) may be reduced to the maximization of $f_y(y^* \mid z; u)$ with respect to u for the given z, or the minimization of $\mathrm{E}_s(\tilde{s} \mid z; u)$ with respect to u for the given z, respectively. Similarly, finding the algorithm $u = \Psi(z)$ such that E_y defined by (3.23) is equal to y^* may be reduced to solving the equation

$$\mathrm{E}_y(\widetilde{y} \mid z; u) = y^* \tag{3.25}$$

with respect to u for the given z.

Consequently, the interpretations of u_a, u_b, u_c for z_n are the same as for $z = const$, i.e. the frequencies and mean values concern the set of successive values y_n, described by (3.21). Note that knowledge of $f_z(z)$ is not needed. It is sufficient to assume that z_n is changing randomly, i.e. that $f_z(z)$ exists.

Remark 3.1. It is important to note that the results u_{ad} and u_{bd} obtained via the determinization of KD have no empirical interpretations analogous to those for u_a and u_b, i.e. the interpretations concerning the output y for which a user's requirement is formulated. □

3.5 Relational Plants with Random Parameters

Let us consider a plant described by a relation

$$R(u, y; x) \subset U \times Y \tag{3.26}$$

where x denotes an unknown vector parameter which is assumed to be a value of a random variable $\widetilde{x}$ described by the probability density $f_x(x)$ (see Sect. 3.1). Now, we have the case of *two-level uncertainty* or second-order uncertainty. The first level denotes the uncertainty concerning the non-deterministic plant, described by a relation which is not a function. The second level denotes the expert's uncertainty, and is described by the probability distribution $f_x(x)$. The probabilistic description of the expert's uncertainty means that the parameter x has been chosen randomly from the set of values X described by the probability density $f_x(x)$, or that the plant has been chosen randomly from a set of plants with different values x in the relation R. The expert does not know the value x in the plant but knows the probabilistic description characterizing the set under consideration. The relational plant described by (3.26) with a random parameter x may be called a *relational and random plant.*

If the relation R is reduced to the function $y = \Phi(u, x)$ then, for the fixed u, the output y is the value of a random variable $\widetilde{y} = \Phi(u, \widetilde{x})$, and the analysis problem may consist in the determination of the probability density $f_y(y; u)$ for the fixed u (see Sect. 3.2). If R is not a function, such a formulation of the analysis problem is not possible. Now, for the fixed u it is possible to determine a set of possible outputs (see Sect. 2.2)

$$D_y(u; x) = \{y \in Y : (u, y) \in R(u, y; x)\}$$

i.e. a *random set* $D_y(u; \widetilde{x})$. Then the analysis problem may consist in the determination for the given y the probability that $y \in D_y(u;\tilde{x})$, or shortly speaking – the probability that y is a possible output. The problem may be extended to a set of output values $D_y \subset Y$, given by a user.

Analysis problem: For the given $\text{KP} = \langle R(u,y;x), f_x(x) \rangle$, u and D_y find the probability

$$P[D_y \subseteq D_y(u;\tilde{x})] \triangleq p(D_y,u), \tag{3.27}$$

i.e. the probability that every y belonging to the set D_y may occur at the output (may be a possible output).

Note that

$$P[D_y \subseteq D_y(u;\tilde{x})] = P[\tilde{x} \in D_x(D_y,u)]$$

where

$$D_x(D_y,u) = \{x \in X : D_y \subseteq D_y(u;x)\}.$$

Then

$$p(D_y,u) = \int\limits_{D_x(D_y,u)} f_x(x)dx\,. \tag{3.28}$$

In particular, for $D_y = \{y\}$ (a singleton), we obtain the following probability that y is a possible output

$$p(y,u) = \int\limits_{D_x(y,u)} f_x(x)dx$$

where

$$D_x(y,u) = \{x \in X : y \in D_y(u;x)\}.$$

Decision problem: For $\text{KP} = \langle R(u,y;x), f_x(x) \rangle$ given by an expert and the set $D_y \subset Y$ given by a user, one should determine the optimal decision u^* maximizing the probability

$$P[D_y(u;\tilde{x}) \subseteq D_y] \triangleq p(u)\,. \tag{3.29}$$

It is one of the possible formulations of the decision making problem, consisting in the determination of the decision u^* maximizing the probability that the set of possible outputs belongs to D_y, i.e. that it is not possible to obtain an output value

y not belonging to D_y.

Since

$$P[D_y(u;\widetilde{x}) \subseteq D_y] = P[\widetilde{x} \in D_x(D_y,u)]$$

where

$$D_x(D_y,u) = \{x \in X : D_y(u;x) \subseteq D_y\},$$

then

$$u^* = \arg\max_{u \in D_u} \int_{D_x(D_y,u)} f_x(x)dx.$$

The considerations may be easily extended for the plant with the vector of external disturbances z, described by the relation

$$R(u,y,z;x) \subset U \times Y \times Z.$$

The analysis problem consists in finding the probability

$$P[D_y \subseteq D_y(u,z;\widetilde{x})] \stackrel{\Delta}{=} p(D_y,u,z) \tag{3.30}$$

where $D_y(u,z;x)$ is the set of possible outputs for the fixed u and z:

$$D_y(u,z;x) = \{y \in Y : (u,y,z) \in R(u,y,z;x)\}. \tag{3.31}$$

The result is analogous to (3.28):

$$p(D_y,u,z) = \int_{D_x(D_y,u,z)} f_x(x)dx \tag{3.32}$$

where

$$D_x(D_y,u,z) = \{x \in X : D_y \in D_y(u,z;x)\}.$$

The decision problem consists in finding for the given z the optimal decision u^* maximizing the probability

$$P[D_y(u,z;\widetilde{x}) \subseteq D_y] \stackrel{\Delta}{=} p(u,z). \tag{3.33}$$

Since

$$p(u,z) = P[\widetilde{x} \in D_x(D_y,u,z)]$$

where

$$D_x(D_y,u,z) = \{x \in X : D_y(u,z;x) \subseteq D_y\},$$

then

$$u^* = \arg\max_{u \in D_u} \int_{D_x(D_y,u,z)} f_x(x)dx \triangleq \Psi(z) . \tag{3.34}$$

The result u^* is a function of z if u^* is a unique value maximizing $p(u,z)$ for the given z. In this case we obtain the deterministic decision algorithm $u^* = \Psi(z)$ in an open-loop decision system (Fig. 3.4).

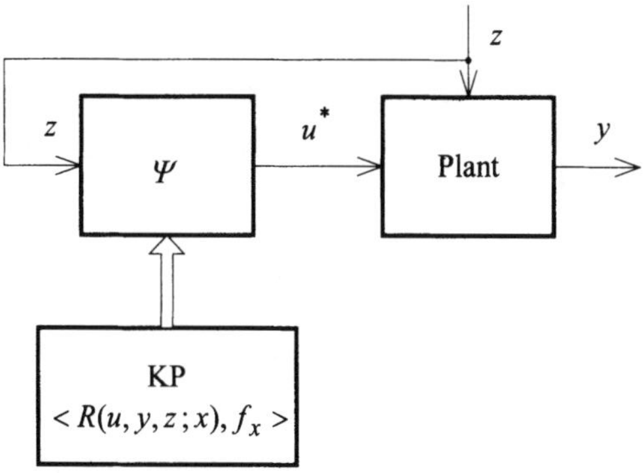

Figure 3.4. Open-loop knowledge-based decision system

Example 3.3.

Let us determine the optimal decision u^* for one-dimensional plant and the following data:

$$xu \le y \le 2xu , \quad u \ge 0 , \quad D_y = [y_1, y_2] , \quad 0 < y_1 < \frac{y_2}{2}$$

$$f_x(x) = \begin{cases} \lambda e^{-\lambda x} & \text{for} \quad x \ge 0 \\ 0 & \text{for} \quad x \le 0 . \end{cases}$$

In this case, the set $D_x(D_y,u)$ is determined by the inequality

$$\frac{y_1}{u} \le x \le \frac{y_2}{2u} .$$

Then

$$p(u) = \int_{\frac{y_1}{u}}^{\frac{y_2}{u}} f_x(x)dx = \exp(-\lambda \frac{y_1}{u}) - \exp(-\lambda \frac{y_2}{2u}) .$$

From the equation $\frac{dp(u)}{du} = 0$ we obtain

$$u^* = \arg\max_u p(u) = \frac{\lambda(\frac{y_2}{2} - y_1)}{\ln y_2 - \ln 2y_1}. \qquad \square$$

3.6 Determinization

The function $p(u,z)$ in (3.33) may be considered as a result of the determinization of the relational and random plant under consideration. For the fixed x and z we can determine the largest set $D_u(z;x) \subset U$ such that the implication

$$u \in D_u(z;x) \rightarrow y \in D_y$$

is satisfied. The set $D_u(z)$ of all possible decisions such that $y \in D_y$ for the plant described by $R(u,y,z)$ has been introduced in Sect. 2.3 and has been called a relational knowledge of the decision making KD, or a relational decision algorithm. Now the set of all possible decisions $D_u(z;x)$ depends on x and

$$\text{KD} = < D_u(z;x), f_x > \tag{3.35}$$

may be called a *relational and random decision algorithm*. Having KD in the form (3.35) we can formulate the decision problem in the following way: find the decision u_d^* maximizing

$$P[u \in D_u(z;\tilde{x})] \triangleq p_d(u,z). \tag{3.36}$$

Since

$$p_d(u,z) = P[\tilde{x} \in D_{xd}(u,z)]$$

where

$$D_{xd}(u,z) = \{x \in X : u \in D_u(z;x)\}, \tag{3.37}$$

then

$$u_d^* = \arg\max_{u \in D_u} \int_{D_{xd}(u,z)} f(x)dx \triangleq \Psi_d(z). \tag{3.38}$$

The function $p_d(u,z)$ in (3.36) may be considered as a result of the determinization of the relational and random decision algorithm. Two concepts of the determination of the deterministic decision algorithm, analogous to those described in Sect. 3.2, are illustrated in Figs. 3.5 and 3.6. In the first case (Fig. 3.5) the decision algorithm $\Psi(z)$ is obtained via the determinization of the knowledge of the plant KP. In the second case (Fig. 3.6) the decision algorithm $\Psi_d(z)$ is obtained via the determinization of KD obtained from KP for the given D_y.

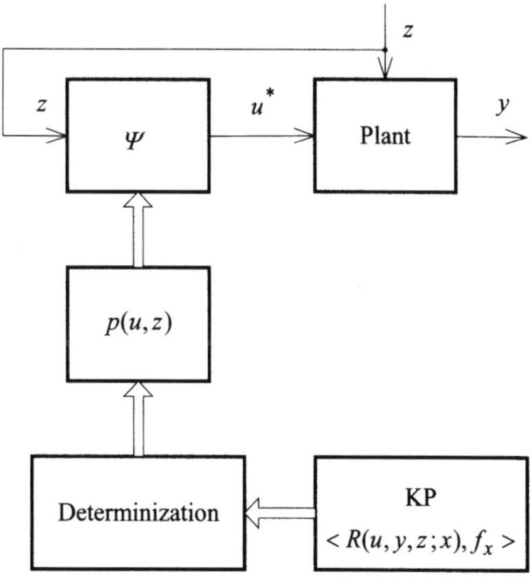

Figure 3.5. Decision system with determinization for relational plant – the first case

Theorem 3.1. Two ways of the determinization described above are equivalent, i.e.

$$\bigwedge_{z \in Z} [\Psi(z) = \Psi_d(z)]. \tag{3.39}$$

Proof: According to the definition of $D_u(z;x)$

$$u \in D_u(z;x) \to y \in D_y \quad \text{and} \quad u \notin D_u(z;x) \to y \notin D_y.$$

Then

$$D_u(z;x) = \{u \in U : D_y(u,z;x) \subseteq D_y\}$$

where $D_y(u,z;x)$ is determined by (3.31). Thus, the properties $u \in D_u(z;x)$ and $D_y(u,z;x) \subseteq D_y$ are equivalent, i.e.

$$\bigwedge_{z \in Z} [u \in D_u(z;x) \leftrightarrow D_y(u,z;x) \subseteq D_y].$$

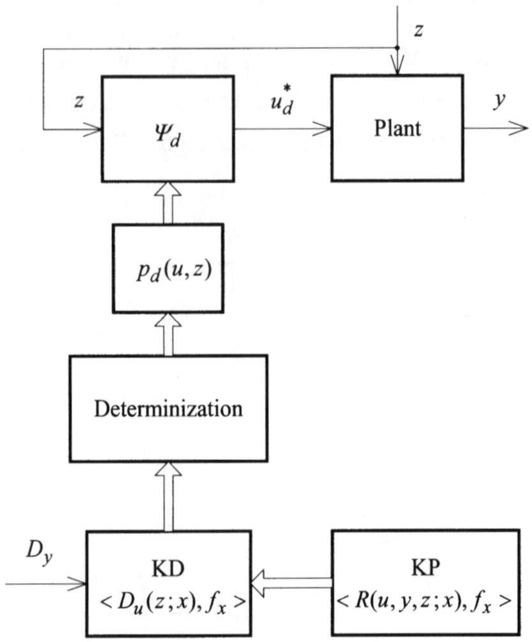

Figure 3.6. Decision system with determinization for relational plant – the second case

Consequently, according to (3.33) and (3.36)

$$\bigwedge_{z \in Z} [\, p(u,z) = p_d(u,z)]$$

and the property (3.39) is satisfied. □

We can consider other approaches to the determinization of KP or KD under consideration, consisting in two successive determinizations: the determinization at the relational level analogous to that described in Sect. 2.4 and the determinization concerning uncertainty described by the probability distribution of $\widetilde{x}$ at the second level.

A. Determinization for the given D_y

In this case we may use the mean value u_M in the determinization of the relational and random decision algorithm $< D_u(z;x), f_x > \stackrel{\Delta}{=} \text{KD}_\text{I}$ at the first level:

$$u_M = \int\limits_{D_u(z;x)} u\, du \cdot [\int\limits_{D_u(z;x)} du\,]^{-1} \stackrel{\Delta}{=} \Phi_d(z,x)\,. \tag{3.40}$$

As a result we obtain the random decision algorithm $< \Phi_d(z;x), f_x > \stackrel{\Delta}{=} \text{KD}_{\text{II}}$ (see

(3.11) and (3.12) in Sect. 3.2). The determinization at the second level presented in (3.13) and (3.14) gives the deterministic decision algorithms Ψ_{ad} and Ψ_{bd}. The idea is illustrated in Fig. 3.7 where the determinizations at the first and second levels are denoted by I and II, respectively.

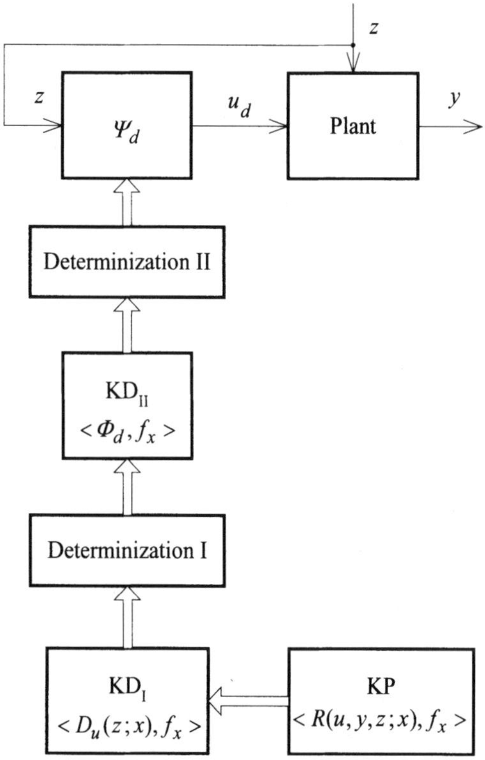

Figure 3.7. Decision system with the determinization at two levels

B. Determinization for the given y^*

Using the mean value at the first level, we obtain

$$y_M = \int_{D_y(u,z;x)} y\, dy \cdot [\int_{D_y(u,z;x)} dy\,]^{-1} \triangleq \Phi(u,z,x).$$

Thus, the relational plant has been replaced by the functional plant described by Φ. Then, at the second level it is possible to apply two concepts of the determination of decision algorithms Ψ, presented in Sect. 3.2. In the first case (Fig. 3.8) the algorithms $\Psi_a(z)$, $\Psi_b(z)$, $\Psi_c(z)$ are obtained via the determinization of $\text{KP}_\text{I} = < \Phi, f_x >$ at the second level. In the second case (Fig.3.9) the algorithms $\Psi_{ad}(z), \Psi_{bd}(z)$ are the results of the determinization of $\text{KD}_\text{I} = < \Phi_d(z,x), f_x >$.

The function $u = \Phi_d(z,x)$ is obtained as a solution of the equation

$$\Phi(u,z,x) = y^*,$$

on the assumption that the unique solution exists (see (3.11)).

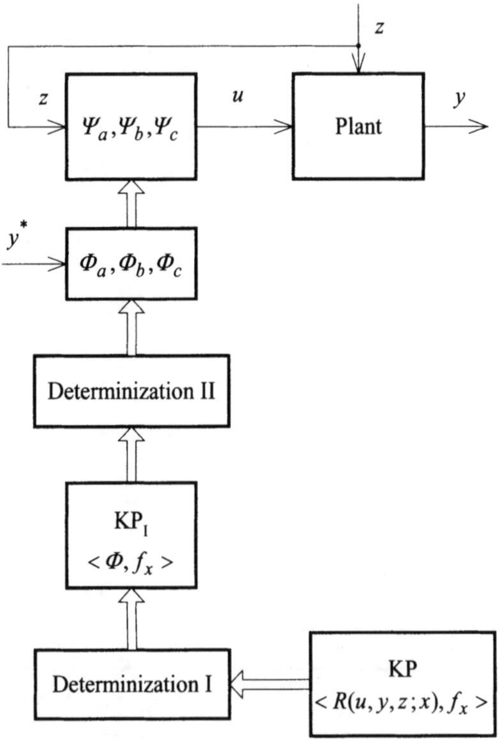

Figure 3.8. Decision system with two-level determinization for the given y^* – the first case

Remark 3.2. It is worth noting that we may obtain different decision algorithms Ψ using the different ways of the determinization described above. It is also important to note that the results Ψ_{ad} and Ψ_{bd} in points A and B have no evident empirical interpretation. □

Example 3.4.

Let us determine the optimal decision u_d^* for a one-dimensional plant and the following data:

$$\frac{u}{x} + z \le y \le \frac{2u}{x} + z, \quad u \ge 0, \quad D_y = [y_1, y_2],$$

$$f_x(x) = \begin{cases} \lambda e^{-\lambda x} & \text{for} \quad x \geq 0 \\ 0 & \text{for} \quad x \leq 0 \,. \end{cases}$$

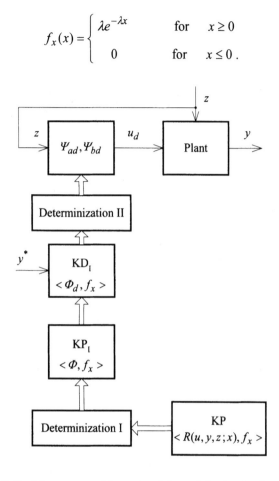

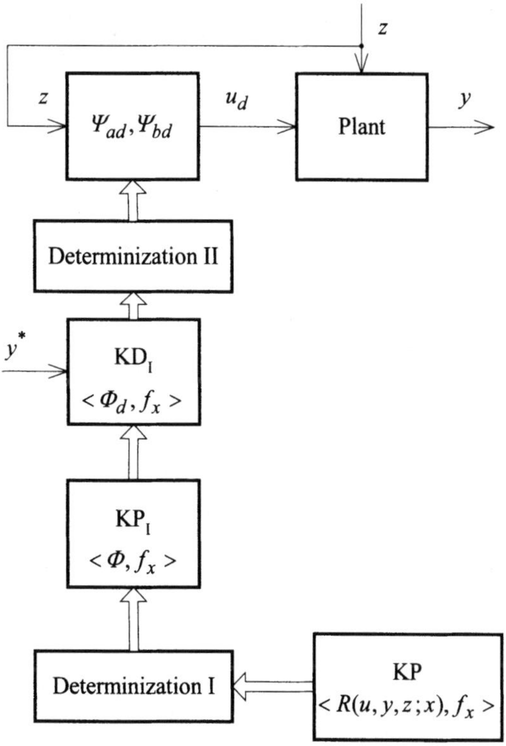

Figure 3.9. Decision system with two-level determinization for the given y^* – the second case

In this case the set $D_{xd}(u,z)$ defined by (3.37) is determined by the inequality

$$\frac{2u}{y_2 - z} \leq x \leq \frac{u}{y_1 - z} \,. \tag{3.41}$$

Then

$$p_d(u,z) = \lambda \int_{\frac{2u}{y_2 - z}}^{\frac{u}{y_1 - z}} e^{-\lambda x} dx = \exp(-\lambda \frac{2u}{y_2 - z}) - \exp(-\lambda \frac{u}{y_1 - z}) \,.$$

From the equation

$$\frac{dp_d(u,z)}{du} = 0$$

we obtain the deterministic decision algorithm

$$u_d^* = \Psi_d(z) = \frac{(y_1 - z)(y_2 - z)\{\ln(y_2 - z) - \ln[2(y_1 - z)]\}}{\lambda(y_2 - 2y_1 + z)}.$$

It is easy to see that, according to Theorem 3.1,

$$u^* = \arg\max_u P[D_y(u,z;\tilde{x}) \subseteq D_y] = u^*.$$

From (3.41) we have the following condition of the existence of the solution, i.e. the existence of u for which $p_d(u,z) \neq 0$:

$$0 \le \frac{2}{y_2 - z} \le \frac{1}{y_1 - z}.$$

□

Example 3.5.
For the plant considered in Example 3.4 let us apply the way described in point A. In our case the set $D_u(z;x)$ is described by the inequality

$$(y_1 - z)x \le u \le \frac{1}{2}(y_2 - z)x.$$

Then

$$u_M = \frac{1}{2}[(y_1 - z)x + \frac{1}{2}(y_2 - z)x] = \frac{2(y_1 - z) + (y_2 - z)}{4}x = \Phi_d(z,x).$$

Using the determinization of $\Phi_d(z,x)$ in the form presented in (3.14), it is easy to obtain

$$u_{bd} = \frac{2(y_1 - z) + (y_2 - z)}{4}\int_0^\infty \lambda x e^{-\lambda x} dx = \frac{2(y_1 - z) + (y_2 - z)}{4\lambda} = \Psi_{bd}(z) \neq \Psi_d(z).$$

□

3.7 Non-parametric Uncertainty. Continuous Case

Let us consider a plant with input vector u and output vector y, described by KP $= < f_y(y \mid u) >$ (see Sect. 3.1). If the value u is chosen randomly according to the known probability density $f_u(u)$, the analysis problem may consist in the determination of $f_y(y)$ characterizing the set of values y. It corresponds to the analysis problem considered in Sect. 2.2 in which one should find D_y for the given $R(u,y)$ and D_y. Now $R(u,y)$, D_u and D_y are replaced by $f_y(y\,|\,u)$, $f_u(u)$ and $f_y(y)$, respectively.

Analysis problem: For the given $\mathrm{KP} = < f_y(y\,|\,u) >$ and $f_u(u)$ find the probability density $f_y(y)$.

According to (3.1)

$$f(u,y) = f_u(u) f_y(y\,|\,u)$$

and

$$f_y(y) = \int\limits_U f(u,y)du\,.$$

Then

$$f_y(y) = \int\limits_U f_u(u) f_y(y\,|\,u)du\,. \tag{3.42}$$

The decision problem is an inverse problem consisting in finding $f_u(u)$ for a desirable probability distribution $f_y(y)$ given by a user. The user's requirement in the form $y \in D_y$ introduced in Sect. 2.2 is now replaced by $f_y(y)$, i.e. the requirement is greater, expressed more precisely.

Decision problem: For the given $\mathrm{KP} = < f_y(y\,|\,u) >$ and $f_y(y)$ find the probability density $f_u(u)$.

To find the solution one should solve Equation (3.42) with respect to the function $f_u(u)$ satisfying the conditions for a probability density, i.e.

$$\bigwedge_{u \in U} f_u(u) \geq 0\,, \qquad \int\limits_U f_u(u)du = 1\,.$$

The probability distribution $f_u(u)$ may be called a *random decision*. A final particular decision u may be chosen randomly from U according to $f_u(u)$, which requires a generator of random numbers. Another way consists in applying a determinization and using the value u_a maximizing $f_u(u)$ or the mean value $u_b = \mathrm{E}(\tilde{u})$.

The analysis and decision problems may be easily extended to the plant with the vector of external disturbances $z \in Z$, on the assumption that z is a value of a random variable $\tilde{z}$ described by a probability density $f_z(z)$. The equation analogous to (3.42) is now as follows:

$$f_y(y) = \int\limits_U \int\limits_Z f_z(z) f_u(u \mid z) f_y(y \mid u,z) du dz \tag{3.43}$$

and the analysis problem consists in finding $f_y(y)$ according to (3.43), for the given $f_y(y \mid u,z)$, $f_u(u \mid z)$ and $f_z(z)$.

Decision problem: For the given $\mathrm{KP} = < f_y(y | u,z) >$ and $f_y(y)$ required by a user one should determine $f_u(u | z)$.
The determination of $f_u(u | z)$ may be decomposed into two steps. In the first step one should find the function $f_{uz}(u,z)$ satisfying the equation

$$f_y(y) = \int\limits_U \int\limits_Z f_{uz}(u,z) f_y(y | u,z) du \, dz$$

and the conditions for a probability density:

$$\bigwedge_{u \in U} \bigwedge_{z \in Z} f_{uz}(u,z) \geq 0, \qquad \int\limits_U \int\limits_Z f_{uz}(u,z) du \, dz = 1 .$$

In the second step, one should determine the function $f_u(u | z)$:

$$f_u(u | z) = \frac{f_{uz}(u,z)}{\int\limits_U f_{uz}(u,z) du} .$$

The function $f_u(u \mid z)$ may be considered as the knowledge of the decision making $\mathrm{KD} = < f_u(u \mid z) >$ or a *random decision algorithm* (the description of a random controller in the open-loop control system). The determination of $f_u(u \mid z)$ may be connected with great computational difficulties. In having $f_u(u \mid z)$, one can obtain the deterministic decision algorithm $\Psi(z)$ as a result of the determinization of the random decision algorithm $f_u(u \mid z)$. Two versions corresponding to versions I and II in Sect. 3.2 are the following:
Version I

$$u_a = \arg\max_{u \in U} f_u(u \mid z) \stackrel{\Delta}{=} \Psi_a(z) .$$

Version II

$$u_b = \mathrm{E}_u(\widetilde{u} \mid z) = \int\limits_U u \, f_u(u \mid z) du \stackrel{\Delta}{=} \Psi_b(z) .$$

The deterministic decision algorithms $\Psi_a(z)$ or $\Psi_b(z)$ are based on the knowledge of the decision making $\text{KD} = < f_u(u \mid z) >$, which is determined from the knowledge of the plant KP for the given $f_y(y)$ (Fig. 3.10).

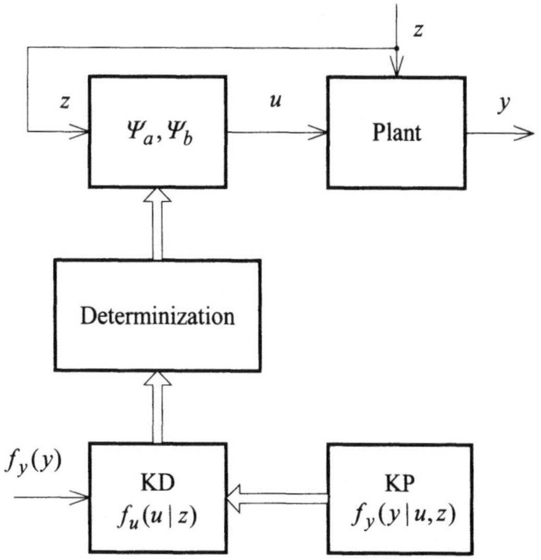

Figure 3.10. Open-loop decision system with the determinization in the non-parametric case

It is worth noting that for the non-parametric description of the plant in the form $\text{KP} = < f_y(y \mid u,z) >$ we can state three versions of the decision problem with the deterministic requirement $y = y^*$, analogous to those considered in Sect. 3.2:

Version I. To find the decision $u \stackrel{\Delta}{=} u_a$ maximizing $f_y(y^* \mid u,z)$.

Version II. To find the decision $u \stackrel{\Delta}{=} u_b$ such that $\mathrm{E}_y(\tilde{y} \mid u,z) = y^*$.

Version III. To find $u \stackrel{\Delta}{=} u_c$ minimizing $\mathrm{E}_s(\tilde{s} \mid u,z)$ where $s = \varphi(y, y^*)$ is a quality index. Consequently, we obtain the functions Ψ_a, Ψ_b, Ψ_c as three versions of the deterministic decision algorithm $u = \Psi(z)$ in an open-loop decision system (Fig. 3.1), corresponding to three different ways of the determinization of the random plant.

Example 3.6.

Let us illustrate the analysis problem for a one-dimensional plant $(u, y, z \in R^1)$ described by

$$f_y(y \mid u,z) = \frac{1}{\sqrt{\pi}} e^{-(y - \frac{u+z}{2})^2} \stackrel{\Delta}{=} \frac{1}{\sqrt{\pi}} \exp[-(y - \frac{u+z}{2})^2] ,$$

and the following data concerning the random decision algorithm and the disturbances:

$$f_u(u \mid z) = \frac{1}{2\sqrt{\pi}} \exp[-\frac{(u-2z)^2}{4}], \tag{3.44}$$

$$f_z(z) = \frac{\sqrt{7}}{2\sqrt{2\pi}} \exp(-\frac{7z^2}{8}).$$

Putting $f_y(y \mid u,z)$, $f_u(u \mid z)$ and $f_z(z)$ into (3.43) we shall obtain

$$f_y(y) = \frac{\sqrt{7}}{4\sqrt{2\pi}} \exp(-\frac{7y^2}{32}). \tag{3.45}$$

Then $\mathrm{E}_y(\widetilde{y}) = 0$ and the variance $\sigma_y^2 = \frac{16}{7}$.

In the decision problem, for the required probability density (3.45) we may obtain the probability density (3.44) as one of possible solutions, i.e. one of the possible descriptions of the random decision algorithm. As a result of the determinization we have

$$u_a = u_b = \Psi_{a,b}(z) = 2z \ . \qquad \square$$

3.8 Non-parametric Uncertainty. Discrete Case

In the discrete case

$$u \in U = \{\overline{u}_1, \overline{u}_2, ..., \overline{u}_\alpha\}, \qquad y \in Y = \{\overline{y}_1, \overline{y}_2, ..., \overline{y}_\beta\},$$

$$z \in Z = \{\overline{z}_1, \overline{z}_2, ..., \overline{z}_\gamma\}.$$

Let us start with a plant without disturbances, described by KP in the form of a matrix of conditional probabilities (see Sect. 3.1)

$$P_y = [p_{ij}]_{\substack{i=1,2,\dots,\beta \\ j=1,2,\dots,\alpha}}$$

where

$$p_{ij} = P(\widetilde{y} = \overline{y}_i \mid \widetilde{u} = \overline{u}_j).$$

Now, instead of the densities $f_u(u)$ and $f_y(y)$, we have the vectors of respective probabilities:

$$p_u = \begin{bmatrix} p_{u1} \\ p_{u2} \\ \vdots \\ p_{u\alpha} \end{bmatrix}, \qquad p_y = \begin{bmatrix} p_{y1} \\ p_{y2} \\ \vdots \\ p_{y\beta} \end{bmatrix}$$

where

$$p_{uj} = P(\tilde{u} = \bar{u}_j), \qquad p_{yi} = P(\tilde{y} = \bar{y}_i).$$

Analysis problem: For the given $\mathrm{KP} = < P_y >$ and p_u find p_y.

According to (3.3), where $p_j = p_{uj}$ and $p_i = p_{yi}$, we have

$$p_y = P_y \cdot p_u . \tag{3.46}$$

Decision problem: For the given $\mathrm{KP} = < P_y >$ and the desirable probability distribution p_y find the probability distribution p_u.

To find the solution one should solve Equation (3.46) with respect to the vector p_u satisfying the following conditions:

$$\bigwedge_j p_{uj} \geq 0, \qquad \sum_{j=1}^{\alpha} p_{uj} = 1 .$$

A final particular decision u may be chosen randomly from U according to the probability distribution p_u (which requires a generator of random numbers) or may be obtained via the determinization using a mean value:

$$u_d = \mathrm{E}_u(\tilde{u}) = \sum_{j=1}^{\alpha} \bar{u}_j p_{uj} .$$

For the plane with external disturbances $z \in Z$ let us introduce the following notations:

$$p_z = \begin{bmatrix} p_{z1} \\ p_{z2} \\ \vdots \\ p_{z\gamma} \end{bmatrix} \quad \text{where} \quad p_{zl} = P(\tilde{z} = \bar{z}_l),$$

$$p_u(\bar{z}_l) = \begin{bmatrix} p_{u1}(\bar{z}_l) \\ p_{u2}(\bar{z}_l) \\ \vdots \\ p_{u\alpha}(\bar{z}_l) \end{bmatrix} \quad \text{where} \quad p_{uj}(\bar{z}_l) = P(\tilde{u} = \bar{u}_j \mid \tilde{z} = \bar{z}_l),$$

$$p_y(\bar{z}_l) = \begin{bmatrix} p_{y1}(\bar{z}_l) \\ p_{y2}(\bar{z}_l) \\ \vdots \\ p_{y\beta}(\bar{z}_l) \end{bmatrix} \quad \text{where} \quad p_{yi}(\bar{z}_l) = P(\tilde{y} = \bar{y}_i \mid \tilde{z} = \bar{z}_l),$$

$$P_y(\bar{z}_l) = [p_{ij}(\bar{z}_l)]_{\substack{i = 1,2,\ldots,\beta \\ j = 1,2,\ldots,\alpha}}$$

where

$$p_{ij}(\bar{z}_l) = P(\tilde{y} = \bar{y}_i \mid \tilde{u} = \bar{u}_j, \tilde{z} = \bar{z}_l).$$

Decision problem: For the given

$$\text{KP} = < P_y(\bar{z}_1), P_y(\bar{z}_2), \ldots, P_y(\bar{z}_\gamma);\ p_z >$$

and p_y, one should determine

$$[p_u(\bar{z}_1)\ p_u(\bar{z}_2) \ldots p_u(\bar{z}_\gamma)] \overset{\Delta}{=} P_u. \tag{3.47}$$

In (3.47) $p_u(\bar{z}_1), \ldots, p_u(\bar{z}_\gamma)$ denote the columns of the matrix P_u, and P_u is the matrix of the conditional probabilities:

$$P = [\bar{p}_{jl}]_{\substack{j = 1,2,\ldots,\alpha \\ l = 1,2,\ldots,\gamma}}$$

where

$$\bar{p}_{jl} = P(\tilde{u} = \bar{u}_j \mid \tilde{z} = \bar{z}_l).$$

In this notation p_y, p_z and P_u correspond to $f_y(y)$, $f_z(z)$ and $f_u(u \mid z)$ in the continuous case, respectively, and the set of matrices $P_y(\bar{z}_1),..., P_y(\bar{z}_\gamma)$ corresponds to $f_y(y \mid u,z)$. Using our notation we can write the following relationships:

$$p_y(\bar{z}_l) = P_y(\bar{z}_l) \cdot p_u(\bar{z}_l),$$

$$p_y = [\, p_y(\bar{z}_1) \;\; p_y(\bar{z}_2) \;\; \cdots \;\; p_y(\bar{z}_\gamma)\,] p_z$$

and finally

$$p_y = [P_y(\bar{z}_1) p_u(\bar{z}_1) \;\; P_y(\bar{z}_2) p_u(\bar{z}_2) \; \dots \; P_y(\bar{z}_\gamma) p_u(\bar{z}_\gamma)] p_z . \quad (3.48)$$

In (3.48) $P_y(\bar{z}_l) \cdot p_u(\bar{z}_l)$ are the columns of the matrix. The relationship (3.48) is analogous to the relationship (3.43) for the continuous case. Any matrix P_u (3.47), i.e. any sequence of vectors $p_u(\bar{z}_1),..., p_u(\bar{z}_\gamma)$, satisfying Equation (3.48) is a solution of our decision problem. It is worth noting that the solution of Equation (3.48) with respect to $p_u(\bar{z}_1),..., p_u(\bar{z}_\gamma)$ may not be unique. Note that (3.48) denotes β scalar equations with $\alpha\gamma$ unknowns to be determined. The matrix P_u may be considered as a knowledge of the decision making KD or a random decision algorithm. In having P_u, we can obtain the deterministic decision algorithm using two versions of the determinization analogous to versions I and II for the continuous case.

Version I. For the given $z = \bar{z}_l$, the decision

$$u_a(\bar{z}_l) = u_{j^*}$$

where

$$j^* = \arg\max_j p_{uj}(\bar{z}_l) .$$

Version II. For the given $z = \bar{z}_l$ the decision $u_b(\bar{z}_l)$ is obtained by rounding off the expected value $\mathrm{E}_u(\tilde{u} \mid \bar{z}_l)$ to the nearest value $\bar{u}_j$. The rounding off is reasonable if α is sufficiently large and the distance between the different $\bar{u}_j$ is sufficiently small.

Example 3.7.

Assume that in the plant under consideration $\alpha = \gamma = 2$, $\beta = 4$ and KP is the following:

$$P_y(\bar{z}_1) = \begin{bmatrix} 0.5 & 0.3 \\ 0.1 & 0.2 \\ 0.2 & 0.1 \\ 0.2 & 0.4 \end{bmatrix}, \quad P_y(\bar{z}_2) = \begin{bmatrix} 0.1 & 0.5 \\ 0.6 & 0.1 \\ 0.1 & 0.1 \\ 0.2 & 0.3 \end{bmatrix}, \quad p_z = \begin{bmatrix} 0.1 \\ 0.9 \end{bmatrix}.$$

For example, the first row of the matrix $P_y(\bar{z}_1)$ means: if $\tilde{z} = \bar{z}_1$ then

$$P(\tilde{y} = \bar{y}_1 \mid \tilde{u} = \bar{u}_1) = 0.5, \qquad P(\tilde{y} = \bar{y}_1 \mid \tilde{u} = \bar{u}_2) = 0.3.$$

The requirement concerning y is as follows:

$$P(\tilde{y} = \bar{y}_1) = 0.2, \quad P(\tilde{y} = \bar{y}_2) = 0.4, \quad P(\tilde{y} = \bar{y}_3) = 0.1, \quad P(\tilde{y} = \bar{y}_4) = 0.3.$$

For our numerical data Equation (3.48) takes the form:

$$\begin{bmatrix} 0.2 \\ 0.4 \\ 0.1 \\ 0.3 \end{bmatrix} = \begin{bmatrix} 0.5p_{u1}(\bar{z}_1) + 0.3p_{u2}(\bar{z}_1) & 0.1p_{u1}(\bar{z}_2) + 0.5p_{u2}(\bar{z}_2) \\ 0.1p_{u1}(\bar{z}_1) + 0.2p_{u2}(\bar{z}_1) & 0.6p_{u1}(\bar{z}_2) + 0.1p_{u2}(\bar{z}_2) \\ 0.2p_{u1}(\bar{z}_1) + 0.1p_{u2}(\bar{z}_1) & 0.1p_{u1}(\bar{z}_2) + 0.1p_{u2}(\bar{z}_2) \\ 0.2p_{u1}(\bar{z}_1) + 0.4p_{u2}(\bar{z}_1) & 0.2p_{u1}(\bar{z}_2) + 0.3p_{u2}(\bar{z}_2) \end{bmatrix} \begin{bmatrix} 0.1 \\ 0.9 \end{bmatrix}$$

which may be presented as a set of four equations with the unknowns $p_{u1}(\bar{z}_1)$, $p_{u2}(\bar{z}_1)$, $p_{u1}(\bar{z}_2)$, $p_{u2}(\bar{z}_2)$. Solving this set of equations we obtain

$$p_{u1}(\bar{z}_1) = P(\tilde{u} = \bar{u}_1 \mid \tilde{z} = \bar{z}_1) \approx 0.2, \qquad p_{u2}(\bar{z}_1) = P(\tilde{u} = \bar{u}_2 \mid \tilde{z} = \bar{z}_1) \approx 0.8,$$

$$p_{u1}(\bar{z}_2) = P(\tilde{u} = \bar{u}_1 \mid \tilde{z} = \bar{z}_2) \approx 0.6, \qquad p_{u2}(\bar{z}_2) = P(\tilde{u} = \bar{u}_2 \mid \tilde{z} = \bar{z}_2) \approx 0.4.$$

□

4 Uncertain Logics and Variables

In this chapter, we present basic theoretical foundations of uncertain variables – as a comparatively new tool for analysis and decision problems in uncertain systems [44, 46, 54]. The uncertain variable is described by a certainty distribution given by an expert and characterizing his or her knowledge on approximate values of the variable. The uncertain variables are related to random and fuzzy variables, but there are also essential differences. The comparison with random and fuzzy variables will be presented in Chapter 6. The definitions of the uncertain variables are based on uncertain logics described in Sects. 4.1 and 4.2. It is worth noting that the uncertain logics and variables are defined for any metric space X (the set of values). Starting from Sect. 4.4, the considerations are concerned with a real number vector space X.

4.1 Uncertain Logic

Our considerations are based on multi-valued logic. To introduce terminology and notation employed in our presentation of uncertain logic and uncertain variables, let us recall that multi-valued (exactly speaking – infinite-valued) propositional logic deals with propositions $(\alpha_1, \alpha_2, ...)$ whose logic values $w(\alpha) \in [0, 1]$ and

$$w(\neg\alpha) = 1 - w(\alpha),$$

$$w(\alpha_1 \vee \alpha_2) = \max\{w(\alpha_1), w(\alpha_2)\}, \tag{4.1}$$

$$w(\alpha_1 \wedge \alpha_2) = \min\{w(\alpha_1), w(\alpha_2)\}.$$

Multi-valued predicate logic deals with predicates $P(x)$ defined on a set X, i.e. properties concerning x, which for the fixed value of x form propositions in multi-valued propositional logic, i.e.

$$w[P(x)] \triangleq \mu_p(x) \in [0, 1] \quad \text{for each } x \in X. \tag{4.2}$$

For the fixed x, $\mu_p(x)$ denotes the *degree of truth*, i.e. the value $\mu_p(x)$ shows to what degree P is satisfied. If for each $x \in X$ $\mu_p(x) \in \{0, 1\}$ then $P(x)$ will be called here a *crisp* or a *well-defined property*, and $P(x)$ which is not well-defined will be called a *soft property*. The crisp property defines a set

$$D_x = \{x \in X\colon w[P(x)] = 1\} \triangleq \{x \in X\colon P(x)\}. \tag{4.3}$$

Consider now a universal set Ω, $\omega \in \Omega$, a set X which is assumed to be a metric space, a function $g\colon \Omega \to X$, and a crisp property $P(x)$ in the set X. The property P and the function g generate the crisp property $\Psi(\omega, P)$ in Ω: "For the value $\bar{x} = g(\omega) \triangleq \bar{x}(\omega)$ assigned to ω the property P is satisfied", i.e.

$$\Psi(\omega, P) = P[\bar{x}(\omega)].$$

Let us introduce now the property $G(\bar{x}, x) =$ "$\bar{x} \cong x$" for $\bar{x}, x \in X$, which means: "$\bar{x}$ is approximately equal to x". The equivalent formulations are "x is the approximate value of $\bar{x}$" or "x belongs to a small neighbourhood of $\bar{x}$" or "the value of the metric $d(x, \bar{x})$ is small". Note that $G(\bar{x}, x)$ is a reflexive, symmetric and transitive relation in $X \times X$. For the fixed ω, $G[\bar{x}(\omega), x] \triangleq G_\omega(x)$ is a soft property in X. The properties $P(x)$ and $G_\omega(x)$ generate the soft property $\overline{\Psi}(\omega, P)$ in Ω: "the approximate value of $\bar{x}(\omega)$ satisfies P" or "$\bar{x}(\omega)$ approximately satisfies P", i.e.

$$\overline{\Psi}(\omega, P) = G_\omega(x) \wedge P(x) = [\bar{x}(\omega) \cong x] \wedge P(x) \tag{4.4}$$

where x is a free variable. The property $\overline{\Psi}$ may be denoted by

$$\overline{\Psi}(\omega, P) = \text{“}\bar{x}(\omega) \,\tilde{\in}\, D_x\text{”} \tag{4.5}$$

where D_x is defined by (4.3) and "$\bar{x} \,\tilde{\in}\, D_x$" means: "the approximate value of $\bar{x}$ belongs to D_x" or "$\bar{x}$ approximately belongs to D_x". Denote by $h_\omega(x)$ the logic value of $G_\omega(x)$

$$w[G_\omega(x)] \triangleq h_\omega(x), \qquad \bigwedge_{x \in X} (h_\omega(x) \ge 0), \tag{4.6}$$

$$\max_{x \in X} h_\omega(x) = 1. \tag{4.7}$$

Definition 4.1 (*uncertain logic*). The uncertain logic is defined by a universal set Ω, a metric space X, crisp properties (predicates) $P(x)$, the properties $G_\omega(x)$ and the corresponding functions (4.6) for $\omega \in \Omega$. In this logic we consider soft properties (4.4) generated by P and G_ω. The logic value of $\overline{\Psi}$ is defined in the following way:

$$w[\overline{\Psi}(\omega, P)] \triangleq v[\overline{\Psi}(\omega, P)] = \begin{cases} \max\limits_{x \in D_x} h_\omega(x) & \text{for } D_x \neq \varnothing \\ 0 & \text{for } D_x = \varnothing \end{cases} \tag{4.8}$$

and is called a degree of certainty or *certainty index*. The operations for the certainty indexes are defined as follows:

$$v[\neg \overline{\Psi}(\omega, P)] = 1 - v[\overline{\Psi}(\omega, P)], \tag{4.9}$$

$$v[\Psi_1(\omega, P_1) \vee \Psi_2(\omega, P_2)] = \max\{v[\Psi_1(\omega, P_1)], v[\Psi_2(\omega, P_2)]\}, \tag{4.10}$$

$$v[\Psi_1(\omega, P_1) \wedge \Psi_2(\omega, P_2)] = \begin{cases} 0 & \text{if } w(P_1 \wedge P_2) = 0 \text{ for each } x \\ \min\{v[\Psi_1(\omega, P_1)], v[\Psi_2(\omega, P_2)]\} & \text{otherwise} \end{cases} \tag{4.11}$$

where Ψ_1 is $\overline{\Psi}$ or $\neg\overline{\Psi}$, and Ψ_2 is $\overline{\Psi}$ or $\neg\overline{\Psi}$. □

Using the notation (4.5) we have

$$v[\bar{x}(\omega) \mathrel{\tilde{\notin}} D_x] = 1 - v[\bar{x}(\omega) \mathrel{\tilde{\in}} D_x], \tag{4.12}$$

$$v[\bar{x}(\omega) \mathrel{\tilde{\in}} D_1 \vee \bar{x}(\omega) \mathrel{\tilde{\in}} D_2] = \max\{v[\bar{x}(\omega) \mathrel{\tilde{\in}} D_1], v[\bar{x}(\omega) \mathrel{\tilde{\in}} D_2]\}, \tag{4.13}$$

$$v[\bar{x}(\omega) \mathrel{\tilde{\in}} D_1 \wedge \bar{x}(\omega) \mathrel{\tilde{\in}} D_2] = \min\{v[\bar{x}(\omega) \mathrel{\tilde{\in}} D_1], v[\bar{x}(\omega) \mathrel{\tilde{\in}} D_2]\} \tag{4.14}$$

for $D_1 \cap D_2 \neq \varnothing$ and 0 for $D_1 \cap D_2 = \varnothing$ – where $\tilde{\in} D_1$ and $\tilde{\in} D_2$ may be replaced by $\tilde{\notin} D_1$ and $\tilde{\notin} D_2$, respectively.

From (4.7) and (4.8), $v[\bar{x} \mathrel{\tilde{\in}} X] = 1$. Let $D_{x,t}$ for $t \in T$ be a family of sets D_x. Then, according to (4.13) and (4.14)

$$v[\bigvee_{t \in T} \bar{x}(\omega) \mathrel{\tilde{\in}} D_{x,t}] = \max_{t \in T} v[\bar{x}(\omega) \mathrel{\tilde{\in}} D_{x,t}], \tag{4.15}$$

$$v[\bigwedge_{t \in T} \bar{x}(\omega) \mathrel{\tilde{\in}} D_{x,t}] = \min_{t \in T} v[\bar{x}(\omega) \mathrel{\tilde{\in}} D_{x,t}]. \tag{4.16}$$

One can note that $G_\omega(x) =$ " $\bar{x}(\omega) \cong x$ " is a special case of $\overline{\Psi}$ for $D_x = \{x\}$ (a singleton) and

$$v[\bar{x}(\omega) \cong x] = h_\omega(x), \qquad v[\bar{x}(\omega) \ncong x] = 1 - h_\omega(x). \tag{4.17}$$

According to (4.4), (4.5), (4.17), (4.15), (4.16)

$$v[\bar{x}(\omega) \,\tilde{\in}\, D_x] = v[\bigvee_{x \in D_x} \bar{x}(\omega) \cong x] = \max_{x \in D_x} h_\omega(x)$$

which coincides with (4.8), and

$$v[\bar{x}(\omega) \,\tilde{\notin}\, D_x] = v[\bigwedge_{x \in D_x} \bar{x}(\omega) \ncong x] = \min_{x \in D_x} [1 - h_\omega(x)] = 1 - \max_{x \in D_x} h_\omega(x)$$

which coincides with (4.8) and (4.12). From (4.8) one can immediately deliver the following property: if $P_1 \to P_2$ for each x (i.e. $D_1 \subseteq D_2$) then

$$v[\overline{\Psi}(\omega, P_1)] \le v[\overline{\Psi}(\omega, P_2)]$$

or

$$v[\bar{x}(\omega) \,\tilde{\in}\, D_1] \le v[\bar{x}(\omega) \,\tilde{\in}\, D_2]. \tag{4.18}$$

Theorem 4.1.

$$v[\overline{\Psi}(\omega, P_1 \vee P_2)] = v[\overline{\Psi}(\omega, P_1) \vee \overline{\Psi}(\omega, P_2)], \tag{4.19}$$

$$v[\overline{\Psi}(\omega, P_1 \wedge P_2)] \le \min\{v[\overline{\Psi}(\omega, P_1)], v[\overline{\Psi}(\omega, P_2)]\}. \tag{4.20}$$

Proof: From (4.8) and (4.10)

$$v[\overline{\Psi}(\omega, P_1) \vee \overline{\Psi}(\omega, P_2)] = \max\{\max_{x \in D_1} h_\omega(x), \max_{x \in D_2} h_\omega(x)\}$$
$$= \max_{x \in D_1 \cup D_2} h_\omega(x) = v[\overline{\Psi}(\omega, P_1 \vee P_2)].$$

Inequality (4.20) follows immediately from $D_1 \cap D_2 \subseteq D_1$, $D_1 \cap D_2 \subseteq D_2$ and (4.18). □

Theorem 4.2.

$$v[\overline{\Psi}(\omega, \neg P)] \ge v[\neg \overline{\Psi}(\omega, P)]. \tag{4.21}$$

Proof: Let $P_1 = P$ and $P_2 = \neg P$ in (4.19). Since $w(P \vee \neg P) = 1$ for each x ($D_x = X$ in this case),

$$1 = v[\overline{\Psi}(\omega, P) \vee \overline{\Psi}(\omega, \neg P)] = \max\{v[\overline{\Psi}(\omega, P)], v[\overline{\Psi}(\omega, \neg P)]\}$$

and

$$v[\overline{\Psi}(\omega, \neg P)] \ge 1 - v[\overline{\Psi}(\omega, P)] = v[\neg \overline{\Psi}(\omega, P)]. \quad □$$

Inequality (4.21) may be written in the form

$$v[\bar{x}(\omega) \,\tilde{\in}\, \overline{D}_x] \ge v[\bar{x}(\omega) \,\tilde{\notin}\, D_x] = 1 - v[\bar{x}(\omega) \,\tilde{\in}\, D_x] \tag{4.22}$$

where $\overline{D}_x = X - D_x$.

As was said in Sect.1.2, the definition of uncertain logic should contain two parts: a mathematical model (which is described above) and its interpretation (semantics). The semantics are provided in the following: the uncertain logic operates with crisp predicates $P[\overline{x}(\omega)]$, but for the given ω it is not possible to state whether $P(\overline{x})$ is true or false because the function $\overline{x} = g(\omega)$ and consequently the value $\overline{x}$ corresponding to ω is unknown. The exact information, i.e. the knowledge of g is replaced by $h_\omega(x)$, which for the given ω characterizes the different possible approximate values of $\overline{x}(\omega)$. If we use the terms knowledge, information, data etc., it is necessary to determine the subject (who knows?, who gives the information?).

In our considerations this subject is called an *expert*. So the expert does not know exactly the value $\overline{x}(\omega)$, but by "looking at" ω he obtains some information concerning $\overline{x}$, which he does not express in an explicit form but uses it to formulate $h_\omega(x)$. Hence, the expert is the source of $h_\omega(x)$ which for particular x evaluates his opinion that $\overline{x} \cong x$. That is why $h_\omega(x)$ and consequently $v[\overline{\Psi}(\omega,P)]$ are called degrees of certainty. For example, Ω is a set of persons, $\overline{x}(\omega)$ denotes the age of ω and the expert looking at the person ω gives the function $h_\omega(x)$ whose value for the particular x is his degree of certainty that the age of this person is approximately equal to x. The predicates $\overline{\Psi}(\omega, P)$ are soft because of the uncertainty of the expert. The result of including $h_\omega(x)$ into the definition of uncertain logic is that for the same (Ω, X) we may have the different logics specified by different experts.

The logic introduced by Definition 4.1 will be denoted by L-logic. In the next section we shall consider other versions of uncertain logic which will be denoted by L_p, L_n and L_c.

4.2 Other Versions of Uncertain Logic

Definition 4.2 (*L_p-logic*). The first part is the same as in Definition 4.1. The certainty index

$$v_p[\overline{\Psi}(\omega, P)] = v[\overline{\Psi}(\omega, P)] = \max_{x \in D_x} h_\omega(x). \tag{4.23}$$

The operations are defined in the following way:

$$\neg\overline{\Psi}(\omega, P) = \overline{\Psi}(\omega, \neg P), \tag{4.24}$$

$$\overline{\Psi}(\omega, P_1) \vee \overline{\Psi}(\omega, P_2) = \overline{\Psi}(\omega, P_1 \vee P_2), \tag{4.25}$$

$$\overline{\Psi}(\omega, P_1) \wedge \overline{\Psi}(\omega, P_2) = \overline{\Psi}(\omega, P_1 \wedge P_2) . \qquad \square \quad (4.26)$$

Consequently, we have the same equalities for v_p , i.e.

$$v_p[\neg\overline{\Psi}(\omega, P)] = v_p[\overline{\Psi}(\omega, \neg P)] , \qquad (4.27)$$

$$v_p[\overline{\Psi}(\omega, P_1) \vee \overline{\Psi}(\omega, P_2)] = v_p[\overline{\Psi}(\omega, P_1 \vee P_2)] , \qquad (4.28)$$

$$v_p[\overline{\Psi}(\omega, P_1) \wedge \overline{\Psi}(\omega, P_2)] = v_p[\overline{\Psi}(\omega, P_1 \wedge P_2)] . \qquad (4.29)$$

In a similar way as for L-logic it is easy to prove that

$$\text{If} \quad P_1 \to P_2 \quad \text{then} \quad v_p[\overline{\Psi}(\omega, P_1)] \le v_p[\overline{\Psi}(\omega, P_2)] , \qquad (4.30)$$

$$v_p[\overline{\Psi}(\omega, P_1 \vee P_2)] = \max \{v_p[\overline{\Psi}(\omega, P_1)], v_p[\overline{\Psi}(\omega, P_2)]\} , \qquad (4.31)$$

$$v_p[\overline{\Psi}(\omega, P_1 \wedge P_2)] \le \min \{v_p[\overline{\Psi}(\omega, P_1)], v_p[\overline{\Psi}(\omega, P_2)]\} , \qquad (4.32)$$

$$v_p[\overline{\Psi}(\omega, \neg P)] \ge 1 - v_p[\overline{\Psi}(\omega, P)] . \qquad (4.33)$$

Definition 4.3 (*L_n-logic*). The certainty index of $\overline{\Psi}$ is defined as follows:

$$v_n[\overline{\Psi}(\omega, P)] = 1 - v_p[\overline{\Psi}(\omega, \neg P)] = 1 - \max_{x \in \overline{D}_x} h_\omega(x) . \qquad (4.34)$$

The operations are the same as for v_p in L_p-logic, i.e. (4.24)–(4.26) and (4.27)–(4.29) with v_n in place of v_p . □

It may be proved that

$$\text{If} \quad P_1 \to P_2 \quad \text{then} \quad v_n[\overline{\Psi}(\omega, P_1)] \le v_n[\overline{\Psi}(\omega, P_2)] , \qquad (4.35)$$

$$v_n[\overline{\Psi}(\omega, P_1 \vee P_2)] \ge \max \{v_n[\overline{\Psi}(\omega, P_1)], v_n[\overline{\Psi}(\omega, P_2)]\} , \qquad (4.36)$$

$$v_n[\overline{\Psi}(\omega, P_1 \wedge P_2)] = \min \{v_n[\overline{\Psi}(\omega, P_1)], v_n[\overline{\Psi}(\omega, P_2)]\} \qquad (4.37)$$

for $w(P_1 \wedge P_2) > 0$, and

$$v_n[\overline{\Psi}(\omega, \neg P)] \le 1 - v_n[\overline{\Psi}(\omega, P)] . \qquad (4.38)$$

The statement (4.35) follows immediately from (4.30) and (4.34). Property (4.36) follows from $D_1 \cup D_2 \supseteq D_1$, $D_1 \cup D_2 \supseteq D_2$ and (4.35). From (4.34) we have

$$v_n[\overline{\Psi}(\omega, P_1 \wedge P_2)] = 1 - \max_{x \in \overline{D}_1 \cup \overline{D}_2} h_\omega(x) = 1 - \max\{\max_{x \in \overline{D}_1} h_\omega(x), \max_{x \in \overline{D}_2} h_\omega(x)\}$$
$$= 1 - \max\{1 - v_n[\overline{\Psi}(\omega, P_1)], 1 - v_n[\overline{\Psi}(\omega, P_2)]\},$$

which proves (4.37). Substituting (4.34) into (4.33) we obtain (4.38).

In Definition 4.2 the certainty index is defined in "a positive way", so we may use the term "positive" logic (L_p). In Definition 4.3 the certainty index is defined in "a negative way" and consequently we may use the term "negative" logic (L_n). In (4.23) the shape of the function $h_\omega(x)$ in $\overline{D}_x$ is not taken into account, and in (4.34) the function $h_\omega(x)$ in D_x is neglected. They are the known disadvantages of these definitions. To avoid them consider the *combined logic* denoted by L_c.

Definition 4.4 (*L_C-logic*). The certainty index of $\overline{\Psi}$ and the negation $\neg\overline{\Psi}$ are defined as follows:

$$v_c[\overline{\Psi}(\omega, P)] = \frac{v_p[\overline{\Psi}(\omega, P)] + v_n[\overline{\Psi}(\omega, P)]}{2} = \frac{1}{2}[\max_{x \in D_x} h_\omega(x) + 1 - \max_{x \in \overline{D}_x} h_\omega(x)], \tag{4.39}$$

$$\neg\overline{\Psi}(\omega, P) = \overline{\Psi}(\omega, \neg P). \tag{4.40}$$

The operations for v_c are determined by the operations for v_p and v_n. □

According to (4.40)

$$v_c[\neg\overline{\Psi}(\omega, P)] = v_c[\overline{\Psi}(\omega, \neg P)] .$$

Using (4.39) and (4.28), (4.29) for v_p and v_n, it is easy to show that

$$v_c[\overline{\Psi}(\omega, P_1) \vee \overline{\Psi}(\omega, P_2)] = v_c[\overline{\Psi}(\omega, P_1 \vee P_2)], \tag{4.41}$$

$$v_c[\overline{\Psi}(\omega, P_1) \wedge \overline{\Psi}(\omega, P_2)] = v_c[\overline{\Psi}(\omega, P_1 \wedge P_2)]. \tag{4.42}$$

L_C-logic may be defined independently of v_p and v_n, with the right-hand side of (4.39) and the definitions of operations (4.40), (4.41), (4.42). The operations may be rewritten in the following form:

$$\bar{x} \,\tilde{\notin}\, D_x = \bar{x} \,\tilde{\in}\, \overline{D}_x, \tag{4.43}$$

$$v_c[\bar{x}(\omega) \,\tilde{\in}\, D_1 \vee \bar{x}(\omega) \,\tilde{\in}\, D_2] = v_c[\bar{x}(\omega) \,\tilde{\in}\, D_1 \cup D_2], \tag{4.44}$$

$$v_c[\bar{x}(\omega) \,\tilde{\in}\, D_1 \wedge \bar{x}(\omega) \,\tilde{\in}\, D_2] = v_c[\bar{x}(\omega) \,\tilde{\in}\, D_1 \cap D_2]. \tag{4.45}$$

From (4.8), (4.34) and (4.39)

$$v_c[\overline{x}(\omega) \tilde{\in} X] = 1, \qquad v_c[\overline{x}(\omega) \tilde{\in} \varnothing] = 0. \tag{4.46}$$

One can note that $G_\omega(x) =$ "$\overline{x} \cong x$" is a special case of $\overline{\Psi}$ for $D_x = \{x\}$ and according to (4.39)

$$v_c[\overline{x}(\omega) \cong x] = \frac{1}{2}[h_\omega(x) + 1 - \max_{\breve{x} \in X - \{x\}} h_\omega(\breve{x})], \tag{4.47}$$

$$v_c[\overline{x}(\omega) \ncong x] = \frac{1}{2}[\max_{\breve{x} \in X - \{x\}} h_\omega(\breve{x}) + 1 - h_\omega(x)]. \tag{4.48}$$

It is worth noting that if $h_\omega(x)$ is a continuous function then

$$v_c[\overline{x}(\omega) \cong x] = \frac{1}{2} h_\omega(x) .$$

Using (4.30) and (4.35), we obtain the following property: if for each x $P_1 \to P_2$ (i.e. $D_1 \subseteq D_2$) then

$$v_c[\overline{\Psi}(\omega, P_1)] \le v_c[\overline{\Psi}(\omega, P_2)]$$

or

$$v_c[\overline{x}(\omega) \tilde{\in} D_1] \le v_c[\overline{x}(\omega) \tilde{\in} D_2]. \tag{4.49}$$

Theorem 4.3.

$$v_c[\overline{\Psi}(\omega, P_1 \vee P_2)] \ge \max\{v_c[\overline{\Psi}(\omega, P_1)], v_c[\overline{\Psi}(\omega, P_2)]\}, \tag{4.50}$$

$$v_c[\overline{\Psi}(\omega, P_1 \wedge P_2)] \le \min\{v_c[\overline{\Psi}(\omega, P_1)], v_c[\overline{\Psi}(\omega, P_2)]\}. \tag{4.51}$$

Proof: Inequality (4.50) may be obtained from $D_1 \cup D_2 \supseteq D_1$, $D_1 \cup D_2 \supseteq D_2$ and (4.49). Inequality (4.51) follows from $D_1 \cap D_2 \subseteq D_1$, $D_1 \cap D_2 \subseteq D_2$ and (4.49). The property (4.50) can also be delivered from (4.39), (4.31), (4.36), and the property (4.51) – from (4.39), (4.32), (4.37). □

Theorem 4.4.

$$v_c[\neg \overline{\Psi}(\omega, P)] = 1 - v_c[\overline{\Psi}(\omega, P)]. \tag{4.52}$$

Proof: From (4.34) and (4.39)

$$v_c[\overline{\Psi}(\omega, P)] = \frac{1}{2}\{v_p[\overline{\Psi}(\omega, P)] + 1 - v_p[\overline{\Psi}(\omega, \neg P)]\} .$$

Then

$$v_c[\neg\overline{\Psi}(\omega,P)] = \frac{1}{2}\{v_p[\overline{\Psi}(\omega,\neg P)] + 1 - v_p[\overline{\Psi}(\omega,P)]\} = 1 - v_c[\overline{\Psi}(\omega,P)]\,. \square$$

It is worth noting that in *L*-logic

$$v[\overline{\Psi}(\omega,\neg P)] \geq v[\neg\overline{\Psi}(\omega,P)]$$

(see (4.21)), and in L_C-logic

$$v_c[\overline{\Psi}(\omega,\neg P)] = v_c[\neg\overline{\Psi}(\omega,P)]\,.$$

Consequently, in *L*-logic

$$v[\overline{x}(\omega)\,\tilde{\in}\,\overline{D}_x] \geq v[\overline{x}(\omega)\,\tilde{\notin}\,D_x] = 1 - v[\overline{x}(\omega)\,\tilde{\in}\,D_x]$$

and in L_C-logic

$$v_c[\overline{x}(\omega)\,\tilde{\in}\,\overline{D}_x] = v_c[\overline{x}(\omega)\,\tilde{\notin}\,D_x] = 1 - v_c[\overline{x}(\omega)\,\tilde{\in}\,D_x]\,.$$

Untill now it has been assumed that $\overline{x}(\omega), x \in X$. The considerations can be extended for the case $\overline{x}(\omega) \in \overline{X}$ and $x \in X \subset \overline{X}$. It means that the set of approximate values X evaluated by an expert may be a subset of the set of the possible values of $\overline{x}(\omega)$. In a typical case $X = \{x_1, x_2, ..., x_m\}$ (a finite set), $x_i \in \overline{X}$ for $i \in \overline{1,m}$. In our example with persons and age an expert may give the values $h_\omega(x)$ for natural numbers, e.g. $X = \{18, 19, 20, 21, 22\}$.

4.3 Uncertain Variables

The variable $\overline{x}$ for a fixed ω will be called an uncertain variable. Four versions of uncertain variables will be defined. The precise definition will contain: $h(x)$ given by an expert, the definition of the certainty index $w(\overline{x}\,\tilde{\in}\,D_x)$ and the definitions of $w(\overline{x}\,\tilde{\notin}\,D_x)$, $w(\overline{x}\,\tilde{\in}\,D_1 \vee \overline{x}\,\tilde{\in}\,D_2)$, $w(\overline{x}\,\tilde{\in}\,D_1 \wedge \overline{x}\,\tilde{\in}\,D_2)$.

Definition 4.5 (*uncertain variable*). An uncertain variable $\overline{x}$ is defined by the set of values *X*, the function $h(x) = v(\overline{x} \cong x)$ (i.e. the certainty index that $\overline{x} \cong x$, given by an expert) and the following definitions:

$$v(\overline{x}\,\tilde{\in}\,D_x) = \begin{cases} \max\limits_{x \in D_x} h(x) & \text{for } D_x \neq \varnothing \\ 0 & \text{for } D_x = \varnothing\,, \end{cases} \tag{4.53}$$

$$v(\overline{x}\,\tilde{\notin}\,D_x) = 1 - v(\overline{x}\,\tilde{\in}\,D_x)\,, \tag{4.54}$$

$$v(\overline{x}\,\tilde{\in}\,D_1 \vee \overline{x}\,\tilde{\in}\,D_2) = \max\{v(\overline{x}\,\tilde{\in}\,D_1), v(\overline{x}\,\tilde{\in}\,D_2)\}\,, \tag{4.55}$$

$$v(\bar{x} \tilde{\in} D_1 \wedge \bar{x} \tilde{\in} D_2) = \begin{cases} \min\{v(\bar{x} \tilde{\in} D_1), v(\bar{x} \tilde{\in} D_2)\} & \text{for } D_1 \cap D_2 \neq \varnothing \\ 0 & \text{for } D_1 \cap D_2 = \varnothing . \end{cases} \tag{4.56}$$

The function $h(x)$ will be called a *certainty distribution*. □

The definition of the uncertain variable is based on the uncertain logic, i.e. *L*-logic (see Definition 4.1). Then the properties (4.17)–(4.20) and (4.22) are satisfied. The properties (4.19) and (4.20) may be presented in the following form:

$$v(\bar{x} \tilde{\in} D_1 \cup D_2) = \max\{v(\bar{x} \tilde{\in} D_1), v(\bar{x} \tilde{\in} D_2)\}, \tag{4.57}$$

$$v(\bar{x} \tilde{\in} D_1 \cap D_2) \leq \min\{v(\bar{x} \tilde{\in} D_1), v(\bar{x} \tilde{\in} D_2)\}. \tag{4.58}$$

Example 4.1.
$X = \{1, 2, 3, 4, 5, 6, 7\}$ and the corresponding values of $h(x)$ are $(0.5, 0.8, 1, 0.6, 0.5, 0.4, 0.2)$, i.e. $h(1) = 0.5$, $h(2) = 0.8$ etc.
Let

$$D_1 = \{1, 2, 4, 5, 6\}, \qquad D_2 = \{3, 4, 5\}.$$

Then $D_1 \cup D_2 = \{1, 2, 3, 4, 5, 6\}$, $D_1 \cap D_2 = \{4, 5\}$, $v(\bar{x} \tilde{\in} D_1) = \max\{0.5, 0.8, 0.6, 0.5, 0.4\} = 0.8$, $v(\bar{x} \tilde{\in} D_2) = \max\{1, 0.6, 0.5\} = 1$, $v(\bar{x} \tilde{\in} D_1 \cup D_2) = \max\{0.5, 0.8, 1, 0.6, 0.5, 0.4\} = 1$, $v(\bar{x} \tilde{\in} D_1 \vee \bar{x} \tilde{\in} D_2) = \max\{0.8, 1\} = 1$, $v(\bar{x} \tilde{\in} D_1 \cap D_2) = \max\{0.6, 0.5\} = 0.6$, $v(\bar{x} \tilde{\in} D_1 \wedge \bar{x} \tilde{\in} D_2) = \min\{0.8, 1\} = 0.8$.

□

Example 4.2.
The certainty distribution is shown in Fig. 4.1. Let $D_x = [0, 4]$. Then

$$v(\bar{x} \tilde{\in} D_x) = v(\bar{x} \tilde{\in} [0, 4]) = 0.8,$$
$$v(\bar{x} \tilde{\in} \overline{D}_x) = v(\bar{x} \tilde{\in} (4, 16]) = 1,$$
$$v(\bar{x} \tilde{\notin} \overline{D}_x) = v(\bar{x} \tilde{\notin} (4, 16]) = 1 - 1 = 0 < v(\bar{x} \tilde{\in} D_x),$$
$$v(\bar{x} \tilde{\notin} D_x) = v(\bar{x} \tilde{\notin} [0, 4]) = 1 - 0.8 = 0.2 < v(\bar{x} \tilde{\in} \overline{D}_x).$$ □

Definition 4.6 (*P-uncertain variable*): A *P*-uncertain variable $\bar{x}$ is defined by the set of values *X*, the function $h(x) = v(\bar{x} \cong x)$ given by an expert (the same as in Definition 4.5) and the following definitions:

$$v_p(\bar{x} \tilde{\in} D_x) = \max_{x \in D_x} h(x) = v(\bar{x} \cong x),$$

$$v_p(\bar{x} \tilde{\notin} D_x) = v_p(\bar{x} \tilde{\in} \overline{D}_x),$$

$$v_p(\bar{x} \tilde{\in} D_1 \vee \bar{x} \tilde{\in} D_2) = v_p(\bar{x} \tilde{\in} D_1 \cup D_2),$$

$$v_p(\bar{x} \tilde{\in} D_1 \wedge \bar{x} \tilde{\in} D_2) = v_p(\bar{x} \tilde{\in} D_1 \cap D_2).$$ □

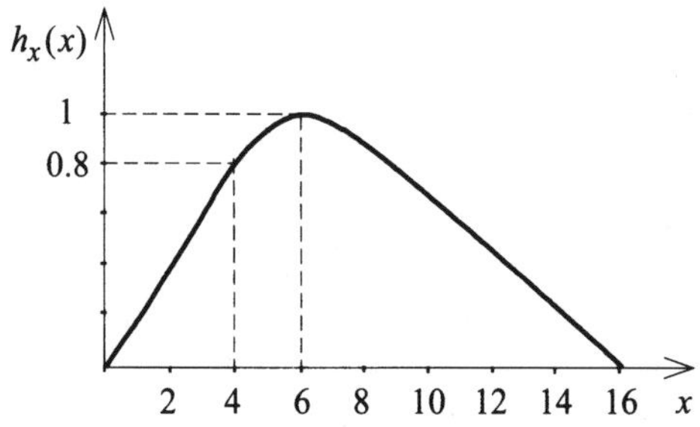

Figure 4.1. Example of certainty distribution

The definition of a *P*-uncertain variable is based on L_p-logic (see Definition 4.2). Then the properties (4.31), (4.32) and (4.33) are satisfied. These may be presented in the following form:

$$v_p(\bar{x} \,\tilde{\in}\, D_1 \cup D_2) = \max \{v_p(\bar{x} \,\tilde{\in}\, D_1), v_p(\bar{x} \,\tilde{\in}\, D_2)\},$$

$$v_p(\bar{x} \,\tilde{\in}\, D_1 \cap D_2) \le \min \{v_p(\bar{x} \,\tilde{\in}\, D_1), v_p(\bar{x} \,\tilde{\in}\, D_2)\},$$

$$v_p(\bar{x} \,\tilde{\in}\, \overline{D}_x) \ge 1 - v_p(\bar{x} \,\tilde{\in}\, D_x).$$

Definition 4.7 (*N-uncertain variable*). An *N*-uncertain variable $\bar{x}$ is defined by the set of values *X*, the function $h(x) = v(\bar{x} \cong x)$ given by an expert and the following definitions:

$$v_n(\bar{x} \,\tilde{\in}\, D_x) = 1 - v_p(\bar{x} \,\tilde{\in}\, \overline{D}_x) = 1 - \max_{x \in \overline{D}_x} h(x),$$

$$v_n(\bar{x} \,\tilde{\notin}\, D_x) = v_n(\bar{x} \,\tilde{\in}\, \overline{D}_x),$$

$$v_n(\bar{x} \,\tilde{\in}\, D_1 \vee \bar{x} \,\tilde{\in}\, D_2) = v_n(\bar{x} \,\tilde{\in}\, D_1 \cup D_2),$$

$$v_n(\bar{x} \,\tilde{\in}\, D_1 \wedge \bar{x} \,\tilde{\in}\, D_2) = v_n(\bar{x} \,\tilde{\in}\, D_1 \cap D_2). \qquad \square$$

The definition of an *N*-uncertain variable is based on L_n-logic and the properties corresponding to (4.36), (4.37) and (4.38) are as follows:

$$v_n(\bar{x} \,\tilde{\in}\, D_1 \cup D_2) \ge \max \{v_n(\bar{x} \,\tilde{\in}\, D_1), v_n(\bar{x} \,\tilde{\in}\, D_2)\},$$

$$v_n(\bar{x} \,\tilde{\in}\, D_1 \cap D_2) = \min \{v_n(\bar{x} \,\tilde{\in}\, D_1), v_n(\bar{x} \,\tilde{\in}\, D_2)\},$$

$$v_n(\bar{x} \,\tilde{\in}\, \overline{D}_x) \le 1 - v_n(\bar{x} \,\tilde{\in}\, D_x).$$

Definition 4.8 (*C-uncertain variable*). A *C*-uncertain variable $\bar{x}$ is defined by the set of values *X*, the function $h(x) = v(\bar{x} \cong x)$ given by an expert and the following definitions:

$$v_c(\bar{x} \,\tilde{\in}\, D_x) = \frac{1}{2}[\max_{x \in D_x} h(x) + 1 - \max_{x \in \overline{D}_x} h(x)], \tag{4.59}$$

$$v_c(\bar{x} \,\tilde{\notin}\, D_x) = 1 - v_c(\bar{x} \,\tilde{\in}\, D_x), \tag{4.60}$$

$$v_c(\bar{x} \,\tilde{\in}\, D_1 \vee \bar{x} \,\tilde{\in}\, D_2) = v_c(\bar{x} \,\tilde{\in}\, D_1 \cup D_2)\ , \tag{4.61}$$

$$v_c(\bar{x} \,\tilde{\in}\, D_1 \wedge \bar{x} \,\tilde{\in}\, D_2) = v_c(\bar{x} \,\tilde{\in}\, D_1 \cap D_2)\ . \qquad \square \tag{4.62}$$

The definition of a *C*-uncertain variable is based on L_C-logic (see Definition 4.4). Then the properties (4.50), (4.51), (4.52) are satisfied. According to (4.40) and (4.52)

$$v_c(\bar{x} \,\tilde{\notin}\, D_x) = v_c(\bar{x} \,\tilde{\in}\, \overline{D}_x). \tag{4.63}$$

Inequalities (4.50) and (4.51) may be presented in the following form:

$$v_c(\bar{x} \,\tilde{\in}\, D_1 \cup D_2) \geq \max\{v_c(\bar{x} \,\tilde{\in}\, D_1), v_c(\bar{x} \,\tilde{\in}\, D_2)\}, \tag{4.64}$$

$$v_c(\bar{x} \,\tilde{\in}\, D_1 \cap D_2) \leq \min\{v_c(\bar{x} \,\tilde{\in}\, D_1), v_c(\bar{x} \,\tilde{\in}\, D_2)\}. \tag{4.65}$$

The function $v_c(\bar{x} \cong x) \stackrel{\Delta}{=} h_c(x)$ expressed by (4.47) may be called a *C-certainty distribution*. Note that the certainty distribution $h(x)$ is given by an expert and the *C*-certainty distribution may be determined according to (4.47), by using $h(x)$. The *C*-certainty distribution does not determine the certainty index $v_c(\bar{x} \,\tilde{\in}\, D_x)$. To determine v_c, it is necessary to know $h(x)$ and to use (4.59). According to (4.64)

$$\max_{x \in D_x} h_c(x) \leq v_c(\bar{x} \,\tilde{\in}\, D_x)\ .$$

The formula (4.59) may be presented in the following way:

$$v_c(\bar{x} \,\tilde{\in}\, D_x) = \begin{cases} \dfrac{1}{2} \max\limits_{x \in D_x} h(x) = \dfrac{1}{2} v(\bar{x} \,\tilde{\in}\, D_x) & \text{if } \max\limits_{x \in \overline{D}_x} h(x) = 1 \\ 1 - \dfrac{1}{2} \max\limits_{x \in \overline{D}_x} h(x) = v(\bar{x} \,\tilde{\in}\, D_x) - \dfrac{1}{2} v(\bar{x} \,\tilde{\in}\, \overline{D}_x) & \text{otherwise.} \end{cases} \tag{4.66}$$

The formula (4.66) shows the relation between the certainty indexes v and v_c for the same D_x : if $D_x \neq X$ and $D_x \neq \varnothing$ then $v_c < v$. In particular, (4.47) becomes

$$h_c(x) = \begin{cases} \frac{1}{2} h(x) & \text{if} \quad \max\limits_{\breve{x} \in X-\{x\}} h(\breve{x}) = 1 \\ 1 - \frac{1}{2} \max\limits_{\breve{x} \in X-\{x\}} h(\breve{x}) & \text{otherwise}. \end{cases} \tag{4.67}$$

In the continuous case

$$h_c(x) = \frac{1}{2} h(x)$$

and in the discrete case

$$h_c(x_i) = \begin{cases} \frac{1}{2} h(x_i) & \text{if} \ \max\limits_{x \neq x_i} h(x) = 1 \\ 1 - \frac{1}{2} \max\limits_{x \neq x_i} h(x) & \text{otherwise}. \end{cases}$$

Example 4.3.
The set X and $h(x)$ are the same as in Example 4.1. Using (4.67) we obtain $h_c(1) = 0.25$, $h_c(2) = 0.4$, $h_c(3) = 1 - \frac{0.8}{2} = 0.6$, $h_c(4) = 0.3$, $h_c(5) = 0.25$, $h_c(6) = 0.2$, $h_c(7) = 0.1$.

Let D_1 and D_2 be the same as in Example 4.1. Using (4.66) and the values v obtained in Example 4.1 we have: $v_c(\bar{x} \,\tilde{\in}\, D_1) = \frac{1}{2} v = 0.4$, $v_c(\bar{x} \,\tilde{\in}\, D_2) = 1 - \frac{0.8}{2} = 0.6$, $v_c(\bar{x} \,\tilde{\in}\, D_1 \vee \bar{x} \,\tilde{\in}\, D_2) = v_c(\bar{x} \,\tilde{\in}\, D_1 \cup D_2) = 1 - \frac{0.2}{2} = 0.9$, $v_c(\bar{x} \,\tilde{\in}\, D_1 \wedge \bar{x} \,\tilde{\in}\, D_2) = v_c(\bar{x} \,\tilde{\in}\, D_1 \cap D_2) = \frac{0.6}{2} = 0.3$. In this case, for both D_1 and D_2, $v_c(\bar{x} \,\tilde{\in}\, D) = \max h_c(x)$ for $x \in D$. Let $D = \{2, 3, 4\}$. Now $v_c(\bar{x} \,\tilde{\in}\, D) = 1 - \frac{0.5}{2} = 0.75$ and $\max h_c(x) = \max\{0.4, 0.6, 0.3\} = 0.6 < v_c$. □

Example 4.4.
The certainty distribution and D_x are the same as in Example 4.2.

$$v_c(\bar{x} \,\tilde{\in}\, D_x) = \frac{1}{2}[v(\bar{x} \,\tilde{\in}\, D_x) + v(\bar{x} \,\tilde{\notin}\, \overline{D}_x)] = \frac{1}{2}[0.8 + 0] = 0.4,$$

$$v_c(\bar{x} \,\tilde{\in}\, \overline{D}_x) = v_c(\bar{x} \,\tilde{\notin}\, D_x) = 1 - v_c(\bar{x} \,\tilde{\in}\, D_x) = 0.6,$$

$$v_c(\bar{x} \,\tilde{\notin}\, \overline{D}_x) = v_c(\bar{x} \,\tilde{\in}\, D_x) = 0.4,$$

$$v(\bar{x} \,\tilde{\notin}\, D_x) = v_c(\bar{x} \,\tilde{\in}\, \overline{D}_x) = 0.6.$$ □

In the further considerations we shall use uncertain variables (defined by

Definition 4.5) and C-uncertain variables, because of the advantages of these formulations. In both cases the logic value of the negation is $w(\bar{x} \tilde{\notin} D_x) = 1 - w(\bar{x} \tilde{\in} D_x)$ (see (4.54) and (4.60)). In the first case it is easy to determine the certainty indexes for $\bar{x} \tilde{\in} D_1 \vee \bar{x} \tilde{\in} D_2$ and $\bar{x} \tilde{\in} D_1 \wedge \bar{x} \tilde{\in} D_2$, and all operations are the same as in (4.1) for multi-valued logic. In the second case, in the definition of the certainty index $v_c(\bar{x} \tilde{\in} D_x)$, the values of $h(x)$ for $\overline{D}_x$ are also taken into account and the logic operations (negation, disjunction and conjunction) correspond to the operations in the family of subsets D_x (complement, union and intersection). On the other hand, the calculations of the certainty indexes for disjunction and conjunction are more complicated than in the first case and are not determined by the certainty indexes for $\bar{x} \tilde{\in} D_1$, $\bar{x} \tilde{\in} D_2$, i.e. they cannot be reduced to operations in the set of certainty indexes $v_c(\bar{x} \tilde{\in} D)$. These features should be taken into account when making a choice between the application of the uncertain variable or C-uncertain variable in particular cases.

4.4 Additional Description of Uncertain Variables

For the further considerations we assume $X \subseteq R^k$ (k-dimensional real number vector space) and we shall consider two cases: the discrete case with $X = \{x_1, x_2, ..., x_m\}$ and the continuous case in which $h(x)$ is a continuous function.

Definition 4.9. In the discrete case

$$\bar{h}(x_i) = \frac{h(x_i)}{\sum_{j=1}^{m} h(x_j)}, \qquad i \in \overline{1,m} \tag{4.68}$$

will be called a *normalized certainty distribution*. The value

$$\mathrm{M}(\bar{x}) = \sum_{i=1}^{m} x_i \, \bar{h}(x_i) \tag{4.69}$$

will be called a *mean value* of the uncertain variable $\bar{x}$. In the continuous case the normalized certainty distribution and the mean value are defined as follows:

$$\bar{h}(x) = \frac{h(x)}{\int_X h(x)\,dx}, \qquad \mathrm{M}(\bar{x}) = \int_X x\,\bar{h}(x)\,dx, \tag{4.70}$$

on the condition that the integrals in (4.70) exist.

For a C-uncertain variable the normalized C-certainty distribution $\bar{h}_c(x)$ and the mean value $\mathrm{M}_c(\bar{x})$ are defined in the same way, with h_c in place of h in (4.68), (4.69) and (4.70). □

In the continuous case $h_c(x) = \frac{1}{2}h(x)$, then $\bar{h}_c(x) = \bar{h}(x)$ and $\mathrm{M}_c = \mathrm{M}$. In the discrete case, if x^* is a unique value for which $h(x^*) = 1$ and

$$\max_{x \neq x^*} h(x) \approx 1$$

then $\mathrm{M}_c \approx \mathrm{M}$. As a value characterizing $h(x)$ or $h_c(x)$, one can also use

$$x^* = \arg \max_{x \in X} h(x) \qquad \text{or} \qquad x_c^* = \arg \max_{x \in X} h_c(x) \, .$$

Replacing the uncertain variable $\bar{x}$ by its deterministic representation $\mathrm{M}(\bar{x})$ or x^* may be called *determinization* (analogous to defuzzification for fuzzy numbers).

Let us now consider a pair of uncertain variables $(\bar{x}, \bar{y}) = < X \times Y, h(x, y) >$ where $h(x, y) = v[(\bar{x}, \bar{y}) \cong (x, y)]$ is given by an expert and is called a *joint certainty distribution*. Then, using (4.1) for the disjunction in multi-valued logic, we have the following *marginal certainty distributions*:

$$h_x(x) = v(\bar{x} \cong x) = \max_{y \in Y} h(x, y), \tag{4.71}$$

$$h_y(y) = v(\bar{y} \cong y) = \max_{x \in X} h(x, y) \, . \tag{4.72}$$

If the certainty index $v[\bar{x}(\omega) \cong x]$ given by an expert depends on the value of y for the same ω (i.e. if the expert changes the value $h_x(x)$ when he obtains the value y for the element ω "under observation") then $h_x(x \mid y)$ may be called a *conditional certainty distribution*. The variables $\bar{x}$, $\bar{y}$ are called independent when

$$h_x(x \mid y) = h_x(x) \, , \qquad h_y(y \mid x) = h_y(y) \; .$$

Using (4.1) for the conjunction in multi-valued logic we obtain

$$h(x, y) = v(\bar{x} \cong x \wedge \bar{y} \cong y) = \min\{h_x(x), h_y(y \mid x)\} = \min\{h_y(y), h_x(x \mid y)\} \, . \tag{4.73}$$

In the discrete case, where $X = \{x_1, x_2, ..., x_m\}$ and $Y = \{y_1, y_2, ..., y_n\}$, the conditional certainty distribution may be presented as a matrix of conditional certainty indexes

$$\begin{bmatrix} h_x(x_1 \mid y_1) & h_x(x_1 \mid y_2) & \dots & h_x(x_1 \mid y_n) \\ h_x(x_2 \mid y_1) & h_x(x_2 \mid y_2) & \dots & h_x(x_2 \mid y_n) \\ \dots & \dots & \dots & \dots \\ h_x(x_m \mid y_1) & h_x(x_m \mid y_2) & \dots & h_x(x_m \mid y_n) \end{bmatrix}.$$

According to (4.71) and (4.73)

$$h_x(x_i) = \max\{\min[h_x(x_i \mid y_1), h_y(y_1)], \min[h_x(x_i \mid y_2), h_y(y_2)], \\ \dots, \min[h_x(x_i \mid y_2\, y_n), h_y(y_n)]\}, \qquad i = 1, 2, \dots, m.$$

In this way, by knowing the matrix of conditional certainty indexes and the certainty distribution $h_y(y_1), h_y(y_2), \dots, h_y(y_n)$, one can determine the certainty distribution $h_x(x_1), h_x(x_2), \dots, h_x(x_m)$.

By taking into account the relationships between the certainty distributions one can see that they cannot be given independently by an expert. If the expert gives $h_x(x)$ and $h_y(y \mid x)$ or $h_y(y)$ and $h_x(x \mid y)$ then $h(x, y)$ is already determined by (4.73). The joint distribution $h(x, y)$ given by an expert determines $h_x(x)$ (4.71) and $h_y(y)$ (4.72) but does not determine $h_x(x \mid y)$ and $h_y(y \mid x)$. In such a case only sets of functions $h_x(x \mid y)$ and $h_y(y \mid x)$ satisfying (4.73) are determined.

4.5 Functions of Uncertain Variables

Consider now a function $\Phi : X \rightarrow Y$, $Y \subseteq R^k$, i.e. $y = \Phi(x)$. We say that the uncertain variable $\bar{y} = < Y, h_y(y) >$ is a function of the uncertain variable $\bar{x} = < X, h_x(x) >$, i.e. $\bar{y} = \Phi(\bar{x})$ where the certainty distribution $h_y(y)$ is determined by $h_x(x)$ and Φ:

$$h_y(y) = v(\bar{y} \cong y) = \max_{x \in D_x(y)} h_x(x) \qquad (4.74)$$

where

$$D_x(y) = \{x \in X : \Phi(x) = y\} .$$

If $y = \Phi(x)$ is one-to-one mapping and $x = \Phi^{-1}(y)$ then

$$D_x(y) = \{\, \Phi^{-1}(y) \,\}$$

and

$$h_y(y) = h_x[\, \Phi^{-1}(y) \,].$$

In this case, according to (4.68) and (4.69),

$$\mathrm{M}_y(\bar{y}) = \sum_{i=1}^{m} \Phi(x_i) h_x(x_i) [\sum_{j=1}^{m} h_x(x_j)]^{-1} .$$

For C-uncertain variables, C-certainty distribution $h_{cy}(y) = v_c(\bar{y} \cong y)$ may be determined in two ways:

1. According to (4.67)

$$h_{cy}^{\mathrm{I}}(y) = \begin{cases} \frac{1}{2} h_y(y) & \text{if} \quad \max\limits_{\breve{y} \in Y - \{y\}} h_y(\breve{y}) = 1 \\ 1 - \frac{1}{2} \max\limits_{\breve{y} \in Y - \{y\}} h_y(\breve{y}) & \text{otherwise}. \end{cases} \tag{4.75}$$

where $h_y(y)$ is determined by (4.74).

2. According to (4.66)

$$h_{cy}^{\mathrm{II}}(y) = \begin{cases} \frac{1}{2} \max\limits_{x \in D_x(y)} h_x(x) & \text{if} \quad \max\limits_{x \in \overline{D}_x} h_x(x) = 1 \\ 1 - \frac{1}{2} \max\limits_{x \in \overline{D}_x(y)} h_x(x) & \text{otherwise} . \end{cases} \tag{4.76}$$

Theorem 4.5.

$$h_{cy}^{\mathrm{I}}(y) = h_{cy}^{\mathrm{II}}(y) .$$

Proof: It is sufficient to prove that

$$\max_{\breve{y} \in Y - \{y\}} h_y(\breve{y}) = \max_{x \in \overline{D}_x} h_x(x) .$$

From (4.74)

$$\max_{\breve{y} \in Y - \{y\}} h_y(\breve{y}) = \max_{\breve{y} \in Y - \{y\}} [\max_{x \in D_x(\breve{y})} h_x(x)] .$$

Note that if $y_1 \neq y_2$ then $D_x(y_1) \cap D_x(y_2) = \varnothing$. Consequently, $D_x(\breve{y}) \cap D_x(y) = \varnothing$ and

$$\bigcup_{\breve{y} \in Y - \{y\}} D_x(\breve{y}) = \overline{D}_x(y) .$$

Therefore

$$\max_{\breve{y} \in Y - \{y\}} [\max_{x \in D_x(\breve{y})} h_x(x)] = \max_{x \in \overline{D}_x(y)} h_x(x) . \qquad \square$$

It is important to note that $h_{cy}(y)$ is not determined by $h_{cx}(x)$. To determine $h_{cy}(y)$ it is necessary to know $h_x(x)$ and to use (4.76), or (4.74) and (4.75).

Example 4.5.

Determine the certainty distribution of the uncertain variable $\bar{y} = \bar{x}^2$ where $\bar{x}$ has a triangular certainty distribution (Fig. 4.2)

$$h_x(x) = \begin{cases} \frac{1}{d}(x-c)+1 & \text{for } c-d \le x \le c \\ -\frac{1}{d}(x-c)+1 & \text{for } c \le x \le c+d \\ 0 & \text{otherwise} \end{cases}$$

and $d > 0$. If $c-d \ge 0$ or $c+d \le 0$ then $y = x^2$ is one-to-one mapping for x such that $h_x(x) > 0$. In this case

$$h_y(y) = \begin{cases} \frac{1}{d}(\sqrt{y}-c)+1 & \text{for } (c-d)^2 \le y \le c^2 \\ -\frac{1}{d}(\sqrt{y}-c)+1 & \text{for } c^2 \le y \le (c+d)^2 \\ 0 & \text{otherwise .} \end{cases}$$

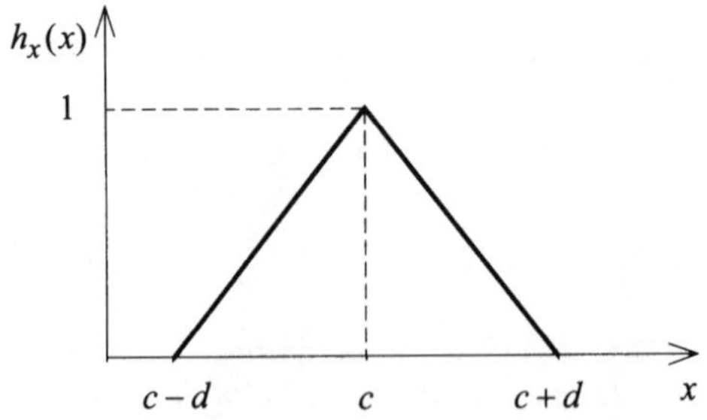

Figure 4.2. Example of certainty distribution $h_x(x)$

Assume that $c-d < 0$ and $c+d > 0$. Now $D_x(y) = \{-\sqrt{y}, \sqrt{y}\}$ and according to (4.74)

$$h_y(y) = \max\{h_x(-\sqrt{y}), h_x(\sqrt{y})\}.$$

It is easy to see that for $c \ge 0$ $h_x(\sqrt{y}) \ge h_x(-\sqrt{y})$ and for $c \le 0$ $h_x(\sqrt{y}) \le h_x(-\sqrt{y})$. The result is then as follows:

1. For $c \geq 0$

$$h_y(y) = \begin{cases} \frac{1}{d}(\sqrt{y} - c) + 1 & \text{for} \quad 0 \leq y \leq c^2 \\ -\frac{1}{d}(\sqrt{y} - c) + 1 & \text{for} \quad c^2 \leq y \leq (c+d)^2 \\ 0 & \text{otherwise} \ . \end{cases}$$

2. For $c \leq 0$

$$h_y(y) = \begin{cases} \frac{1}{d}(-\sqrt{y} - c) + 1 & \text{for} \quad c^2 \leq y \leq (c-d)^2 \\ -\frac{1}{d}(-\sqrt{y} - c) + 1 & \text{for} \quad 0 \leq y \leq c^2 \\ 0 & \text{otherwise} \ . \end{cases}$$

The distributions $h_y(y)$ in the two cases are illustrated in Fig. 4.3. □

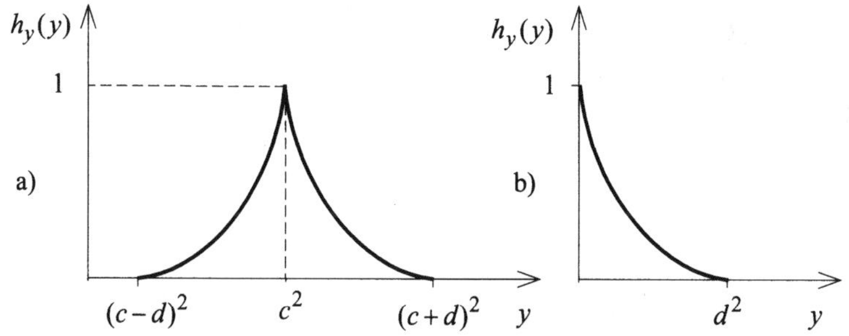

Figure 4.3. Certainty distribution $h_y(y)$: a) for $c - d \geq 0$, b) for $c = 0$

For the function $\bar{y} = \Phi(\bar{x})$ where $\bar{x}$ is a pair of variables $(\bar{x}_1, \bar{x}_2)$, $x_{1,2} \in X$, according to (4.74)

$$h_y(y) = \max_{(x_1, x_2) \in D(y)} h(x_1, x_2) \ ,$$

$h(x_1, x_2)$ is determined by (4.73) for $x = x_1$, $y = x_2$, and

$$D(y) = \{(x_1, x_2) \in X \times X : \Phi(x_1, x_2) = y\} \ .$$

Consider one-dimensional variables $\bar{x}_1, \bar{x}_2 \in R^1$ and the function

$$\bar{y} = k_1 \bar{x}_1 + k_2 \bar{x}_2$$

where k_1, k_2 are real numbers. In this case

$$D(y) = \{(x_1, x_2) : k_1 x_1 + k_2 x_2 = y\} .$$

Assume that $\bar{x}_1$, $\bar{x}_2$ are independent variables described by the certainty distributions $h_{x1}(x_1)$ and $h_{x2}(x_2)$, respectively. Then

$$h_y(y) = \max_{x_1} \min\{h_{x1}(x_1), h_{x2}(\frac{y - k_1 x_1}{k_2})\}$$

$$= \max_{x_2} \min\{h_{x1}(\frac{y - k_2 x_2}{k_1}), h_{x2}(x_2)\} . \tag{4.77}$$

Example 4.6.
Determine the certainty distribution for $\bar{y} = \bar{x}_1 + \bar{x}_2$ where $\bar{x}_1, \bar{x}_2$ are independent variables with parabolic certainty distributions (Fig. 4.4):

$$h_{x1}(x_1) = \begin{cases} -(x_1 - c_1)^2 + 1 & \text{for } c_1 - 1 \le x_1 \le c_1 + 1 \\ 0 & \text{otherwise} \end{cases} .$$

and $h_{x2}(x_2)$ is the same with x_2 and c_2 in place of x_1 and c_1, respectively; $c_1 \ge 1$, $c_2 \ge 1$. Let us use the first part of (4.77). It is easy to show that $h_y(y) = h_{x1}(x_1^*)$ where x_1^* is the solution of the equation

$$-(x_1 - c_1)^2 + 1 = -(y - x_1 - c_2)^2 + 1 .$$

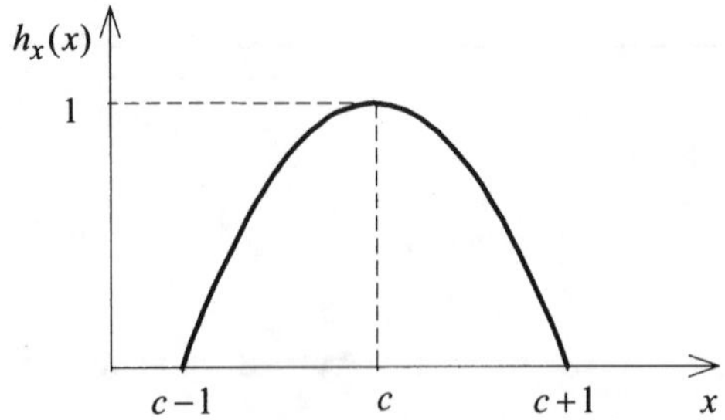

Figure 4.4. Parabolic certainty distribution

For $y \ne c_1 + c_2$ we obtain

$$x_1^* = \frac{y + c_1 - c_2}{2}$$

and $h_{x1}(x_1^*) \ge 0$ for $c_1 - 1 \le x_1^* \le c_1 + 1$, i.e. for

$$c_1 + c_2 - 2 \le y \le c_1 + c_2 + 2 .$$

For $y = c_1 + c_2$

$$\bigwedge_{-\infty \le x_1 \le \infty} [h_{x1}(x_1) = h_{x2}(y - x_1)].$$

Then

$$h_y(y) = \begin{cases} -[\frac{y-(c_1+c_2)}{2}]^2 + 1 & \text{for} \quad c_1 + c_2 - 2 \le y \le c_1 + c_2 + 2 \\ 0 & \text{otherwise} \end{cases}.$$

The distribution $h_y(y)$ is presented in Fig. 4.5.

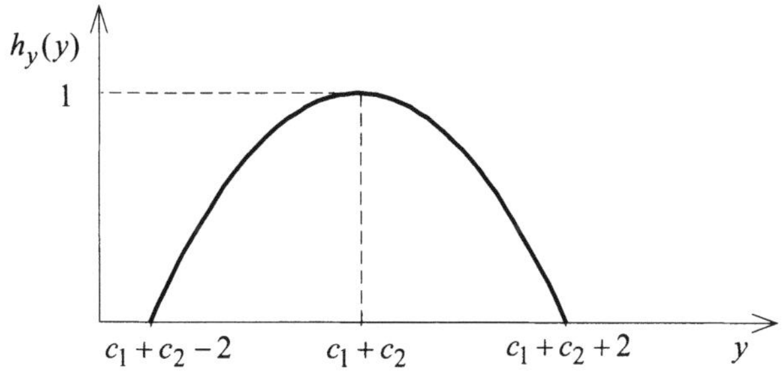

Figure 4.5. Certainty distribution of $\bar{y} = \bar{x}_1 + \bar{x}_2$ in Example 4.6

Example 4.7.
The uncertain variable $\bar{x}_1$ is described by $X_1 = \{0, 1, 4, 5, 6\}$ and the corresponding values of $h_{x1}(x_1)$: $(0.2, 0.3, 1, 0.4, 0.6)$, i.e. $h_{x1}(0) = 0.2$, $h_{x1}(1) = 0.3$, etc.

The uncertain variable $\bar{x}_2$ is described by $X_2 = \{-3, -2, 2\}$ and the corresponding values of $h_{x2}(x_2)$: $(0.5, 1, 0.8)$. The set of values of $\bar{y} = \bar{x}_1 + \bar{x}_2$ is as follows:

$$Y = \{-3, -2, -1, 1, 2, 3, 4, 6, 7, 8\}.$$

By using (4.77) we obtain the corresponding values of $h_y(y)$ (Fig. 4.6):

$$(0.2, 0.3, 0.3, 0.5, 1, 0.5, 0.6, 0.8, 0.4, 0.6).$$

For example,

1. $-3 = 0 - 3$. Then $h_y(-3) = \min(0.2, 0.5) = 0.2$.
2. $-2 = 0 - 2$ or $-2 = 1 - 3$. Then

$$h_y(-2) = \max\{\min(0.2, 1), \min(0.3, 0.5)\} = 0.3.$$

3. $2 = 0 + 2$ or $2 = 4 - 2$ or $2 = 5 - 3$. Then

$$h_y(2) = \max\{\min(0.2, 0.8), \min(1, 1), \min(0.4, 0.5)\} = 1.$$

□

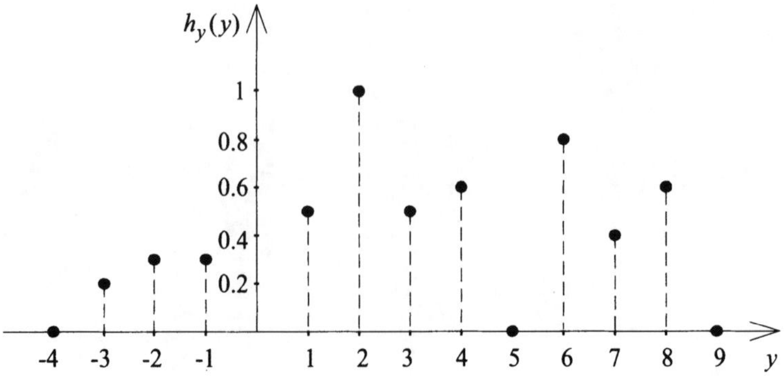

Figure 4.6. Certainty distribution of $\bar{y} = \bar{x}_1 + \bar{x}_2$ in Example 4.7

5 Application of Uncertain Variables

The purpose of this chapter is to show how uncertain variables may be applied to analysis and decision problems for a static plant. In the parametric case, we assume that the unknown parameters in the function or the relation describing the plant are values of uncertain variables described by certainty distributions given by an expert. In the non-parametric case, the plant is described by the conditional certainty distribution characterizing the expert's knowledge of the plant. The considerations are analogous to those for the random variables in Chapter 3.

5.1 Analysis Problem for a Functional Plant

Let us consider a static plant with input vector $u \in U$ and output vector $y \in Y$, where U and Y are real number vector spaces. When the plant is described by a function $y = \Phi(u)$, the analysis problem consists in finding the value y for the given value u. Consider now the plant described by $y = \Phi(u, x)$ where $x \in X$ is an unknown vector parameter which is assumed to be a value of an uncertain variable $\bar{x}$ with the certainty distribution $h_x(x)$ given by an expert. Then y is a value of an uncertain variable $\bar{y}$ and for the fixed u, $\bar{y}$ is the function of $\bar{x}$: $\bar{y} = \Phi(u, \bar{x})$.

Analysis problem: For the given Φ, $h_x(x)$ and u find the certainty distribution $h_y(y)$ of the uncertain variable $\bar{y}$. Having $h_y(y)$, one can determine M_y and

$$\hat{y} = \arg\max_{y \in Y} h_y(y), \qquad \text{i.e.} \qquad h_y(\hat{y}) = 1 .$$

According to (4.74)

$$h_y(y;u) = v(\bar{y} \cong y) = \max_{x \in D_x(y;u)} h_x(x) \tag{5.1}$$

where $D_x(y;u) = \{x \in X : \Phi(u, x) = y\}$. If Φ as a function of x is one-to-one mapping and $x = \Phi^{-1}(u, y)$ then

$$h_y(y;u) = h_x[\Phi^{-1}(u, y)]$$

and $\hat{y} = \Phi(u, \hat{x})$ where $\hat{x} = \arg\max h_x(x)$. From the definition of the certainty distributions h and h_c it is easy to note that in both the continuous and discrete cases $\hat{y} = \hat{y}_c$ where $\hat{y}_c = \arg\max h_{cy}(y)$ and $h_{cy}(y)$ is a certainty distribution of $\overline{y}$ considered as a C-uncertain variable.

Example 5.1.

Let $u, x \in R^2$, $u = (u^{(1)}, u^{(2)})$, $x = (x^{(1)}, x^{(2)})$, $y \in R^1$,

$$y = x^{(1)}u^{(1)} + x^{(2)}u^{(2)},$$

$x^{(1)} \in \{ 3, 4, 5, 6 \}$, $x^{(2)} \in \{ 5, 6, 7 \}$ and the corresponding values of h_{x1}, h_{x2} given by an expert are (0.3, 0.5, 1, 0.6) for $x^{(1)}$ and (0.8, 1, 0.4) for $x^{(2)}$. Assume that $\overline{x}^{(1)}$ and $\overline{x}^{(2)}$ are independent, i.e.

$$h_x(x_i^{(1)}, x_j^{(2)}) = \min\{h_{x1}(x_i^{(1)}), h_{x2}(x_j^{(2)})\}.$$

Then for $x = (x^{(1)}, x^{(2)}) \in \{(3, 5), (3, 6), (3, 7), (4, 5), (4, 6), (4, 7), (5, 5), (5, 6), (5, 7), (6, 5), (6, 6), (6, 7)\}$ the corresponding values of h_x are (0.3, 0.3, 0.3, 0.5, 0.5, 0.4, 0.8, 1, 0.4, 0.6, 0.6, 0.4).

Let $u^{(1)} = 2$, $u^{(2)} = 1$. The values of $y = 2x^{(1)} + x^{(2)}$ corresponding to the set of pairs $(x^{(1)}, x^{(2)})$ are the following: {11, 12, 13, 13, 14, 15, 15, 16, 17, 17, 18, 19}. Then $h_y(11) = h_x(3, 5) = 0.3$, $h_y(12) = h_x(3, 6) = 0.3$, $h_y(13) = \max\{h_x(3, 7), h_x(4, 5)\} = 0.5$, $h_y(14) = h_x(4, 6) = 0.5$, $h_y(15) = \max\{h_x(4, 7), h_x(5, 5)\} = 0.8$, $h_y(16) = h_x(5, 6) = 1$, $h_y(17) = \max\{h_x(5, 7), h_x(6, 5)\} = 0.6$, $h_y(18) = h_x(6, 6) = 0.6$, $h_y(19) = h_x(6, 7) = 0.4$.

For $h_y(y)$ we have $\hat{y} = 16$. Using (4.68) and (4.69) for $\overline{y}$ we obtain $\overline{h}_y = 5$, $\mathrm{M}_y = \frac{77}{5} = 15.40$.

Using (4.75), we obtain the corresponding values of $h_{yc}(y)$: (0.15, 0.15, 0.25, 0.25, 0.4, $1 - \frac{0.8}{2} = 0.6$, 0.3, 0.3, 0.2). Then $\hat{y}_c = \hat{y} = 16$, $v_c(\overline{y} \cong 16) = 0.6$, $\overline{h}_{yc} = 2.6$, $\mathrm{M}_{yc} = 15.43 \approx \mathrm{M}_y$. □

5.2 Decision Making Problem for a Functional Plant

For the functional system $y = \Phi(u)$ the basic decision problem consists in finding the decision u^* for the given desirable value y^*. Consider now the system with the unknown parameter x, described in Sect. 5.1.

Decision problem:

Version I. To find the decision u^* maximizing $v(\bar{y} \cong y^*)$.

Version II. To find u^* such that $\mathrm{M}_y(\bar{y};u) = y^*$ where M_y is the mean value of $\bar{y} = \Phi(u, \bar{x})$ for the fixed u.

Version III. To find u^* minimizing $\mathrm{M}_s(\bar{s};u)$ where $s = \varphi(y, y^*)$ is a quality index, e.g. $s = (y - y^*)^{\mathrm{T}}(y - y^*)$ where y and y^* are column vectors.

When $\bar{x}$ is assumed to be a *C*-uncertain variable, the formulations are the same with v_c, M_{cy}, M_{cs} instead of v, M_y, M_s. It is worth noting that the decision problem statements are analogous to those in the probabilistic approach where x is assumed to be a value of a random variable with the known probability distribution. In each version $h_y(y;u)$ should be determined according to (5.1). Then, in version I u^* is the value of u maximizing $h_y(y^*;u)$, i.e. u^* is the solution of the equation $\varepsilon(u) = y^*$ where $\hat{y} = \varepsilon(u)$ is a value of y maximizing $h_y(y;u)$. In version II, u^* is obtained as a solution of the equation $\mathrm{M}_y(\bar{y};u) = \hat{y}$.

In version III, for the determination of $\mathrm{M}_s(\bar{s};u)$ one should find

$$h_s(s;u) = \max_{y \in D_y(s)} h_y(y;u) \tag{5.2}$$

where $D_y(s) = \{y \in Y : \varphi(y, y^*) = s\}$.

When $\bar{x}$ is considered as a *C*-uncertain variable, it is necessary to determine h_{cy} using (4.75) or (4.76) and in version II to find $\mathrm{M}_{cy}(\bar{y};u)$. In version III it is necessary to find h_{cs} according to (4.75) or (4.76) with (s, x) instead of (y, x) and then to determine $\mathrm{M}_{cs}(\bar{s};u)$. Using (4.75) it is easy to see that if $\hat{y}$ is the only value for which $h_y = 1$ then $\hat{y}_c = \hat{y}$ where $\hat{y}$ is the value maximizing h_{cy}, and $\mathrm{M}_{cy} \approx \mathrm{M}_y$. Consequently, in this case the results u^* in version I are the same and in version II are approximately the same for the uncertain and *C*-uncertain variable, and for $u = u^*$ in version I $v(\bar{y} \cong y^*) = 1$, $v_c(\bar{y} \cong y^*) < 1$.

Example 5.2.

Let $u, y, x \in R^1$, $y = xu$, $X = \{3, 4, 5, 6, 7\}$ and the corresponding values of $h_x(x)$ are $(0.5, 1, 0.3, 0.1, 0.1)$. Using (4.67) or (4.75) for x we obtain the corresponding values of h_{cx}: $(0.25, 0.75, 0.15, 0.05, 0.05)$.

In version I $\hat{y} = 4u$, i.e. $u^* = 0.25y^*$, $v(\bar{y} \cong y^*) = 1$, $v_c(\bar{y} \cong y^*) = 0.75$.

In our example $Y = \{3u, 4u, 5u, 6u, 7u\}$, the values of $h_y(y;u)$ are the same as h_x, the values of $\bar{h}_y$ are (0.25, 0.5, 0.15, 0.05, 0.05) and the values of $h_{cy} = 1.25\bar{h}_{cy}$ are the same as h_{cx}. In version II, using (4.69) for y we obtain $\mathrm{M}_y = 4.15u$ and $\mathrm{M}_{cy} = 4.12u$. Then in both cases the result is approximately the same: $u^* \approx 0.24y^*$.

Let in version III $s = (y - y^*)^2$. Then $M_s(\bar{s};u)$ is equal to $0.25(3u - y^*)^2 + 0.5(4u - y^*)^2 + 0.15(5u - y^*)^2 + 0.05(6u - y^*)^2 +$

$0.05(7u - y^*)^2 \triangleq \overline{\mathrm{M}}_s(u)$ except for

$$u = \frac{2y^*}{x_i + x_j} \triangleq u_d, \quad x_i \neq x_j, \quad x_i, x_j \in \{3, 4, 5, 6, 7\}. \tag{5.3}$$

For example, for $x_1 = 3$, $x_2 = 4$ and $u = u_d$ we have $(3u - y^*)^2 = (4u - y^*)^2$. Consequently, $s \in \{(4u - y^*)^2, (5u - y^*)^2, (6u - y^*)^2, (7u - y^*)^2\}$, according to (5.2) the values of h_s are $[\max(0.5, 1), 0.3, 0.1, 0.1]$ and

$$\mathrm{M}_s(\bar{s};u) = \frac{1}{1.5}[(4u_d - \hat{y})^2 + 0.3(5u_d - \hat{y})^2 + 0.1(6u_d - \hat{y})^2 + 0.1(7u_d - \hat{y})^2].$$

Then $\mathrm{M}_s(\bar{s};u)$ is a discontinuous function of u. The value of u minimizing $\overline{\mathrm{M}}_s(u)$ is

$$u_{\min} = \frac{4.15}{15.8} y^* \approx 0.26 y^*.$$

From the sensitivity point of view it is reasonable not to take into account the points of discontinuity u_d (5.3) and to accept $u^* = u_{\min}$. □

5.3 External Disturbances

The considerations may be easily extended for a plant with external disturbances, described by a function

$$y = \Phi(u, z, x) \tag{5.4}$$

where $z \in Z$ is a vector of the disturbances which can be measured.

Analysis problem: For the given Φ, $h_x(x)$, u and z find the certainty distribution

$h_y(y)$.

According to (4.74)

$$h_y(y;u,z) = v(\bar{y} \cong y) = v[\bar{x} \,\tilde{\in}\, D_x(y;u,z)] = \max_{x \in D_x(y;u,z)} h_x(x) \quad (5.5)$$

where

$$D_x(y;u,z) = \{x \in X : \Phi(u,z,x) = y\}.$$

Having $h_y(y;u,z)$ one can determine the mean value

$$\mathrm{M}_y(\bar{y};u,z) = \int_Y y h_y(y;u,z)dy \cdot [\int_Y h_y(y;u,z)dy]^{-1} \triangleq \Phi_b(u,z) \quad (5.6)$$

(for the continuous case) and

$$\hat{y} = \arg\max_{y \in Y} h_y(y;u,z),$$

i.e. such a value $\hat{y}$ that $h_y(\hat{y};u,z) = 1$. If Φ as a function of x is one-to-one mapping and $x = \Phi^{-1}(u,z,y)$ then

$$h_y(y;u,z) = h_x[\Phi^{-1}(u,z,y)] \quad (5.7)$$

and $\hat{y} = \Phi(u,z,\hat{x})$ where $\hat{x}$ satisfies the equation $h_x(x) = 1$. It is easy to note that $\hat{y} = \hat{y}_c$ where

$$\hat{y}_c = \arg\max_{y \in Y} h_{cy}(y;u,z)$$

and h_{cy} is a certainty distribution for the *C*-uncertain variable.

Decision problem: For the given Φ, $h_x(x)$, z and y^*

I. One should find $u \triangleq u_a$ maximizing $v(\bar{y} \cong y^*)$.

II. One should find $u \triangleq u_b$ such that $\mathrm{M}_y(\bar{y}) = y^*$.

Versions I, II correspond to versions I, II of the decision problem presented in 5.2. The third version presented in Sect. 5.2 is much more complicated and will not be considered. In version I

$$u_a = \arg\max_{u \in U} \Phi_a(u,z) \triangleq \Psi_a(z) \quad (5.8)$$

where $\Phi_a(u,z) = h_y(y^*;u,z)$ and h_y is determined according to (5.5). The result u_a is a function of z if u_a is a unique value maximizing Φ_a for the given z.

In version II one should solve the equation

$$\Phi_b(u,z) = y^* \tag{5.9}$$

where the function Φ_b is determined by (5.6). If Equation (5.9) has a unique solution with respect to u for the given z then as a result one obtains $u_b = \Psi_b(z)$. The functions Ψ_a and Ψ_b are two versions of the decision algorithm $u = \Psi(z)$ in an open-loop decision system (Fig. 5.1). It is worth noting that u_a is a decision for which $v(\bar{y} \cong y^*) = 1$.

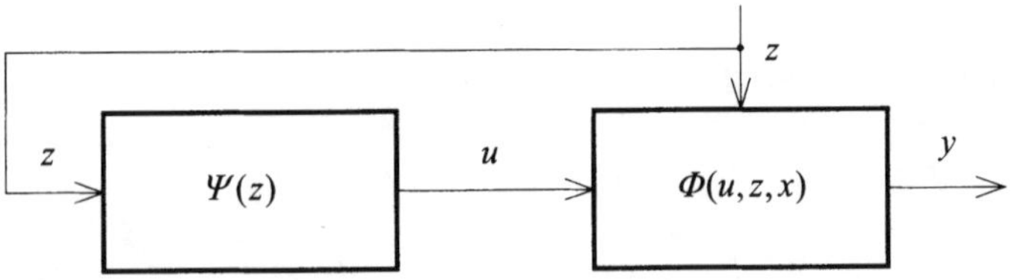

Figure 5.1. Open-loop decision system

The functions Φ_a, Φ_b are the results of two different ways of *determinization* of the uncertain plant, and the functions Ψ_a, Ψ_b are the respective decision algorithms based on the *knowledge of the plant* (KP):

$$\text{KP} = < \Phi, h_x >. \tag{5.10}$$

Assume that the equation

$$\Phi(u,z,x) = y^*$$

has a unique solution with respect to u:

$$u \stackrel{\Delta}{=} \Phi_d(z,x). \tag{5.11}$$

The relationship (5.11) together with the certainty distribution $h_x(x)$ may be considered as a *knowledge of the decision making* (KD):

$$\text{KD} = < \Phi_d, h_x >, \tag{5.12}$$

obtained by using KP and y^*. Equation (5.11) together with h_x may also be called an *uncertain decision algorithm* in the open-loop decision system. The determinization of this algorithm leads to two versions of the deterministic decision algorithm Ψ_d, corresponding to versions I and II of the decision problem:

Version I.

$$u_{ad} = \arg\max_{u \in U} h_u(u\,;z) \triangleq \Psi_{ad}(z) \tag{5.13}$$

where

$$h_u(u\,;z) = \max_{x \in D_x(u\,;z)} h_x(x) \tag{5.14}$$

and

$$D_x(u\,;z) = \{x \in X : u = \Phi_d(z,x)\}\,.$$

Version II.

$$u_{bd} = \mathrm{M}_u(\bar{u}\,;z) \triangleq \Psi_{bd}(z)\,. \tag{5.15}$$

The decision algorithms Ψ_{ad} and Ψ_{bd} are based directly on the knowledge of the decision making. Two concepts of the determination of deterministic decision algorithms are illustrated in Figs. 5.2 and 5.3. In the first case (Fig. 5.2) the decision algorithms $\Psi_a(z)$ and $\Psi_b(z)$ are obtained via the determinization of the knowledge of the plant KP. In the second case (Fig. 5.3) the decision algorithms $\Psi_{ad}(z)$ and $\Psi_{bd}(z)$ are based on the determinization of the knowledge of the decision making KD obtained from KP for the given y^*. The results of these two approaches may be different.

Theorem 5.1. For the plant described by KP in the form (5.10) and for KD in the form (5.12), if there exists an inverse function $x = \Phi^{-1}(u,z,y)$ then

$$\Psi_a(z) = \Psi_{ad}(z)\,.$$

Proof: According to (5.7) and (5.13)

$$h_y(y^*;u,z) = h_x[\Phi^{-1}(u,z,y^*)]\,,$$

$$h_u(u\,;z) = h_x[\Phi^{-1}(u,z,y^*)]\,.$$

Then, by using (5.8) and (5.13) we obtain $\Psi_a(z) = \Psi_{ad}(z)$. □

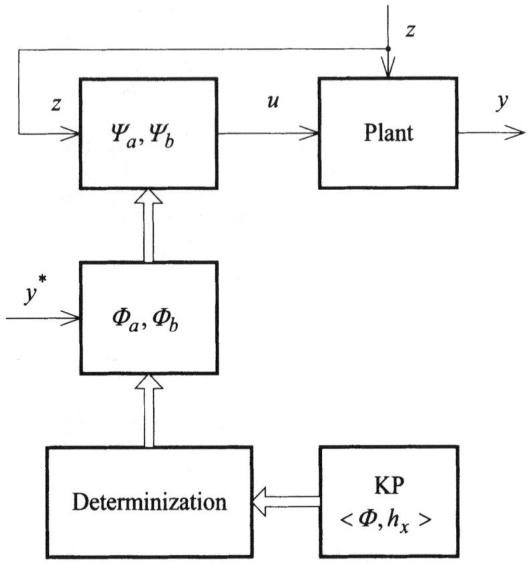

Figure 5.2. Decision system with determinization – the first case

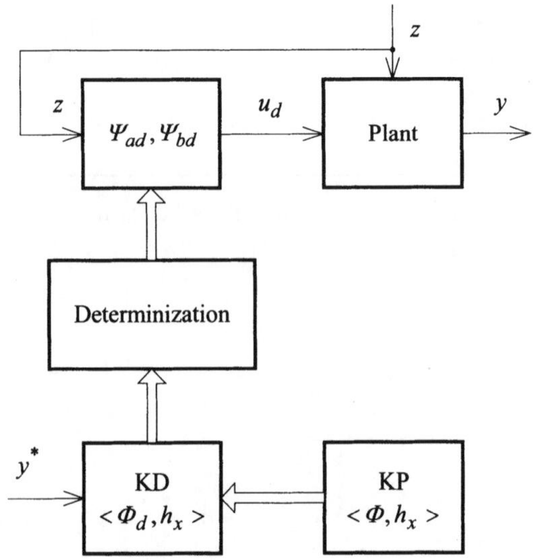

Figure 5.3. Decision system with determinization – the second case

It is worth noting that the considerations using uncertain variables are analogous to those using random variables, presented in Sect. 3.2. In particular, the formulas (5.5), (5.6), (5.8), (5.9), (5.13) and (5.15) correspond to the formulas (3.5), (3.7), (3.8), (3.9), (3.13) and (3.14), respectively. Consequently, Figs. 5.2 and 5.3 are analogous to Figs. 3.2 and 3.3, respectively.

Example 5.3.
Let $u, y, x, z \in R^1$ and

$$y = xu + z .$$

Then

$$M_y(\bar{y}) = u\,M_x(\bar{x}) + z$$

and from the equation $M_y(\bar{y}) = \hat{y}$ we obtain

$$u_b = \Psi_b(z) = \frac{y^* - z}{M_x(\bar{x})} .$$

The uncertain decision algorithm is

$$u = \Phi_d(z, x) = \frac{y^* - z}{x}$$

and after the determinization

$$u_{bd} = \Psi_{bd}(z) = (y^* - z) M_x(\bar{x}^{-1}) \neq \Psi_b(z) . \qquad \square$$

This very simple example shows that the deterministic decision algorithm $\Psi_b(z)$ obtained via the determinization of the uncertain plant may differ from the deterministic decision algorithm $\Psi_{bd}(z)$ obtained as a result of the determinization of the uncertain decision algorithm.

5.4 Analysis for Relational Plants with Uncertain Parameters

Let us consider the plant described by a relation $R(u, y; x) \subseteq U \times Y$ where $x \in X$ is an unknown vector parameter which is assumed to be a value of an uncertain variable $\bar{x}$ with the certainty distribution $h_x(x)$ given by an expert. Now the sets of all possible values y in (2.5) and (2.6) depend on x. For the given set of inputs D_u we have

$$D_y(x) = \{ y \in Y : \bigvee_{u \in D_u} (u, y) \in R(u, y; x) \}$$

and for the given value u

$$D_y(u; x) = \{ y \in Y : \; (u, y) \in R(u, y; x) \} .$$

The analysis may consist in evaluating the input with respect to a set $D_y \subset Y$ given by a user. We can consider two formulations with the different practical interpretations: the determination of $v[D_y \tilde{\supseteq} D_y(\bar{x})]$ (version I) or the determination of $v[D_y(\bar{x}) \subseteq D_y]$ (version II). The analogous formulations may be considered for the given u, with $D_y(u\,;\bar{x})$ in place of $D_y(\bar{x})$.

Analysis problem – version I: For the given $R(u,y\,;x)$, $h_x(x)$, u and $D_y \subset Y$ one should determine

$$v[D_y \tilde{\subseteq} D_y(u\,;\bar{x})] \stackrel{\Delta}{=} g(D_y,u)\,. \tag{5.16}$$

The value (5.16) denotes the certainty index of the soft property: "the set of all possible outputs approximately contains the set D_y given by a user" or "the approximate value of $\bar{x}$ is such that $D_y \subseteq D_y(u\,;x)$" or "the approximate set of the possible outputs contains all values from the set D_y". Let us note that

$$v[D_y \tilde{\subseteq} D_y(u\,;\bar{x})] = v[\bar{x} \,\tilde{\in}\, D_x(D_y,u)] \tag{5.17}$$

where

$$D_x(D_y,u) = \{x \in X : \quad D_y \subseteq D_y(u\,;x)\}\,. \tag{5.18}$$

Then

$$g(D_y,u) = \max_{x \in D_x(D_y,u)} h_x(x)\,. \tag{5.19}$$

In particular, for $D_y = \{y\}$ (a singleton), the certainty index that the given value y may occur at the output of the plant is

$$g(y,u) = \max_{x \in D_x(y,u)} h_x(x) \tag{5.20}$$

where

$$D_x(y,u) = \{x \in X : \quad y \in D_y(u\,;x)\}\,. \tag{5.21}$$

When $\bar{x}$ is considered as a *C*-uncertain variable, it is necessary to determine

$$v[\bar{x} \,\tilde{\in}\, \overline{D}_x(D_y,u)] = \max_{x \in \overline{D}_x(D_y,u)} h_x(x)$$

where $\overline{D}_x(D_y,u) = X - D_x(D_y,u)$. Then, according to (4.59)

$$v_c[D_y \,\tilde{\subseteq}\, D_y(u\,;\bar{x})] = \frac{1}{2}\{v[\bar{x} \,\tilde{\in}\, D_x(D_y,u)] + 1 - v[\bar{x} \,\tilde{\in}\, \overline{D}_x(D_y,u)]\}\,.$$

The considerations may be extended for a plant described by a relation $R(u, y, z\,; x)$ where $z \in Z$ is the vector of disturbances which may be measured. For the given z

$$D_y(u, z\,; x) = \{y \in Y: \quad (u, y, z) \in R(u, y, z\,; x)\}$$

and

$$v[D_y \,\tilde{\subseteq}\, D_y(u, z\,; \bar{x})] = \max_{x \in D_x(D_y, u, z)} h_x(x)$$

where

$$D_x(D_y, u, z) = \{x \in X: \quad D_y \subseteq D_y(u, z\,; x)\}\,.$$

Consequently, the certainty index that the approximate set of the possible outputs contains all the values from the set D_y depends on z.

For the given set D_u, the formulas analogous to (5.16)–(5.21) have the following form:

$$v[D_y \,\tilde{\subseteq}\, D_y(\bar{x})] \triangleq g(D_y, D_u)\,,$$

$$v[D_y \,\tilde{\subseteq}\, D_y(\bar{x})] = v[\bar{x} \,\tilde{\in}\, D_x(D_y, D_u)]\,,$$

$$D_x(D_y, D_u) = \{x \in X: \; D_y \subseteq D_y(x)\}\,,$$

$$g(D_y, D_u) = \max_{x \in D_x(D_y, D_u)} h_x(x)\,,$$

$$g(y, D_u) = \max_{x \in D_x(y, D_u)} h_x(x)\,,$$

$$D_x(y, D_u) = \{x \in X: \; y \in D_y(x)\}\,.$$

Analysis problem – version II: For the given $R(u, y; x)$, $h_x(x)$, u and $D_y \subset Y$ one should determine

$$v[D_y(u;\bar{x}) \tilde{\subseteq} D_y] \triangleq g(D_y,u). \tag{5.22}$$

The value (5.22) denotes the certainty index of the soft property: "the set D_y given by a user contains the approximate set of all possible outputs". The formulas corresponding to (5.17), (5.18) and (5.19) are as follows:

$$v[D_y(u;\bar{x}) \tilde{\subseteq} D_y] = v[\bar{x} \in D_x(D_y,u)]$$

where

$$D_x(D_y,u) = \{x \in X: \ D_y(u;x) \subseteq D_y\}, \tag{5.23}$$

$$g(D_y,u) = \max_{x \in D_x(D_y,u)} h_x(x). \tag{5.24}$$

For the given set D_u one should determine

$$v[D_y(\bar{x}) \tilde{\subseteq} D_y] = v[\bar{x} \tilde{\in} D_x(D_y,D_u)] = \max_{x \in D_x(D_y,D_u)} h_x(x) \tag{5.25}$$

where

$$D_x(D_y,D_u) = \{x \in X: \ D_y(x) \subseteq D_y\}. \tag{5.26}$$

In the case where $\bar{x}$ is considered as a *C*-uncertain variable it is necessary to find v (5.25) and

$$v[\bar{x} \tilde{\in} \overline{D}_x(D_y,D_u)] = \max_{x \in \overline{D}_x(D_y,D_u)} h_x(x) \tag{5.27}$$

where

$$\overline{D}_x(D_y,D_u) = X - D_x(D_y,D_u).$$

Then, according to (4.59)

$$v_c[D_y(\bar{x}) \tilde{\subseteq} D_y] = \frac{1}{2}\{v[\bar{x} \tilde{\in} D_x(D_y,D_u)] + 1 - v[\bar{x} \tilde{\in} \overline{D}_x(D_y,D_u)]\}. \tag{5.28}$$

The considerations for the plant described by $R(u,y,z;x)$ are analogous to those in version I.

Example 5.4.

Let u, y, $x \in \mathrm{R}^1$, the relation R is given by the inequality $xu \le y \le 2xu$, $D_u = [u_1, u_2]$, $u_1 > 0$, $D_y = [y_1, y_2]$, $y_1 > 0$.

For these data $D_y(x) = [xu_1, 2xu_2]$ and (5.26) becomes $D_x(D_y, D_u) = [\frac{y_1}{u_1}, \frac{y_2}{2u_2}]$. Assume that x is a value of an uncertain variable $\bar{x}$ with triangular certainty distribution

$$h_x(x) = \begin{cases} 2x & \text{for } 0 \le x \le \frac{1}{2} \\ -2x+2 & \text{for } \frac{1}{2} \le x \le 1 \\ 0 & \text{otherwise} \end{cases}$$

(Fig. 5.4). Using (5.25), we obtain for $u_1 y_2 \ge 2u_2 y_1$

$$v[D_y(\bar{x}) \mathrel{\tilde{\subseteq}} D_y] = v[\bar{x} \mathrel{\tilde{\in}} D_x(D_y, D_u)]$$

$$= \begin{cases} \frac{y_2}{u_2} & \text{when} \quad y_2 \le u_2 \\ 1 & \text{when} \quad 2y_1 \le u_1 \text{ and } y_2 \ge u_2 \\ 2(1 - \frac{y_1}{u_1}) & \text{when} \quad 2y_1 \ge u_1 \text{ and } y_2 \le u_2 \\ 0 & \text{when} \quad y_1 \ge u_1 \,. \end{cases}$$

For $u_1 y_2 < 2u_2 y_1$, $D_x(D_y, D_u) = \varnothing$ and $v[D_y(\bar{x}) \mathrel{\tilde{\subseteq}} D_y] = 0$. For example, for $u_1 = 5$, $u_2 = 6$, $y_1 = 4$, $y_2 = 12$ we have $2y_1 \ge u_1$, $y_1 \le u_1$ and $v(D_y(\bar{x}) \mathrel{\tilde{\subseteq}} [4, 12]) = 0.4$. When $y_1 = 2$ we have $2y_1 \le u_1$, $y_2 \ge u_2$ and $v = 1$.

To apply the description for a *C*-uncertain variable, one should determine $v[\bar{x} \in \overline{D}_x(D_y, D_u)]$ according to (5.27):

$$v[\bar{x} \in \overline{D}_x(D_y, D_u)] = \begin{cases} 1 & \text{when} \quad y_2 \le u_2 \\ \max\left(\frac{2y_1}{u_1}, \frac{-y_2}{u_2} + 2\right) & \text{when} \quad 2y_1 \le u_1 \text{ and } y_2 \ge u_2 \\ 1 & \text{when} \quad 2y_1 \ge u_1 \,. \end{cases}$$

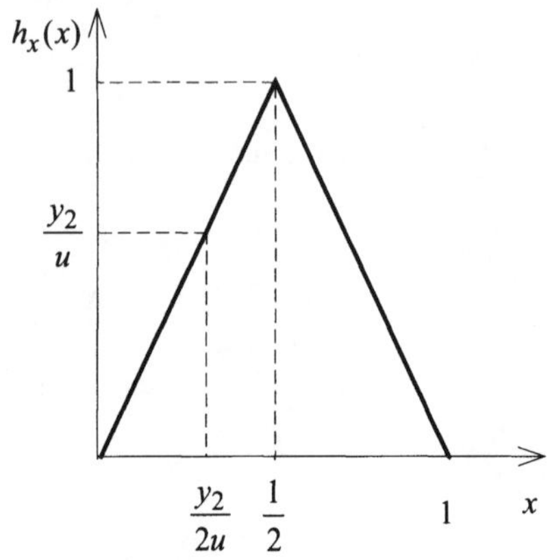

Figure 5.4. Example of certainty distribution

Then, using (5.28) we obtain $v_c[D_y(\bar{x}) \tilde{\subseteq} D_y]$. For the numerical data in the first case ($y_1 = 4$) $v[\bar{x} \in \overline{D}_x(D_y, D_u)] = 1$, $v_c = 0.2$. In the second case ($y_1 = 2$) $v[\bar{x} \in \overline{D}_x(D_y, D_u)] = 0.8$, $v_c = 0.6$. □

5.5 Decision Making for Relational Plants with Uncertain Parameters

We can formulate the different versions of the decision problem with different practical interpretations – corresponding to the formulations of the analysis problem presented in Sect. 5.4.

Decision problem – version I: For the given $R(u, y; x)$, $h_x(x)$ and $D_y \subset Y$ find

$$u^* = \arg\max_{u \in U} v[D_y \tilde{\subseteq} D_y(u; \bar{x})]. \tag{5.29}$$

In this formulation u^* is a decision maximizing the certainty index that the approximate set of the possible outputs contains the set D_y given by a user. To obtain the optimal decision one should determine the function g in (5.19) and to maximize it with respect to u, i.e.

$$u^* = \arg\max_{u \in U} \max_{x \in D_x(D_y, u)} h_x(x) \tag{5.30}$$

where $D_x(D_y,u)$ is defined by (5.18).

Decision problem – version II: For the given $R(u,y;x)$, $h_x(x)$ and $D_y \subset Y$ find

$$u^* = \arg\max_{u\in U} v[D_y(u;\bar{x}) \tilde{\subseteq} D_y]. \tag{5.31}$$

Now u^* is a decision maximizing the certainty index that the approximate set of all possible outputs (i.e. the set of all possible outputs for the approximate value of $\bar{x}$) belongs to the set D_y given by a user. To obtain the optimal decision one should determine the function g in (5.24) and to maximize it with respect to u, i.e. u^* is determined by (5.30) where $D_x(D_y,u)$ is defined by (5.23). It is worth noting that in both versions the solution may not be unique, i.e. we may obtain the set D_u of the decisions (5.29). Denote by $\hat{x}$ the value maximizing $h_x(x)$, i.e. such that $h_x(\hat{x})=1$. Then

$$D_u = \{u \in U : \hat{x} \in D_x(D_y,u)\}$$

and for every $u \in D_u$ the maximal value of the certainty index in (5.29) and (5.31) is equal to 1. To determine the set of optimal decisions D_u it is not necessary to know the form of the function $f_x(x)$. It is sufficient to know the value $\hat{x}$.

In the case where $\bar{x}$ is considered as a C-uncertain variable one should determine

$$v_c[\bar{x} \tilde{\in} D_x(D_y,u)] = \frac{1}{2}\{v[\bar{x} \tilde{\in} D_x(D_y,u)] + 1 - v[\bar{x} \tilde{\in} \overline{D}_x(D_y,u)]\} \tag{5.32}$$

where

$$v[\bar{x} \tilde{\in} \overline{D}_x(D_y,u)] = \max_{x\in \overline{D}_x(D_y,u)} h_x(x). \tag{5.33}$$

Then the optimal decision u_c^* is obtained by maximization of v_c:

$$u_c^* = \max_{u\in U} v_c[\bar{x} \tilde{\in} D_x(D_y,u)]$$

where $D_x(D_y,u)$ is defined by (5.18) in version I or by (5.23) in version II.

In a similar way as in Sect. 5.4, the considerations may be extended for the plant with the vector of external disturbances z, described by $R(u,y,z;x)$. Now the set $D_u(z)$ of the optimal decisions depends on z. In the case of the unique solution u^* for every z, we obtain the deterministic decision algorithm $u^* = \Psi(z)$ in an open-loop decision system. It is the decision algorithm based on the knowledge of the plant

$$\text{KP} = < R(u, y, z\,; x),\ h_x(x) > .$$

For the fixed x and z we may solve the decision problem such as in Sect. 2.3, i.e. determine the largest set $D_u(z\,; x)$ such that the implication

$$u \in D_u(z\,; x) \to y \in D_y$$

is satisfied. According to (2.19)

$$D_u(z\,; x) = \{u \in U : \ D_y(u, z\,; x) \subseteq D_y\} \triangleq \overline{R}(z, u\,; x)$$

where

$$D_y(u, z\,; x) = \{y \in Y : \ (u, y, z) \in R(u, y, z\,; x)\} .$$

Then we can determine the optimal decision

$$u_d = \arg\max_{u \in U} \ v[u \,\tilde{\in}\, D_u(z\,; \overline{x})] \triangleq \Psi_d(z) \tag{5.34}$$

where

$$v[u \,\tilde{\in}\, D_u(z\,; \overline{x})] = v[\overline{x} \,\tilde{\in}\, D_{xd}(D_y, u, z)]$$

and

$$D_{xd}(D_y, u, z) = \{x \in X : \ u \in D_u(z\,; x)\} .$$

Hence

$$v[u \,\tilde{\in}\, D_u(z\,; \overline{x})] = \max_{x \in D_{xd}(D_y, u, z)} h_x(x) . \tag{5.35}$$

In general, we may obtain the set D_{ud} of decisions u_d maximizing the certainty index (5.35). Let us note that the decision algorithm $\Psi_d(z)$ is based on the knowledge of the decision making

$$\text{KD} = < \overline{R}(z, u\,; x),\ h_x(x) > .$$

The relation $\overline{R}(z,u\,;x)$ or the set $D_u(z\,;x)$ may be called an *uncertain decision algorithm* in the case under consideration. It is easy to see that in this case $u_d = u^*$ for every z, i.e. $\Psi_d(z) = \Psi(z)$ where $u^* = \Psi(z)$ is the optimal decision in version II. This follows from the fact that

$$u \in D_u(z\,;x) \leftrightarrow D_y(u,z\,;x) \subseteq D_y,$$

i.e. the properties $u \in D_u(z\,;x)$ and $D_y(u,z\,;x) \subseteq D_y$ are equivalent. The optimal decision in version II $u^* = u_d$ is then the decision which with the greatest certainty index belongs to the set of decisions $D_u(z\,;x)$ for which the requirement $y \in D_y$ is satisfied. The determination of $u^* = u_d$ from (5.34) and (5.35) may be easier than from (5.31) with $D_y(u,z\,;\overline{x})$ in place of $D_y(u\,;\overline{x})$. In the case without z the optimal decision (5.31) may be obtained in the following way:

$$u^* = \arg\max_u v[u \,\tilde{\in}\, D_u(\overline{x})]$$

where

$$v[u \,\tilde{\in}\, D_u(\overline{x})] = v[\overline{x} \,\tilde{\in}\, D_{xd}(D_y,u)] = \max_{x \in D_{xd}(D_y,u)} h_x(x), \qquad (5.36)$$

and

$$\left.\begin{aligned} D_{xd}(D_y,u) &= \{x \in X : \; u \in D_u(x)\}, \\ D_u(x) &= \{u \in U : \; D_y(u\,;x) \subseteq D_y\}, \\ D_y(u\,;x) &= \{y \in Y : \; (u,y) \in R(u,y\,;x)\}. \end{aligned}\right\} \qquad (5.37)$$

Example 5.5 (decision problem – version II).
Let $u, y, x \in \mathrm{R}^1$ and $R(u,y,x)$ be given by the inequality

$$3x - u \le y \le u^2 + x^2 + 1.$$

For $D_y = [0, 2]$ the set $D_u(x)$ (5.37) is determined by

$$u \le 3x \qquad \text{and} \qquad u^2 + x^2 \le 1. \qquad (5.38)$$

Assume that x is a value of an uncertain variable $\overline{x}$ with triangular certainty distribution: $h_x = 2x$ for $0 \le x \le \frac{1}{2}$, $h_x = -2x + 2$ for $\frac{1}{2} \le x \le 1$, $h_x = 0$ otherwise. From (5.38) we have $D_x(u) = [\frac{u}{3}, \sqrt{1-u^2}]$ and the set of all possible u: $\Delta_u = [-1, \frac{3}{\sqrt{10}}]$ (the value $\frac{3}{\sqrt{10}}$ is obtained from the equations $u = 3x$, $u^2 + x^2 = 1$). It is easy to see that $\frac{1}{2} \in D_x(u)$ iff $\sqrt{1-u^2} \ge \frac{1}{2}$. Then, according to (5.36)

$$v[u \mathrel{\tilde{\in}} D_u(\bar{x})] \triangleq v(u) = \begin{cases} 1 & \text{for} \quad -\frac{\sqrt{3}}{2} \le u \le \frac{\sqrt{3}}{2} \\ 2\sqrt{1-u^2} & \text{otherwise in } \Delta_u . \end{cases} \tag{5.39}$$

For example, $v(0.5) = 1$, $v(0.9) \approx 0.88$. As the decision u^* we can choose any value from $[-\frac{\sqrt{3}}{2}, \frac{\sqrt{3}}{2}]$ and the property $D_y(u; \bar{x}) \mathrel{\tilde{\subseteq}} D_y$ is satisfied with the certainty index equal to 1.

To apply the description for a *C*-uncertain variable it is necessary to determine $v[\bar{x} \mathrel{\tilde{\in}} \overline{D}_x(D_y, u)]$. Using (5.33) let us note that for $|u| < \frac{\sqrt{3}}{2}$

$$v[\bar{x} \mathrel{\tilde{\in}} \overline{D}_x(D_y, u)] = \max\{\frac{2u}{3}, \ 2 - 2\sqrt{1-u^2}\}.$$

Then

$$v[\bar{x} \mathrel{\tilde{\in}} \overline{D}_x(D_y, u)] = \begin{cases} \max\{\frac{2u}{3}, \ 2 - 2\sqrt{1-u^2}\} & \text{for} \quad -\frac{\sqrt{3}}{2} \le u \le \frac{\sqrt{3}}{2} \\ 1 & \text{otherwise in} \quad \Delta_u . \end{cases} \tag{5.40}$$

Substituting (5.39) and (5.40) into (5.32) we obtain $v_c(u)$. For example, $v_c(0.5) = \frac{5}{6}$, $v_c(0.9) \approx 0.44$. It is easy to note that in this case $u_c^* = 0$ and $v_c(u_c^*) = 1$. □

Example 5.6 (decision problem – version II).

R and $h_x(x)$ are the same as in Example 5.4, $D_y = [y_1, y_2]$, $y_1 > 0$, $y_2 > 2y_1$. Then $D_u(x) = [\frac{y_1}{x}, \frac{y_2}{2x}]$, $D_x(u) = [\frac{y_1}{u}, \frac{y_2}{2u}]$ and $v(u)$ in (5.36) is the same as $v[D_y(\bar{x}) \mathrel{\tilde{\subseteq}} D_y]$ in Example 5.4, with $u_1 = u_2 = u$. Thus, u^* is any value from $[2y_1, y_2]$ and $v(u^*) = 1$.

In the case of a *C*-uncertain variable $v[\bar{x} \,\tilde{\in}\, \overline{D}_x(D_y, u)]$ is the same as in Example 5.4, with $u_1 = u_2 = u$. Using (5.32), we obtain

$$v_c(u) = \begin{cases} \dfrac{y_2}{2u} & \text{when} \quad u \ge y_1 + 0.5y_2 \\ 1 - \dfrac{y_1}{u} & \text{when} \quad y_1 \le u \le y_1 + 0.5y_2 \\ 0 & \text{when} \quad u \le y_1 \,. \end{cases}$$

It is easy to see that $u_c^* = y_1 + 0.5y_2$ and $v_c(u_c^*) = \dfrac{y_2}{2y_1 + y_2}$. For example, for $y_1 = 2$, $y_2 = 12$ we obtain $u^* \in [4, 12]$ and $v = 1$, $u_c^* = 8$ and $v_c = 0.75$. The function $v_c(u)$ is illustrated in Fig. 5.5. □

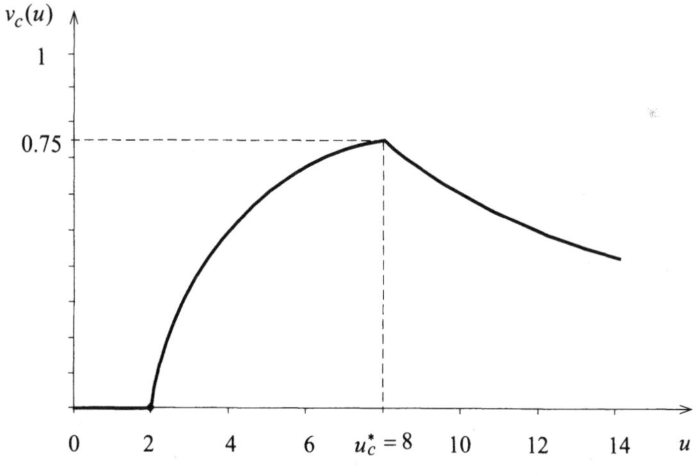

Figure 5.5. Example of the relationship between v_c and u

5.6 Computational Aspects [35]

The application of *C*-uncertain variables with the certainty index v_c instead of v means better using the expert's knowledge, but may be connected with much greater computational difficulties. In the discrete case, when the number of possible values x is small, it may be acceptable to determine all possible values of v_c and then to choose the value u_c^* for which v_c is the greatest. Let us explain it for the decision problem in version II. Assume that X and U are finite discrete sets:

$$X = \{x_1, x_2, \ldots, x_m\}, \quad U = \{u_1, u_2, \ldots, u_p\}.$$

Now the relation $R(u, y; x)$ is reduced to the family of sets

$$D_y(u_i; x_j) \subset Y, \quad i \in \overline{1,p}, \quad j \in \overline{1,m},$$

i.e. the sets of possible outputs for all pairs (u_i, x_j).

The algorithm for the determination of u^* is as follows:

1. For u_i $(i = 1, 2, \ldots, p)$ prove if

$$D_y(u_i; x_j) \subseteq D_y, \quad j = 1, 2, \ldots, m. \tag{5.41}$$

2. Determine

$$v_i = \max_{x \in D_x(D_y, u_i)} h_x(x)$$

where $D_x(D_y, u_i)$ is the set of all x_j satisfying the property (5.41).

3. Choose $u^* = u_i$ for $i = \hat{i}$ where $\hat{i}$ is an index for which v_i is the greatest.

The algorithm for the determination of u_c^* is then the following:

1. For u_i $(i = 1, 2, \ldots, p)$ prove if

$$D_y(u_i; x_j) \subseteq D_y, \quad j = 1, 2, \ldots, m.$$

If yes then $x_j \in D_x(D_y, u_i)$. In this way, for $j = m$ we obtain the set $D_x(D_y, u_i)$ as a set of all x_j satisfying the property (5.41).

2. Determine v_{ci} according to (4.66) and (5.32):

$$v_{ci} = \begin{cases} \dfrac{1}{2} \max\limits_{x \in D_x(D_y, u_i)} h_x(x) & \text{if} \quad \hat{x} \in \overline{D}_x(D_y, u_i) \\ 1 - \dfrac{1}{2} \max\limits_{x \in \overline{D}_x(D_y, u_i)} h_x(x) & \text{otherwise} \end{cases}$$

where $\hat{x} \in X$ is such that $h_x(\hat{x}) = 1$ and $\overline{D}_x(D_y, u_i) = X - D_x(D_y, u_i)$.

3. Choose $i = \hat{i}$ such that v_{ci} is the maximum value in the set of v_{ci} determined in the former steps. Then $u_c^* = u_i$ for $i = \hat{i}$.

Let us consider the relational plant with a one-dimensional output, described by the inequality

$$\Phi_1(u;e) \le y \le \Phi_2(u;d)$$

where $\Phi_1 : U \to \mathrm{R}^1$, $\Phi_2 : U \to \mathrm{R}^1$, e and d are the subvectors of the parameter vector $x = (e, d)$,

$$e \in E = \{e_1, e_2, ..., e_s\}, \quad d \in D = \{d_1, d_2, ..., d_l\}.$$

Now $m = s \cdot l$ where m is a number of the pairs (e_γ, d_δ); $\gamma \in \overline{1,s}$, $\delta \in \overline{1,l}$. If $D_y = [y_{\min}, y_{\max}]$ then the set $D_y(u_i; e_\gamma, d_\delta)$ is described by the inequalities

$$\Phi_1(u_i; e_\gamma) \ge y_{\min} \quad \text{and} \quad \Phi_2(u_i; d_\delta) \le y_{\max}.$$

Assume that $\overline{e}$ and $\overline{d}$ are independent uncertain variables. Then, according to (4.73)

$$h_x(x) = h(e, d) = \min\{h_e(e), h_d(d)\}.$$

Let $\hat{e} = e_\nu$ and $\hat{d} = d_\mu$, i.e. $h_e(e_\nu) = 1$ and $h_d(d_\mu) = 1$.

The algorithm for the determination of the optimal decision u_c^* in this case is as follows:

1. For u_i prove if

$$\Phi_1(u_i; e_\nu) \ge y_{\min} \quad \text{and} \quad \Phi_2(u_i; d_\mu) \le y_{\max}.$$

If yes, go to 2. If not, go to 4.

2. Prove if

$$\Phi_1(u_i; e_\gamma) \le y_{\min} \quad \text{or} \quad \Phi_2(u_i; d_\delta) \ge y_{\max} \tag{5.42}$$

for

$$\gamma = 1, 2, ..., \nu - 1, \nu + 1, ..., s,$$

$$\delta = 1, 2, ..., \mu - 1, \mu + 1, ..., l.$$

3. Determine

$$v_{ci} = 1 - \frac{1}{2} \max_{(e,d) \in \overline{D}_x} \min\{h_e(e_\gamma), h_d(d_\delta)\}$$

where $\overline{D}_x$ is the set of all pairs (e_γ, d_δ) satisfying the property (5.42).

4. Prove if

$$\Phi_1(u_i;e_\gamma) \geq y_{\min} \quad \text{and} \quad \Phi_2(u_i;d_\delta) \leq y_{\max} \tag{5.43}$$

for

$$\gamma = 1, 2, ..., \nu - 1, \nu + 1, ..., s ,$$

$$\delta = 1, 2, ..., \mu - 1, \mu + 1, ..., l .$$

5. Determine

$$v_{ci} = \frac{1}{2} \max_{(c,d)\in D_x} \min\{h_e(e_\gamma), h_d(d_\delta)\}$$

where D_x is the set of all pairs (e_γ, d_δ) satisfying the property (5.43).

6. Execute the points 1–4 for $i = 1, 2, ..., p$.

7. Choose

$$\hat{i} = \arg\max_{i\in\overline{1,p}} v_{ci} .$$

The result (the optimal decision) is: $u^* = u_i$ for $i = \hat{i}$.

The algorithm is illustrated in Fig. 5.6. For the great size of the problem (the great value p) the method of *integer programming* may be used to determine $\hat{i}$.

Example 5.7.

A one-dimensional plant is described by the inequality

$$xu \leq y \leq 2xu ,$$

$$u \in \{1, 2, 3\} , \quad x \in \{3, 4, 5, 6\}$$

and the corresponding values of $h_x(x)$ are $(0.5, 0.6, 1, 0.4)$. The requirement is $y \in D_y = [5, 10]$. Then $D_x(D_y, u)$ is determined by

$$\frac{5}{u} \leq x \leq \frac{10}{u} .$$

For $u = 1$

$$v_1 = v\{\bar{x} \,\tilde{\in}\, [5, 10]\} = \max\{h_x(5), h_x(6)\} = 1 .$$

For $u = 2$

$$v_2 = v\{\bar{x} \,\tilde{\in}\, [2.5, 5]\} = \max\{h_x(3), h_x(4), h_x(5)\} = 1 .$$

For $u = 3$

$$v_3 = v\{\bar{x} \,\tilde{\in}\, [\frac{5}{3}, \frac{10}{3}]\} = h_x(3) = 0.5 .$$

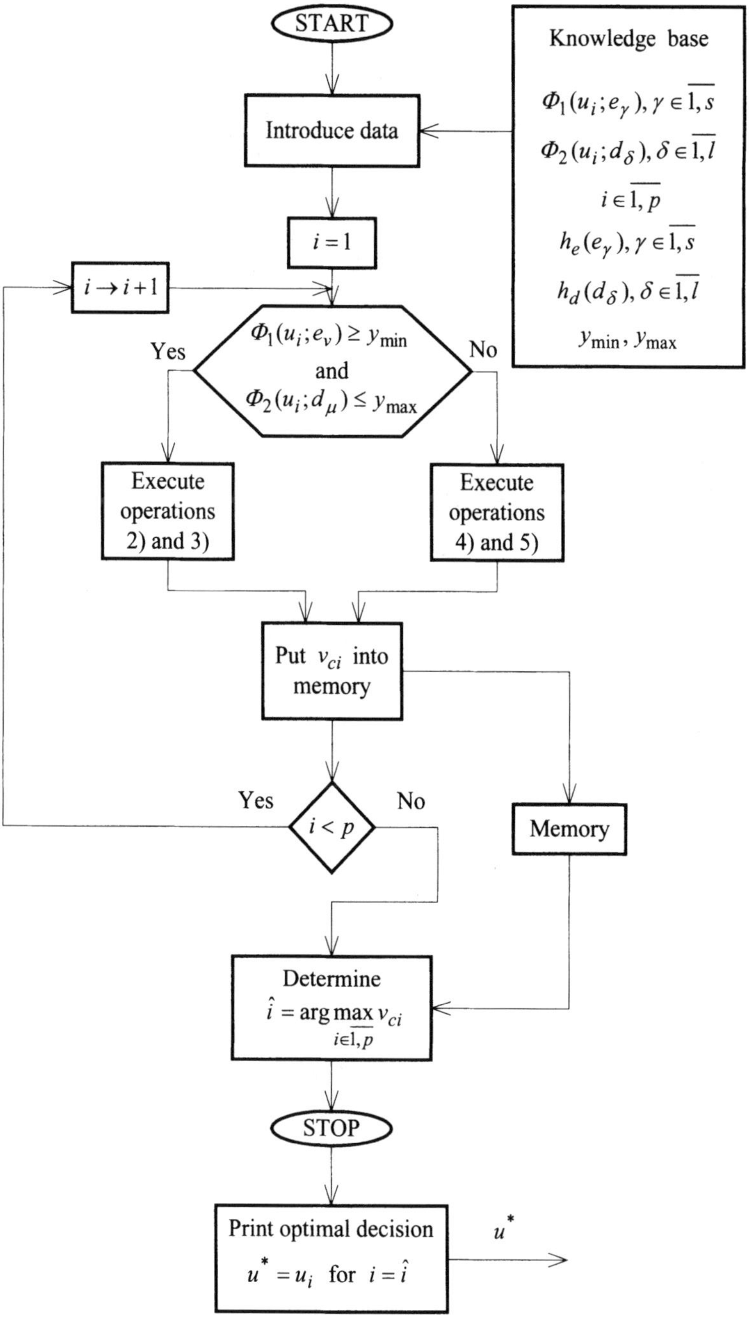

Figure 5.6. Block scheme of decision algorithm

Then $u^* = 1$ or 2, and $v(u^*) = 1$.

Now let us assume that $\bar{x}$ is a C-uncertain variable.

For $u = 1$

$$v_{c1} = v_c\{\bar{x} \,\tilde{\in}\, [5, 10]\} = \frac{1}{2}[1 + 1 - \max\{h_x(3), h_x(4)\}] = 0.7 .$$

For $u = 2$

$$v_{c2} = v_c\{\bar{x} \,\tilde{\in}\, [2.5, 5]\} = \frac{1}{2}[1 + 1 - h_x(6)] = 0.8 .$$

For $u = 3$

$$v_{c3} = v_c\{\bar{x} \,\tilde{\in}\, [\frac{5}{3}, \frac{10}{3}]\} = \frac{1}{2}[0.5 + 1 - \max\{h_x(4), h_x(5), h_x(6)\}] = 0.25 .$$

Then $u_c^* = 2$ and $v_c(u_c^*) = 0.8$, i.e. for $u = 2$ the certainty index that the set of possible outputs belongs to the set [5, 10] is equal to 0.8. □

5.7 Non-parametric Uncertainty

Now we shall present non-parametric problems analogous to those described in Sect. 3.7 for the probabilistic description of the uncertainty. Consider a static plant with the input vector $u \in U$ and the output vector $y \in Y$, and assume that (u, y) are values of uncertain variables $(\bar{u}, \bar{y})$. The plant is described by

$$\mathrm{KP} = < h_y(y \mid u) >$$

where $h_y(y \mid u)$ is a conditional certainty distribution (see Sect. 4.4), given by an expert. If the knowledge of the input is characterized by the certainty distribution $h_u(u)$, the analysis problem may consist in the determination of $h_y(y)$ characterizing the set of possible values y. It corresponds to the analysis problem considered in Sect. 2.2 in which one should find D_y for the given $R(u, y)$ and D_u. Now $R(u, y)$, D_u and D_y are replaced by $h_y(y \mid u)$, $h_u(u)$ and $h_y(y)$, respectively.

Analysis problem: For the given $\mathrm{KP} = < h_y(y \mid u) >$ and $h_u(u)$ find the certainty distribution $h_y(y)$.

According to (4.72) and (4.73)

$$h_y(y) = \max_{u \in U} \min\{h_u(u), h_y(y \mid u)\} . \tag{5.44}$$

The decision problem is an inverse problem consisting in finding $h_u(u)$ for a desirable certainty distribution $h_y(y)$ given by a user. The user's requirement in the form $y \in D_y$ introduced in Sect. 2.2 now is replaced by $h_y(y)$, i.e. the requirement is greater, expressed more precisely. On the other hand, it is not a crisp requirement $y = y^*$, which may be stated and satisfied for a deterministic plant described by a function. In our case

$$y^* = \arg\max_{y \in Y} h_y(y),$$

i.e. $h_y(y^*) = 1$.

Decision problem: For the given $\mathrm{KP} = < h_y(y \mid u) >$ and $h_y(y)$ find the certainty distribution $h_u(u)$.

To find the solution one should solve Equation (5.44) with respect to the function $h_u(u)$ satisfying the conditions for a certainty distribution:

$$\bigwedge_{u \in U} h_u(u) \geq 0, \qquad \max_{u \in U} h_u(u) = 1.$$

The certainty distribution $h_u(u)$ may be called an *uncertain decision.* The deterministic decision may be obtained via determinization using the value u_a maximizing the certainty distribution $h_u(u)$ or the mean value $u_b = \mathrm{M}(\overline{u})$.

Assume that the function

$$h_{uy}(u, y) = \min\{h_u(u), h_y(y \mid u)\} \qquad (5.45)$$

for the given y takes its maximum value in one point

$$\hat{u}(y) = \arg\max_{u \in U} \min\{h_u(u), h_y(y \mid u)\}.$$

Theorem 5.2. For the continuous case (i.e. continuous certainty distributions), assume that:

1. The function $h_u(u)$ has one local maximum for

$$u^* = \arg\max_{u \in U} h_u(u)$$

and it is a unique point such that $h_u(u^*) = 1$.

2. For every $y \in Y$ the distribution $h_y(y \mid u)$ as a function of u has at most one local maximum equal to 1, i.e. the equation

$$h_y(y \mid u) = 1$$

has at most one solution

$$\tilde{u}(y) = \arg\max_{u \in U} h_y(y \mid u).$$

Then

$$\hat{u}(y) = \arg \max_{u \in D_u(y)} h_y(y \mid u)$$

where $D_u(y)$ is a set of values u satisfying the equation

$$h_u(u) = h_y(y \mid u). \tag{5.46}$$

Proof: Let us introduce the following notations

$$\overline{D}_u(y) = \{u \in U : h_u(u) > h_y(y \mid u)\},$$

$$\hat{D}_u(y) = \{u \in U : h_u(u) < h_y(y \mid u)\}.$$

From (5.45) it follows that

$$h_{uy}(u, y) = \begin{cases} h_y(y \mid u) & \text{for } u \in \overline{D}_u(y) \quad (5.47) \\ h_y(y \mid u) = h_u(u) & \text{for } u \in D_u(y) \quad (5.48) \\ h_u(u) & \text{for } u \in \hat{D}_u(y). \quad (5.49) \end{cases}$$

For every u and y the value $h_y(y \mid u)$ is not greater than 1. Then $h_u(u) < 1$ for every $u \in \hat{D}_u(y)$, and according to the assumption 1

$$u^* \notin \hat{D}_u(y).$$

Consequently, according to (5.49), there is no local maximum of $h_y(u, y)$ in $\hat{D}_u(u)$. Then

$$\arg \max_{u \in \hat{D}_u(y) \cap D_u(y)} h_{uy}(u, y) \in D_u(y). \tag{5.50}$$

In a similar way, using the assumption 2 and (5.47) it is easy to show that

$$\tilde{u}(y) \notin \overline{D}_u(y)$$

and

$$\arg \max_{u \in \overline{D}_u(y) \cap D_u(y)} h_{uy}(u, y) \in D_u(y). \tag{5.51}$$

From (5.50) and (5.51) it follows that

$$\hat{u}(y) = \arg\max_{u \in U} h_{uy}(u, y) \in D_u(y)$$

and according to (5.48)

$$\hat{u}(y) = \arg \max_{u \in D_u(y)} h_y(y \mid u) . \qquad \square$$

Let us note that according to (5.44), (5.45) and Theorem 5.2

$$h_u[\hat{u}(y)] = h_y[y \mid \hat{u}(y)] = h_y(y) .$$

Using Theorem 5.2 we may apply the following procedure to the determination of $h_u(u)$ for the fixed u in the continuous case:

1. To solve the equation

$$h_y(y) = h_y(y \mid u)$$

   with respect to y and to obtain a solution $y(u)$ (in general, a set of solutions $D_y(u)$).

2. To determine

$$\bar{h}_u(u) = h_y[y(u)] = h_y[y(u) \mid u] . \tag{5.52}$$

3. To prove whether

$$h_y(y) = \max_{u \in \widetilde{D}_u(y)} h_y(y|u) \tag{5.53}$$

   where $\widetilde{D}_u(y)$ is a set of values u satisfying the equation

$$\bar{h}_u(u) = h_y(y|u). \tag{5.54}$$

4. To accept the solution $\bar{h}_u(u) = h_u(u)$ for which (5.53) is satisfied.

It is worth noting that for the non-parametric description of the plant in the form KP $= \langle h_y(y \mid u) \rangle$ we can state two versions of the decision problem with the deterministic requirement $y = y^*$, analogous to those considered in Sect. 5.3.

Version I. To find the decision $u \triangleq u_a$ maximizing $h_y(y^* \mid u)$.

Version II. To find the decision $u \triangleq u_b$ such that $\mathrm{M}_y(\bar{y} \mid u) = y^*$.

Example 5.8.

Consider a plant with $u, y \in R^1$, described by the conditional certainty distribution given by an expert:

$$h_y(y|u) = \begin{cases} -4(y-u)^2 + 1 & \text{for } u - \dfrac{1}{2} \le y \le u + \dfrac{1}{2} \\ 0 & \text{otherwise} . \end{cases}$$

For the certainty distribution required by a user (Fig. 5.7)

$$h_y(y) = \begin{cases} -(y-c)^2+1 & \text{for } c-1 \le y \le c+1 \\ 0 & \text{otherwise ,} \end{cases} \tag{5.55}$$

one should determine the uncertain decision in the form of the certainty distribution $f_u(u)$.

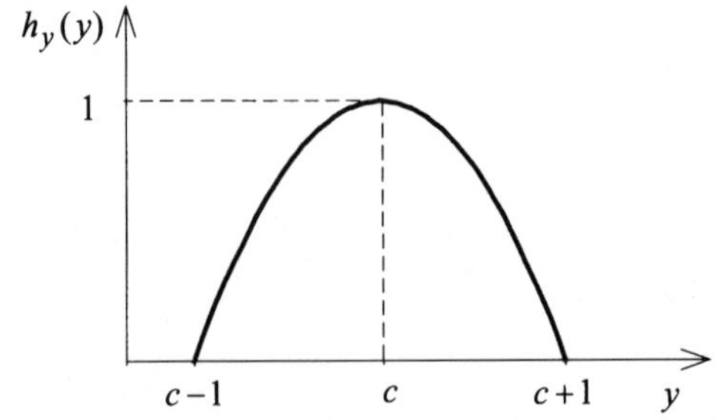

Figure 5.7. Parabolic certainty distribution

The solution of the equation

$$h_y(y) = h_y(y|u) \tag{5.56}$$

has the following form:

1. For

$$c - \frac{1}{2} < u < c + \frac{1}{2}$$

Equation (5.56) has two solutions $y_{1,2}$ such that $h_y(y_{1,2}) > 0$:

$$y_1 = 2u - c, \qquad y_2 = \frac{2u + c}{3}.$$

2. For

$$c + \frac{1}{2} < u < c + \frac{3}{2} \qquad \text{or} \qquad c - \frac{3}{2} < u < c - \frac{1}{2}$$

Equation (5.56) has one solution y such that $h_y(y) > 0$ and $y = y_2$.

3. Otherwise, Equation (5.56) has no solution such that $h_y(y) > 0$.

Then, according to (5.52) and (5.55), we obtain two functions $\overline{h}_u(u)$ corresponding to $y_1(u)$ and $y_2(u)$:

$$\overline{h}_{1u}(u)=\begin{cases}-4(u-c)^2+1 & \text{for} \quad c-\frac{1}{2}\le u\le c+\frac{1}{2} \\ 0 & \text{otherwise}\end{cases}$$

and

$$\overline{h}_{2u}(u)=\begin{cases}-\frac{4}{9}(u-c)^2+1 & \text{for} \quad c-\frac{3}{2}\le u\le c+\frac{3}{2} \\ 0 & \text{otherwise.}\end{cases}$$

In the case $\overline{h}_{1u}(u)$, for $y\neq c$ Equation (5.54) has one solution $u=u_1$ such that $\overline{h}_{1u}(u_1)>0$:

$$u_1=\frac{y+c}{2}$$

and $\overline{h}_{1u}(u_1)=h_y(y)$. Then $\overline{h}_{1u}(u)$ is accepted as a solution of our decision problem. In the case $\overline{h}_{2u}(u)$, for $c-1<y<c+1$ Equation (5.54) has two solutions $u_{1,2}$ such that $\overline{h}_{2u}(u_{1,2})>0$:

$$u_1=\frac{3y+c}{4},\quad u_2=\frac{3y-c}{2}$$

and

$$h_y(y|u_1)=\overline{h}_{2u}(u_1)=-\frac{1}{4}(y-c)^2+1,$$

$$h_y(y|u_2)=\overline{h}_{2u}(u_2)=-(y-c)^2+1.$$

Then

$$\max_{u\in\overline{D}_u(y)} h_y(y|u)=\max\{h_y(y|u_1),\, h_y(y|u_2)\}=-\frac{1}{4}(y-c)^2+1\neq h_y(y).$$

According to (5.53) the function $\overline{h}_{2u}(u)$ should be rejected . The solutions in the cases $\overline{h}_{1u}$ and $\overline{h}_{2u}$ are illustrated in Figs. 5.8 and 5.9, respectively. Finally, the solution of our decision problem is as follows:

$$h_u(u)=\begin{cases}-4(u-c)^2+1 & \text{for} \quad c-\frac{1}{2}\le u\le c+\frac{1}{2} \\ 0 & \text{otherwise.}\end{cases}$$

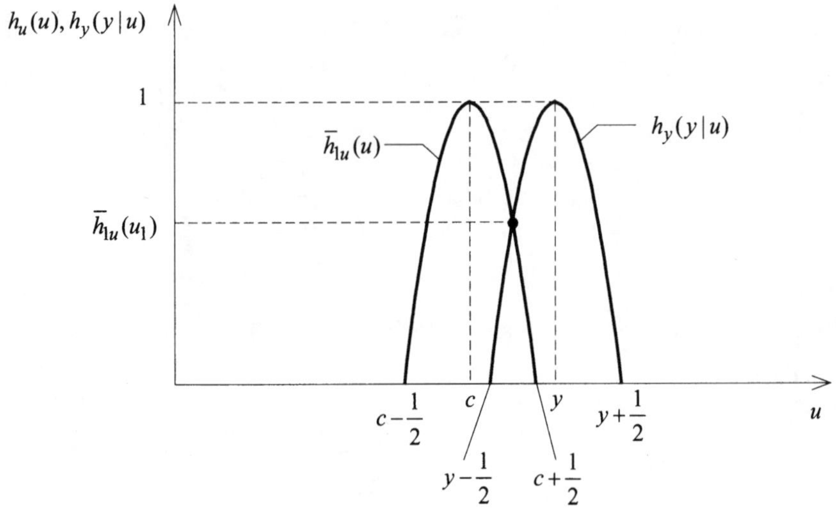

Figure 5.8. Solution of Equation (5.54) for $\overline{h}_{1u}(u)$

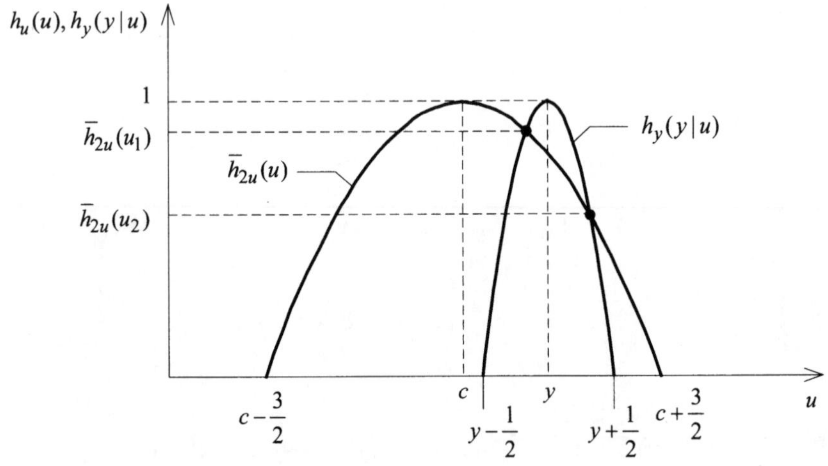

Figure 5.9. Solution of Equation (5.54) for $\overline{h}_{2u}(u)$

It is worth noting that for the deterministic requirement $y^* = c$ we obtain

$$u_a = u_b = c = \mathrm{M}_u(\overline{u}). \qquad \square$$

5.8 Non-parametric Problems for a Plant with External Disturbances

Consider a static plant with input vector $u \in U$, output vector $y \in Y$ and a vector of external disturbances $z \in Z$, and assume that (u, y, z) are values of the uncertain variables $(\overline{u}, \overline{y}, \overline{z})$. The plant is described by

$$\text{KP} = < h_y(y|u, z) >$$

where $h_y(y|u, z)$ is a conditional certainty distribution given by an expert. Now the analysis and decision problems may be formulated as extensions of the problems described in Sect. 5.7.

Analysis problem: For the given $\text{KP} = < h_y(y|u, z) >$, $h_u(u|z)$ and $h_z(z)$, find the certainty distribution $h_y(y)$.

According to (4.72) and (4.73)

$$h_y(y) = \max_{u \in U, z \in Z} h_y(y, u, z) \tag{5.57}$$

where $h_y(y, u, z)$ is the joint certainty distribution for $(\overline{u}, \overline{y}, z)$, i.e.

$$h_y(y, u, z) = \min\{h_{uz}(u, z),\ h_y(y|u, \overline{z})\}. \tag{5.58}$$

Putting

$$h_{uz}(u, z) = \min\{h_z(z),\ h_u(u|\ z)\} \tag{5.59}$$

and (5.58) into (5.57) yields

$$h_y(y) = \max_{u \in U, z \in Z} \min\{h_z(z),\ h_u(u|z),\ h_y(y|u, z)\}. \tag{5.60}$$

Decision problem: For the given $\text{KP} = < h_y(y|u, z) >$ and $h_y(y)$ required by a user one should determine $h_u(u|z)$.

The determination of $h_u(u|z)$ may be decomposed into two steps. In the first step, one should find the function $h_{uz}(u, z)$ satisfying the equation

$$h_y(y) = \max_{u \in U, z \in Z} \min\{h_{uz}(u, z),\ h_y(y|u, z)\} \tag{5.61}$$

and the conditions for a certainty distribution:

$$\bigwedge_{u \in U} \bigwedge_{z \in Z} h_{uz}(u, z) \geq 0, \qquad \max_{u \in U, z \in Z} h_{uz}(u, z) = 1.$$

In the second step, one should determine the function $h_u(u|z)$ satisfying the equation

$$h_{uz}(u,z) = \min\{h_z(z), h_u(u|z)\} \tag{5.62}$$

where

$$h_z(z) = \max_{u \in U} h_{uz}(u,z), \tag{5.63}$$

and the conditions for a certainty distribution:

$$\bigwedge_{u \in U} \bigwedge_{z \in Z} h_u(u|z) \ge 0, \qquad \bigwedge_{z \in Z} \max_{u \in U} h_u(u|z) = 1 .$$

The solution may be not unique. The function $h_u(u|z)$ may be considered as a knowledge of the decision making $\text{KD} = < h_u(u|z) >$ or an *uncertain decision algorithm* (the description of an *uncertain controller* in the open-loop control system). The names and considerations are analogous to those for the probabilistic case presented in Sect. 3.7. In particular, Equation (5.60) is analogous to Equation (3.43). Having $h_u(u|z)$, one can obtain the deterministic decision algorithm $\Psi(z)$ as a result of the determinization of the uncertain decision algorithm $h_u(u|z)$. Two versions corresponding to versions I and II in Sect. 5.3 are the following:

Version I.

$$u_a = \arg\max_{u \in U} h_u(u|z) \triangleq \Psi_a(z) . \tag{5.64}$$

Version II.

$$u_b = M_u(\overline{u}|z) = \int_U u h_u(u|z) du \cdot [\int_U h_u(u|z) du]^{-1} \triangleq \Psi_b(z) . \tag{5.65}$$

The deterministic decision algorithms $\Psi_a(z)$ or $\Psi_b(z)$ are based on the knowledge of the decision making $\text{KD} = < h_u(u|z) >$, which is determined from the knowledge of the plant KP for the given $h_y(y)$ (Fig. 5.10).

To find the solution of the decision problem under consideration, let us note that the problem in the first step is similar to the decision problem in Sect. 5.7 with (u,z) in place of u (see 5.57). Then, to solve this problem in the continuous case, we can use the procedure described in Sect. 5.7.

Theorem 5.3. The set of functions $h_u(u|z)$ satisfying Equation (5.62) is determined as follows:

$$h_u(u \mid z) = \begin{cases} = h_{uz}(u,z) \quad \text{for} \quad (u,z) \notin D(u,z) & (5.66) \\ \geq h_{uz}(u,z) \quad \text{for} \quad (u,z) \in D(u,z) & (5.67) \end{cases}$$

where

$$D(u,z) = \{(u,z) \in U \times Z : \quad h_z(z) = h_{uz}(u,z)\}.$$

Proof: From (5.62) it follows that

$$\bigwedge_{u \in U} \bigwedge_{z \in Z} [h_z(z) \geq h_{uz}(u,z)].$$

If $h_z(z) > h_{uz}(u,z)$ then, according to (5.62), $h_{uz}(u,z) = h_u(u \mid z)$. If $h_z(z) = h_{uz}(u,z)$, i.e. $(u,z) \in D(u,z)$ then $h_u(u \mid z) \geq h_{uz}(u,z)$. □

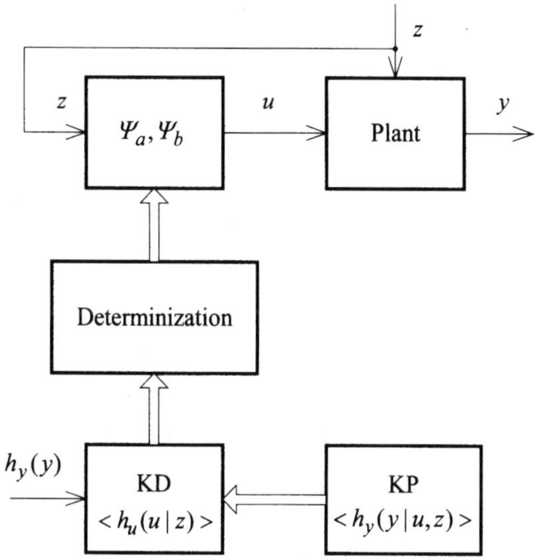

Figure 5.10. Open-loop decision system under consideration

In general, the solution of the problem in the second step is not unique, i.e. we can choose any function $h_u(u|z)$ satisfying the condition (5.67) for $(u,z) \in D(u,z)$, such that

$$\bigwedge_{z \in Z} \max_{u \in U} h_u(u \mid z) = 1.$$

For the fixed z, the set

$$D_u(z) = \{u \in U : (u,z) \in D(u,z)\}$$

is a set of values u maximizing $h_{uz}(u,z)$. If $D_u(z)=\{\hat{u}(z)\}$ (a singleton), then

$$\hat{u}(z)=\arg\max_{u\in U} h_{uz}(u,z)$$

and $h_u(\hat{u}|z)=1$, i.e.

$$h_u(u|z)=\begin{cases} h_{uz}(u,z) & \text{for} \quad u\neq\hat{u}(z) \\ 1 & \text{for} \quad u=\hat{u}(z). \end{cases} \tag{5.68}$$

It is easy to note that $h_u(u|z)$ determined by (5.68) is a continuous function for every $z\in Z$ if and only if

$$\bigwedge_{z\in D_z} [h_z(z)=1],$$

i.e.

$$\bigwedge_{z\in D_z} [\max_{u\in U} h_{uz}(u,z)=1] \tag{5.69}$$

where

$$D_z=\{z\in Z: \bigvee_{u\in U} h_{uz}(u,z)\neq 0\}.$$

If the condition (5.69) is satisfied then $h_u(u|z)=h_{uz}(u,z)$. In this case , according to (5.64) the decision u_a does not depend on z and the decision u_b (5.65) does not depend on z if $u_b=u_a$. It is worth noting that if $D_u(z)$ is a continuous domain, we may obtain a continuous function $h_u(u|z)$ and the decisions u_a, u_b depending on z.

Remark 5.1. The distribution $h_y(y\,|\,u,z)$ given by an expert and / or the result $h_u(u\,|\,z)$ may not satisfy the condition $\max h=1$ (see Example 5.10). The normalization in the form

$$\bar{h}_u(u\,|\,z)=\frac{h_u(u\,|\,z)}{\max\limits_{u\in U} h_u(u\,|\,z)} \tag{5.70}$$

is not necessary if we are interested in the deterministic decisions u_a and u_b, which are the same for h_u and $\bar{h}_u$. □

In a similar way as in Sect. 5.7, we may formulate two versions of the non-parametric decision problem for the deterministic requirement $y=y^*$:

Version I.

$$u_a = \max_{u \in U} h_y(y^* \mid u, z) \triangleq \overline{\Psi}_a(z).$$

Version II. $u_b \triangleq \overline{\Psi}_b(z)$ is a solution of the equation

$$\int_Y y\, h_y(y \mid u, z) dy \cdot [\int_Y h_y(y \mid u, z) dy]^{-1} = y^*.$$

The deterministic algorithms $\overline{\Psi}_a(z)$ and $\overline{\Psi}_b(z)$ are based on the determinization of the plant and, in general, differ from the algorithms $\Psi_a(z)$ and $\Psi_b(z)$ in (5.64) and (5.65), obtained via the determinization of the uncertain decision algorithm $h_u(u \mid z)$.

Example 5.9.

Consider a plant with $u, y, z \in R^1$, described by the conditional certainty distribution given by an expert:

$$h_y(y \mid u, z) = \begin{cases} -z^2(y-u)^2 + 1 & \text{for } u - \frac{1}{z} \le y \le u + \frac{1}{z} \\ & \text{and } 1 < z < 2 \\ 0 & \text{otherwise}. \end{cases}$$

For the certainty distribution $h_y(y)$, the same as in Example 5.8, one should determine the uncertain decision in the form of the conditional certainty distribution $f_u(u \mid z)$.

The solution of the equation

$$h_y(y) = h_y(y \mid u, z) \tag{5.71}$$

has the following form:

1. For

$$c - 1 + \frac{1}{z} < u < c + 1 - \frac{1}{z}$$

Equation (5.71) has two solutions $y_{1,2}$ such that $h_y(y_{1,2}) > 0$:

$$y_1 = \frac{zu - c}{z - 1}, \qquad y_2 = \frac{zu + c}{z + 1}.$$

2. For

$$c + 1 - \frac{1}{z} < u < c + 1 + \frac{1}{z} \quad \text{or} \quad c - 1 - \frac{1}{z} < u < c - 1 + \frac{1}{z}$$

Equation (5.71) has one solution y such that $h_y(y) > 0$ and $y = y_2$.

3. Otherwise, Equation (5.71) has no solution such that $h_y(y) > 0$.

Then, according to (5.52) and (5.55) with (u, z) in place of u, we obtain two functions $\overline{h}_u(u, z)$ corresponding to $y_1(u, z)$ and $y_2(u, z)$:

$$\overline{h}_{1u}(u, z) = \begin{cases} -\dfrac{z^2}{(z-1)^2}(u-c)^2 + 1 & \text{for } c - 1 + \dfrac{1}{z} < u < c + 1 - \dfrac{1}{z} \\ 0 & \text{otherwise} \end{cases}$$

and

$$\overline{h}_{2u}(u, z) = \begin{cases} -\dfrac{z^2}{(z+1)^2}(u-c)^2 + 1 & \text{for } c - 1 - \dfrac{1}{z} < u < c + 1 + \dfrac{1}{z} \\ 0 & \text{otherwise .} \end{cases}$$

In the case $\overline{h}_{1u}(u, z)$ the equation

$$\overline{h}_u(u, z) = h_y(y \mid u, z) \tag{5.72}$$

has two solutions $u_{1,2}$ such that $\overline{h}_u(u_{1,2}, z) > 0$:

$$u_1 = \frac{y(z-1)+c}{z}, \qquad u_2 = \frac{y(z-1)-c}{z-2}$$

and

$$h_y(y \mid u_1, z) = \overline{h}_{1u}(u_1) = -(y-c)^2 + 1,$$

$$h_y(y \mid u_2, z) = \overline{h}_{1u}(u_2) = -\frac{z^2}{(z-2)^2}(y-c)^2 + 1.$$

It is easy to see that for $1 < z < 2$

$$\overline{h}_{1u}(u_1) > \overline{h}_{1u}(u_2)$$

and $\overline{h}_{1u}(u_1) = h_y(y)$. Then $\overline{h}_{1u}(u)$ is accepted as a solution of the problem in the first step. In the case $\overline{h}_{2u}(u)$, for $c - 1 < y < c + 1$ Equation (5.72) has two solutions $u_{1,2}$ such that $\overline{h}_{2u}(u_{1,2}) > 0$:

$$u_1 = \frac{y(z+1)+c}{z+2}, \qquad u_2 = \frac{y(z+1)-c}{z}$$

and

$$h_y(y \mid u_1, z) = \overline{h}_{2u}(u_1) = -\frac{z^2}{(z+2)^2}(y-c)^2 + 1,$$

$$h_y(y \mid u_2, z) = \overline{h}_{2u}(u_2) = -(y-c)^2 + 1.$$

It is easy to note that $h_y(y \mid u_1, z) > h_y(y \mid u_2, z)$ and

$$h_y(y \mid u_1, z) \neq h_y(y).$$

Then, according to (5.53) with (u, z) in place of u, the function $\overline{h}_{2u}(u)$ should be rejected. Finally, the solution of the problem in the first step is as follows:

$$h_{uz}(u, z) = \begin{cases} -\frac{z^2}{(z-1)^2}(u-c)^2 + 1 & \text{for} \quad c - 1 + \frac{1}{z} < u < c + 1 - \frac{1}{z} \\ 0 & \text{otherwise} \ . \end{cases}$$

In this case the condition (5.69) is satisfied. Then $h_{uz}(u, z) = h_{uz}(u \mid z)$ and the decision $u_a = u_b = c$ does not depend on z. □

Example 5.10.

Consider a plant with $u, y, z \in R^1$, described by the conditional certainty distribution given by an expert:

$$h_y(y \mid u, z) = -(y-d)^2 + 1 - u - (b - z) \tag{5.73}$$

for

$$0 \le u \le \frac{1}{2}, \quad b - \frac{1}{2} \le z \le b,$$

$$-\sqrt{1 - u - (b - z)} + d \le y \le \sqrt{1 - u - (b - z)} + d,$$

and $h_y(y \mid u, z) = 0$ otherwise.

For the certainty distribution required by a user (Fig. 5.7):

$$h_y(y) = \begin{cases} -(y-c)^2 + 1 & \text{for} \quad c - 1 \le y \le c + 1 \\ 0 & \text{otherwise} \ , \end{cases}$$

one should determine the uncertain decision algorithm in the form

$$h_u(u \mid z) = h_{uz}(u, z).$$

Let us assume that $b > 0, \ c > 1$ and

$$c+1 \le d \le c+2 \tag{5.74}$$

Then the equation

$$h_y(y) = h_y(y \mid u, z)$$

has a unique solution $y(u,z)$, which is reduced to the solution of the equation

$$-(y-c)^2+1 = -(y-d)^2+1-u-(b-z)$$

and

$$y(u,z) = \frac{d^2-c^2+u+b-z}{2(d-c)} = \frac{1}{2}(d+c+\frac{u+b-z}{d-c}). \tag{5.75}$$

Using (5.75) and (5.74) we obtain

$$h_{uz}(u \mid z) = h_{uz}(u,z) = h_y[y(u,z)]$$

$$= \begin{cases} -[\frac{(d-c)^2+u+b-z}{2(d-c)}]^2+1 & \text{for } u \le 1-[(d-c)-1)]^2-(b-z), \\ & 0 \le u \le \frac{1}{2},\ b-\frac{1}{2} \le z \le b \\ 0 & \text{otherwise} . \end{cases}$$

The values of $h_{uz}(u,z)$ may be greater than zero (i.e. the solution of our decision problem exists) if for every z

$$1-[(d-c)-1]^2-(b-z) > 0 . \tag{5.76}$$

Taking into account the inequality

$$0 \le b-z \le \frac{1}{2},$$

and (5.74), we obtain from (5.76) the following condition:

$$d-c < 1+\frac{1}{\sqrt{2}} .$$

Note that the description (5.73) given by an expert and the solution $h_u(u \mid z) = h_{uz}(u,z)$ do not satisfy the condition $\max h = 1$ (see Remark 5.1). □

6 Fuzzy Variables, Analogies and Soft Variables

The first part of this chapter presents the application of fuzzy variables to non-parametric problems for a static plant, analogous to those described for random and uncertain variables. In Sect. 6.1, a very short description of fuzzy variables (see e.g. [69, 71, 74, 75, 84, 103, 104]) is given in the form needed to formulate our problems and to indicate analogies for non-parametric problems based on random, uncertain and fuzzy variables. These analogies lead to a generalization in the form of soft variables and their applications to non-parametric decision problems.

6.1 Fuzzy Sets and Fuzzy Numbers

Let us consider a universal set *X* and a property (a predicate) $\varphi(x)$ defined on a set *X*, i.e. a property concerning the variable $x \in X$. If $\varphi(x)$ is a crisp property, then for the fixed value *x* the logic value $w[\varphi(x)] \in \{0, 1\}$ and the property $\varphi(x)$ defines a set

$$D_x = \{x \in X : w[\varphi(x)] = 1\} \triangleq \{x \in X : \varphi(x)\}$$

(see Sect. 4.1). If $\varphi(x)$ is a soft property then, for the fixed *x*, $\varphi(x)$ forms a proposition in multi-valued logic and $w[\varphi(x)] \in [0, 1]$. The logic value $w[\varphi(x)]$ denotes the *degree of truth*, i.e. for the fixed *x* the value $w[\varphi(x)]$ shows to what degree the property $\varphi(x)$ is satisfied. The determination of the value $w[\varphi(x)]$ for every $x \in X$ leads to the determination of a function

$$\mu : X \to [0,1], \quad \text{i.e. } w[\varphi(x)] \triangleq \mu(x) .$$

In two-value logic

$$\mu(x) \triangleq I(x) \in \{0,1\}$$

and the set D_x is defined by the pair *X*, *I*(*x*):

$$D_x = < X, I(x) > = \{x \in X : I(x) = 1\} . \tag{6.1}$$

The function $\mu(x)$ is called a *membership function* and the pair $< X, \mu(x) >$ is called a *fuzzy set*. This is a generalization of the function $I(x)$ and the set (6.1), respectively. To every element, the membership function assigns the value $\mu(x)$ from the set $[0,1]$. In practical interpretations it is necessary to determine the property $\varphi(x)$ for which the membership function is given. We assume that the membership function is given by an expert and describes his/her subjective opinions concerning the degree of truth (degree of satisfaction) of the property $\varphi(x)$ for the different elements $x \in X$. For example, let X denote a set of women living in some region. Consider two properties (predicates):

1. $\varphi(x)$= "the age of x is less than 30 years".
2. $\varphi(x)$= "x is beautiful".

The first predicate is a crisp property because for the fixed woman x the sentence $\varphi(x)$ is true or false, i.e. $w[\varphi(x)] \in \{0, 1\}$. The property $\varphi(x)$ determines the set of women (the subset $D_x \subset X$) who are less than 30 years old. The second predicate is a soft property and $w[\varphi(x)] = \mu(x) \in [0, 1]$ may denote a degree of beauty designed to a woman x by an expert. The property $\varphi(x)$ together with the function $\mu(x)$ determines the set of beautiful women. This is a fuzzy set, and for every x the function $\mu(x)$ determines a degree of membership to this set. In the first case (for the crisp property $\varphi(x)$) the expert, not knowing the age of the woman x, may give his/her estimate $\overline{\mu}(x) \in [0,1]$ of the property $\varphi(x)$. Such an estimate is not a membership function of the property $\varphi(x)$ but a value of a certainty index characterizing the expert's uncertainty. Such a difference is important for the proper understanding of fuzzy numbers and their comparison with uncertain variables, presented in Sect. 6.4. We may say that the estimate $\overline{\mu}(x)$ is a membership function of the property "it seems to me that x is less than 30 years old", formulated by the expert.

Let us consider another example: the points x on a plane are red to different degrees: from definitely red via different degrees of pink to definitely white. The value $\mu(x)$ assigned to the point x denotes the degree of red colour of this point. If the definitely red points are concentrated in some domain and the further from this domain they are less red (more white), then the function $\mu(x)$ (the surface upon the plane) reaches its maximum value equal to 1 in this domain and decreases to 0 for the points far from this domain.

According to (4.1), for the determined X and any two functions $\mu_1(x)$, $\mu_2(x)$ (i.e. any two fuzzy sets)

$$\mu_1(x) \vee \mu_2(x) = \max\{\mu_1(x), \mu_2(x)\}, \tag{6.2}$$

$$\mu_1(x) \wedge \mu_2(x) = \min\{\mu_1(x), \mu_2(x)\}, \tag{6.3}$$

$$\neg\mu_1(x) = 1 - \mu_1(x). \tag{6.4}$$

These are definitions of the basic operations in the algebra of fuzzy sets $< X, \mu(x) >$. The relation

$$\mu_1(x) \le \mu_2(x)$$

denotes the inclusion for fuzzy sets, which is a generalization of the inclusion $I_1(x) \le I_2(x)$, i.e. $D_{x1} \subseteq D_{x2}$. It is worth noting that except (4.1) one considers other definitions of the operations $\vee$ and $\wedge$ in the set [0, 1], and consequently – other definitions of the operations (6.2) and (6.3).

If X is a subset of R^1 (the set of real numbers) then the pair $< X, \mu(x) > \triangleq \hat{x}$ is called a *fuzzy number*. In further considerations $\hat{x}$ will be called a *fuzzy variable* to indicate the analogy with random and uncertain variables, and the equation $\hat{x} = x$ will denote that the variable $\hat{x}$ takes a value x. The function $\mu(x)$ is now the membership function of a soft property $\varphi(x)$ concerning a number. The possibilities of the formulation of such properties are rather limited. They may be the formulations concerning the size of the number, e.g. for positive numbers, "x is small", "x is very large" etc. and for real numbers, "x is small positive", "x is large negative" etc. Generally, for the property "$\hat{x}$ is d", the value $\mu(x)$ denotes to what degree this property is satisfied for the value $\hat{x} = x$. For the interpretation of the fuzzy number described by $\mu(x)$ it is necessary to determine the property $\varphi(x)$ for which $\mu(x)$ is given. One assumes that

$$\max_{x \in X} \mu(x) = 1.$$

Usually one considers two cases: the discrete case with $X = \{x_1, x_2, ..., x_m\}$ and the continuous case in which $\mu(x)$ is a continuous function. In the case of fuzzy variables the determinization is called a *defuzzification*. In a way similar to that for random and uncertain numbers, it may consist in replacing the uncertain variable $\hat{x}$ by its deterministic representation

$$x^* = \arg\max_{x \in X} \mu(x)$$

on the assumption that x^* is a unique point such that $\mu(x^*) = 1$, or by the mean value $\mathrm{M}(\hat{x})$. In the discrete case

$$\mathrm{M}(\hat{x}) = \frac{\sum_{i=1}^{m} x_i \mu(x_i)}{\sum_{i=1}^{m} \mu(x_i)} \tag{6.5}$$

and in the continuous case

$$\mathrm{M}(\hat{x}) = \frac{\int\limits_{-\infty}^{\infty} x\mu(x)dx}{\int\limits_{-\infty}^{\infty} \mu(x)dx} \tag{6.6}$$

on the assumption that the respective integrals exist.

Let us consider two fuzzy numbers defined by sets of values $X \subseteq R^1$, $Y \subseteq R^1$ and membership functions $\mu_x(x)$, $\mu_y(y)$, respectively. The membership function $\mu_x(x)$ is the logic value of the soft property $\varphi_x(x)$= "if $\hat{x} = x$ then $\hat{x}$ is d_1" or shortly "$\hat{x}$ is d_1", and $\mu_y(y)$ is the logic value of the soft property $\varphi_y(y)$= "$\hat{y}$ is d_2", i.e.

$$w[\varphi_x(x)] = \mu_x(x), \quad w[\varphi_y(y)] = \mu_y(y)$$

where d_1 and d_2 denote the size of the number, e.g. $\varphi_x(x)$= "$\hat{x}$ is small", $\varphi_y(y)$= "$\hat{y}$ is large". Using the properties φ_x and φ_y we can introduce the property $\varphi_x \to \varphi_y$ (e.g. "if $\hat{x}$ is small, then $\hat{y}$ is large") with the respective membership function

$$w[\varphi_x \to \varphi_y] \stackrel{\Delta}{=} \mu_y(y \mid x),$$

and the properties

$$\varphi_x \vee \varphi_y \quad \text{and} \quad \varphi_x \wedge \varphi_y = \varphi_x \wedge [\varphi_x \to \varphi_y]$$

for which the membership functions are defined as follows:

$$w[\varphi_x \vee \varphi_y] = \max\{\mu_x(x), \mu_y(y)\},$$

$$w[\varphi_x \wedge \varphi_y] = \min\{\mu_x(x), \mu_y(y \mid x)\} \stackrel{\Delta}{=} \mu_{xy}(x, y). \tag{6.7}$$

If we assume that

$$\varphi_x \wedge [\varphi_x \to \varphi_y] = \varphi_y \wedge [\varphi_y \to \varphi_x]$$

then

$$\mu_{xy}(x, y) = \min\{\mu_x(x), \mu_y(y \mid x)\} = \min\{\mu_y(y), \mu_x(x \mid y)\}. \tag{6.8}$$

The properties φ_x, φ_y and the corresponding fuzzy numbers $\hat{x}$, $\hat{y}$ are called *independent* if

$$w[\varphi_x \wedge \varphi_y] = \mu_{xy}(x, y) = \min\{\mu_x(x), \mu_y(y)\}.$$

Using (6.8) it is easy to show that

$$\mu_x(x) = \max_{y \in Y} \mu_{xy}(x, y), \tag{6.9}$$

$$\mu_y(y) = \max_{x \in X} \mu_{xy}(x, y). \tag{6.10}$$

The equations (6.8) and (6.9) describe the relationships between μ_x, μ_y, μ_{xy}, $\mu_x(x \mid y)$ as being analogous to the relationships (4.73), (4.71), (4.72) for uncertain variables, in general defined in the multidimensional sets X and Y. For the given $\mu_{xy}(x, y)$, the set of functions $\mu_y(y \mid x)$ is determined by Equation (6.7) in which

$$\mu_x(x) = \max_{y \in Y} \mu_{xy}(x, y).$$

Theorem 6.1. The set of functions $\mu_y(y \mid x)$ satisfying Equation (6.7) is determined as follows:

$$\mu_y(y \mid x) \begin{cases} = \mu_{xy}(x, y) & \text{for} \quad (x, y) \notin D(x, y) \\ \geq \mu_{xy}(x, y) & \text{for} \quad (x, y) \in D(x, y) \end{cases} \tag{6.11}$$

where

$$D(x, y) = \{(x, y) \in X \times Y : \ \mu_x(x) = \mu_{xy}(x, y)\}.$$

Proof: From (6.7) it follows that

$$\bigwedge_{x \in X} \bigwedge_{y \in Y} [\mu_x(x) \geq \mu_{xy}(x, y)].$$

If $\mu_x(x) > \mu_{xy}(x, y)$ then, according to (6.7), $\mu_{xy}(x, y) = \mu_y(y \mid x)$. If $\mu_x(x) = \mu_{xy}(x, y)$, i.e. $(x, y) \in D(x, y)$ then $\mu_y(y \mid x) \geq \mu_{xy}(x, y)$. □

In particular, as one of the solutions of Equation (6.7), i.e. one of the possible definitions of the membership function for an implication we may accept

$$\mu_y(y \mid x) = \mu_{xy}(x, y). \tag{6.12}$$

If $\mu_{xy}(x, y) = \min\{\mu_x(x), \ \mu_y(y)\}$ then according to (6.12)

$$\mu_y(y \mid x) = \min\{\mu_x(x), \mu_y(y)\}$$

and according to (6.7)

$$\mu_y(y \mid x) = \mu_y(y).$$

Except $\varphi_x(x) \to \varphi_y(y)$ (i.e. the property $\varphi_y(y)$ under the condition φ_x), we can consider the property $\varphi_y(y)$ for the given value $\hat{x} = x$ (i.e. the property $\varphi_y(y)$ under the condition $\hat{x} = x$):

$$\text{“}\hat{x} = x \to \varphi_y(y)\text{”} \triangleq \text{“}\varphi_y(y) \mid x\text{”},$$

and the membership function of this property

$$w[\varphi_y(y) \mid x] = w[\hat{x} = x \to \varphi_y(y)] = w\{[\hat{x} = x \wedge \varphi_x(x)] \wedge [\varphi_x(x) \to \varphi_y(y)]\}$$
$$= \min\{\mu_x(x) \wedge \mu_y(y \mid x)\} = \mu_{xy}(x, y).$$

Then $\mu_{xy}(x, y)$ may be interpreted as a conditional membership function of the property $\varphi_y(y)$ for the given *x*, determined with the help of the property $\varphi_x(x)$. Such an interpretation is widely used in the description of fuzzy controllers in closed-loop systems.

It is worth noting that, according to (6.11), we may use the different functions $\mu_y(y \mid x)$ for the given $\mu_{xy}(x, y)$ and, consequently, for the fixed

$$\mu_x(x) = \max_{y \in Y} \mu_{xy}(x, y)$$

and

$$\mu_y(y) = \max_{x \in X} \mu_{xy}(x, y).$$

In other words, the membership function of the implication

$$w[\varphi_x(x) \to \varphi_y(y)] = \mu_y(y \mid x)$$

may be defined in different ways.

For the fixed *x*, the set

$$D_y(x) = \{y \in Y : (x, y) \in D(x, y)\}$$

is a set of values *x* maximizing $\mu_{xy}(x, y)$. If $D_y(x) = \{y^*(x)\}$ (a singleton), then

$$y^*(x) = \arg\max_{y \in Y} \mu_{xy}(x, y)$$

and $\mu_y(y^* \mid x) = 1$, i.e.

$$\mu_y(y \mid x) = \begin{cases} \mu_{xy}(x,y) & \text{for} \quad y \neq y^*(x) \\ 1 & \text{for} \quad y = y^*(x). \end{cases} \tag{6.13}$$

It is easy to note that $\mu_y(y \mid x)$ determined by (6.13) is a continuous function for every $x \in X$ if and only if

$$\bigwedge_{x \in D_x} [\mu_x(x) = 1],$$

i.e.

$$\bigwedge_{x \in D_x} [\max_{y \in Y} \mu_{xy}(x,y) = 1] \tag{6.14}$$

where

$$D_x = \{x \in X : \bigvee_{y \in Y} \mu_{xy}(x,y) \neq 0\}.$$

If the condition (6.14) is satisfied then $\mu_y(y \mid x) = \mu_{xy}(x,y)$.

6.2 Application of Fuzzy Variables in Analysis and Decision Problems

The description concerning the pair of fuzzy variables may be directly applied to a one-dimensional static plant with single input $u \in U$ and single output $y \in Y$ $(U, Y \subseteq R^1)$. The non-parametric description of uncertainty using fuzzy variables may be formulated by introducing two soft properties $\varphi_u(u)$ and $\varphi_y(y)$. This description (the knowledge of the plant KP) is given by an expert in the form of the membership function

$$w[\varphi_u \to \varphi_y] = \mu_y(y \mid u).$$

For example, the expert says that "if $\hat{u}$ is large then $\hat{y}$ is small" and gives the membership function $\mu_y(y \mid u)$ for this property. In this case the analysis problem may consist in the determination of the membership function $\mu_y(y)$ characterizing the *output property* φ_y for the given membership function $\mu_u(u)$ characterizing the *input property*. The decision problem may be stated as an inverse problem, consisting in finding $\varphi_u(u)$ for a desirable membership function $\mu_y(y)$ given by a user. From a formal point of view, the formulations of these problems and the respective formulas are similar to those for random variables (see Sect. 3.7) and for uncertain variables (see Sects. 5.7 and 5.8).

The essential difference is the following:
The descriptions in the form of $f_u(u)$ or $h_u(u)$, and in the form of $f_y(y)$ or $h_y(y)$, are concerned directly with *values of the input and output*, respectively, and the descriptions in the form of $\mu_u(u)$ and $\mu_y(y)$ are concerned with determined *input and output properties*, respectively. In particular, in the decision problem the functions $f_y(y)$ or $h_y(y)$ describe the user's requirement characterizing directly the *value of the output*, and the function $\mu_y(y)$ required by the user characterizes the determined *output property* $\varphi_u(u)$. Consequently, the solution $\mu_u(u)$ concerns the determined *input property* $\varphi_y(y)$, and not directly the input value u as in the case of $f_u(u)$ or $h_u(u)$.

Analysis problem: For the determined properties $\varphi_u(u)$, $\varphi_y(y)$, the given $\mathrm{KP} = <\mu_y(y \mid u)>$ and $\mu_u(u)$ find the membership function $\mu_y(y)$.
According to (6.10) and (6.7) with u in place of x

$$\mu_y(y) = \max_{u \in U} \min\{\mu_u(u), \mu_y(y \mid u)\}. \tag{6.15}$$

We can also formulate the **analysis problem for the given input**: Find

$$\mu_{uy}(u, y) = w[\varphi_y(y) \mid u] = \min\{\mu_u(u), \mu_y(y \mid u)\}.$$

Having $\mu_{uy}(u, y)$, one can determine the value of y maximizing $\mu_{uy}(u, y)$ or the conditional mean value for the given u:

$$\mathrm{M}(\hat{y} \mid u) = \frac{\int_{-\infty}^{+\infty} y \mu_{uy}(u, y) dy}{\int_{-\infty}^{+\infty} \mu_{uy}(u, y) dy}.$$

Decision problem: For the determined properties $\varphi_u(u)$, $\varphi_y(y)$, the given $\mathrm{KP} = <\mu_y(y \mid u)>$ and $\mu_y(y)$ find the membership function $\varphi_u(u)$.

To find the solution one should solve Equation (6.15) with respect to the function $\mu_u(u)$ satisfying the conditions for a membership function:

$$\bigwedge_{u \in U} \mu_u(u) \geq 0, \qquad \max_{u \in U} \mu_u(u) = 1.$$

The membership function $\mu_u(u)$ may be called a *fuzzy decision*. The deterministic decision may be obtained via a determinization which consists in finding the value u_a maximizing the membership function $\mu_u(u)$ or the mean value $u_b = \mathrm{M}(\hat{u})$. To find the solution $\mu_u(u)$ in the continuous case, we may use the same consideration as for Equation (5.44) concerning uncertain variables.

Assume that the function

$$\mu_{uy}(u,y) = \min\{\mu_u(u), \mu_y(y \mid u)\}$$

for the given y takes its maximum value at one point

$$\hat{u}(y) = \arg\max_{u \in U} \min\{\mu_u(u), \mu_y(y \mid u)\}.$$

Theorem 6.2. For the continuous case (i.e. continuous membership functions), assume that:

1. The function $\mu_u(u)$ has one local maximum for

$$u^* = \arg\max_{u \in U} \mu_u(u)$$

   and it is a unique point such that $\mu_u(u^*) = 1$.
2. For every $y \in Y$ the membership function $\mu_y(y \mid u)$ as a function of u has at most one local maximum equal to 1, i.e. the equation

$$\mu_y(y \mid u) = 1$$

   has at most one solution

$$\widetilde{u}(y) = \arg\max_{u \in U} \mu_y(y \mid u).$$

   Then

$$\hat{u}(y) = \arg\max_{u \in D_u(y)} \mu_y(y \mid u)$$

   where $D_u(y)$ is a set of values u satisfying the equation

$$\mu_u(u) = \mu_y(y \mid u). \qquad \square$$

The proof of Theorem 6.2 and the next considerations are the same as in Sect. 5.7 with $\mu_u(u)$, $\mu_y(y)$, $\mu_{uy}(u,y)$ and $\mu_y(y \mid u)$ in place of $h_u(u)$, $h_y(y)$, $h_{uy}(u,y)$ and $h_y(y \mid u)$, respectively. Then, in the continuous case, the procedure for the determination of $\mu_u(u)$ for the fixed u is the following:

1. To solve the equation

$$\mu_u(u) = \mu_y(y \mid u)$$

with respect to y and to obtain a solution $y(u)$ (in general, a set of solutions $D_y(u)$).

2. To determine

$$\overline{\mu}_u(u) = \mu_y[y(u)] = \mu_y[y(u) \mid u] . \qquad (6.16)$$

3. To prove whether

$$\mu_y(y) = \max_{u \in \widetilde{D}_u(y)} \mu_y(y \mid u) \qquad (6.17)$$

where $\widetilde{D}_u(y)$ is a set of values u satisfying the equation

$$\overline{\mu}_u(u) = \mu_y(y \mid u) .$$

4. To accept the solution $\overline{\mu}_u(u) = \mu_u(u)$ for which (6.17) is satisfied.

Example 6.1.

Consider a plant with $u, y \in R^1$, described by the membership function

$$\mu_y(y \mid u) = \begin{cases} -4(y-u)^2 + 1 & \text{for} \quad u - \frac{1}{2} \le y \le u + \frac{1}{2} \\ 0 & \text{otherwise} . \end{cases}$$

For the membership function required by a user (Fig. 6.1)

$$\mu_y(y) = \begin{cases} -(y-c)^2 + 2 & \text{for} \quad c - \sqrt{2} \le y \le c-1 \\ & \text{or} \quad c+1 \le y \le c + \sqrt{2} \\ 1 & \text{for} \quad c-1 \le y \le c+1 \\ 0 & \text{otherwise} , \end{cases} \qquad (6.18)$$

one should determine the fuzzy decision in the form of the membership function $\mu_u(u)$.

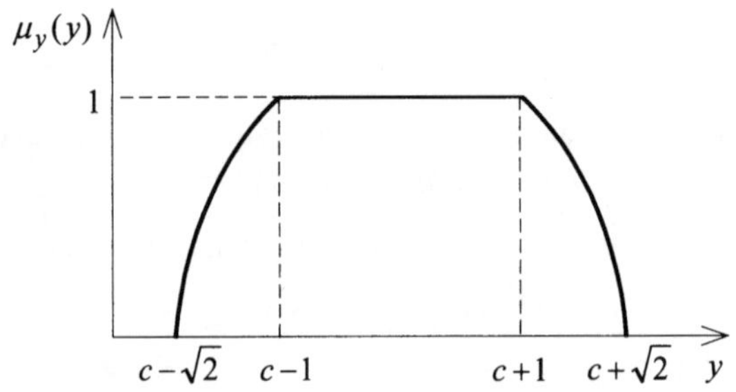

Figure 6.1. Example of the membership function

The solution of the equation

$$\mu_y(y) = \mu_y(y \mid u) \tag{6.19}$$

has the following form:

1. For

$$c-1 \le u \le c+1$$

Equation (6.19) has one solution:

$$y(u) = u\,.$$

2. For

$$c-\sqrt{2}-\frac{1}{2} < u < c-1 \qquad \text{or} \qquad c+1 < u < c+\sqrt{2}+\frac{1}{2}$$

Equation (6.19) is reduced to the equation

$$4(y-u)^2-(y-c)^2+1=0$$

which has one solution such that $h_y(y) > 0$:

$$y(u)=\begin{cases}\dfrac{4u-c+\sqrt{\Delta}}{3} & \text{for} \quad c-\sqrt{2}-\dfrac{1}{2} < u < c-1\\[2ex] \dfrac{4u-c-\sqrt{\Delta}}{3} & \text{for} \quad c+1 < u < c+\sqrt{2}+\dfrac{1}{2}\end{cases}$$

where

$$\Delta = 4(u-c)^2+3\,.$$

3. Otherwise, Equation (6.19) has no solution such that $h_y(y) > 0$.

Then, according to (6.16) and (6.18)

$$\overline{\mu}_u(u)=\mu_y[y(u)]=\begin{cases}(\dfrac{4u-4c+\sqrt{\Delta}}{3})^2+2 & \text{for} \qquad c-\sqrt{2}-\dfrac{1}{2} \le u \le c-1\\[2ex] 1 & \text{for} \qquad c-1 \le u \le c+1\\[2ex] (\dfrac{4u-4c-\sqrt{\Delta}}{3})^2+2 & \text{for} \qquad c+1 \le u \le c+\sqrt{2}+\dfrac{1}{2}\\[2ex] 0 & \qquad\qquad \text{otherwise}\,.\end{cases}$$

□

Remark 6.1. The properties $\varphi_u(u)$ and $\varphi_y(y)$ considered in the example may be introduced by using additional descriptions. For example, if

$$\mu_y(y \mid u) = \begin{cases} -4(y-u)^2 + 1 & \text{for } u - \frac{1}{2} \le y \le u + \frac{1}{2} \\ & \text{and } u > \frac{1}{2} \\ 0 & \text{otherwise} \end{cases}$$

and $c > \sqrt{2} + \frac{1}{2}$, then we can say that

$$\varphi_u(u) = \text{“}u \text{ is medium positive”},$$

$$\varphi_y(y) = \text{“}y \text{ is medium positive”}$$

and $\mu_y(y \mid u)$ is a membership function of the property:

$$\varphi_u(u) \to \varphi_y(y) = \text{“If } u \text{ is medium positive then } y \text{ is medium positive”}.$$

If we introduce a new variable $\overline{u} = u - c$ with the respective constraint then

$$\varphi_u(u) \to \varphi_y(y) = \text{“if } |\overline{u}| \text{ is small then } y \text{ is medium positive”}.$$ □

6.3 Plant with External Disturbances

Consider a static plant with single input $u \in U$, single output $y \in Y$ and single disturbance $z \in Z$ $(U, Y, Z \subseteq R^1)$. Now the non-parametric description of uncertainty using fuzzy variables may be formulated by introducing three soft properties: $\varphi_u(u)$, $\varphi_z(z)$ and $\varphi_y(y)$. This description is given by an expert in the form of the membership function

$$w[\varphi_u \wedge \varphi_z \to \varphi_y] = \mu_y(y \mid u, z),$$

i.e. the knowledge of the plant

$$\text{KP} = < \mu_y(y \mid u, z) > .$$

For example, the expert says that “if $\hat{u}$ is large and $\hat{z}$ is medium then $\hat{y}$ is small” and gives the membership function $\mu_y(y \mid u, z)$ for this property. For such a plant the analysis and decision problems may be formulated as extensions of the problems described in the previous section.

Analysis problem: For the given $\text{KP} = < \mu_y(y \mid u, z) >$, $\mu_u(u \mid z)$ and $\mu_z(z)$ find the membership function $\mu_y(y)$.

According to (6.10) and (6.7)

$$u_y(y) = \max_{u \in U, z \in Z} \mu_y(y,u,z) \tag{6.20}$$

where

$$\mu_y(y,u,z) = w[\varphi_u \wedge \varphi_z \wedge \varphi_y]$$

i.e.

$$\mu_y(y,u,z) = \min\{\mu_{uz}(u,z), \mu_y(y \mid u,z)\}\,. \tag{6.21}$$

Putting

$$\mu_{uz}(u,z) = \min\{\mu_z(z), \mu_u(u \mid z)\} \tag{6.22}$$

and (6.21) into (6.20) yields

$$\mu_y(y) = \arg \max_{u \in U, z \in Z} \min\{\mu_z(z), \mu_u(u \mid z), \mu_y(y \mid u,z)\}\,. \tag{6.23}$$

Decision problem: For the given $\text{KP} = <\mu_y(y \mid u,z)>$ and $\mu_y(y)$ required by a user one should determine $\mu_u(u \mid z)$.

The determination of $\mu_u(u \mid z)$ may be decomposed into two steps. In the first step, one should find the function $\mu_{uz}(u,z)$ satisfying the equation

$$\mu_y(y) = \max_{u \in U, z \in Z} \min\{\mu_{uz}(u,z), \mu_y(y \mid u,z)\} \tag{6.24}$$

and the conditions for a membership function:

$$\bigwedge_{u \in U} \bigwedge_{z \in Z} \mu_{uz}(u,z) \geq 0\,, \qquad \max_{u \in U, z \in Z} \mu_{uz}(u,z) = 1\,.$$

In the second step, one should determine the function $\mu_u(u \mid z)$ satisfying the equation

$$\mu_{uz}(u,z) = \min\{\mu_z(z), \mu_u(u \mid z)\} \tag{6.25}$$

where

$$\mu_z(z) = \max_{u \in U} \mu_{uz}(u,z)\,,$$

and the conditions for a membership function:

$$\bigwedge_{u \in U} \bigwedge_{z \in Z} \mu_u(u \mid z) \geq 0\,, \qquad \bigwedge_{z \in Z} \max_{u \in U} \mu_u(u \mid z) = 1\,.$$

The solution may not be unique. The function $\mu_u(u|z)$ may be considered as a knowledge of the decision making KD $=<\mu_u(u\mid z)>$ or a *fuzzy decision algorithm* (the description of a *fuzzy controller* in the open-loop control system). It is important to remember that the description of the fuzzy controller is concerned with the determined *input and output* properties, i.e.

$$\mu_u(u|z) = w\,[\varphi_z(z) \rightarrow \varphi_u(u)]$$

where the properties $\varphi_z(z)$ and $\varphi_u(u)$ have been used in the description of the plants. Having $\mu_u(u|z)$, one can obtain the deterministic decision algorithm $\Psi(z)$ as a result of the determinization (defuzzification) of the fuzzy decision algorithm $\mu_u(u\mid z)$. Two versions corresponding to versions I and II in Sects. 5.3 and 5.8 are the following:

Version I.

$$u_a = \arg\max_{u\in U} \mu_u(u|z) \stackrel{\Delta}{=} \Psi_a(z).$$

Version II.

$$u_b = \mathrm{M}_u(\hat{u}|z) = \int_U u\mu_u(u|z)du\;[\int_U \mu_u(u|z)du\;]^{-1} \stackrel{\Delta}{=} \Psi_b(z). \qquad (6.26)$$

Using $\mu_u(u\mid z)$ or $\mu_{uz}(u,z)$ with the fixed z in the determination of u_a or u_b, one obtains two versions of $\Psi(z)$. In the second version the fuzzy controller has the form KD $=< w[\varphi_u(u)\mid z]>=<\mu_{uz}(u,z)>$ and the second step with Equation (6.25) is not necessary. Both versions are the same if we assume that $\mu_u(u\mid z) = \mu_{uz}(u,z)$. Let us note that in the analogous problems for random variables (Sect. 3.7) and for uncertain variables (Sect. 5.8) it is not possible to introduce two versions of KD considered here for fuzzy numbers. It is caused by the fact that $\mu_{uz}(u,z)$ and $\mu_u(u\mid z)$ do not concern directly the *values* of the variables (as probability distributions or certainty distributions) but are concerned with the *properties* φ_u, φ_z and

$$\mu_{uz}(u,z) = w[\varphi_u \wedge \varphi_z], \qquad \mu_u(u\mid z) = w[\varphi_z \rightarrow \varphi_u] = w[\varphi_u \mid \varphi_z].$$

The deterministic decision algorithms $\Psi_a(z)$ or $\Psi_b(z)$ are based on the knowledge of the decision making KD $=<\mu_u(u|z)>$, which is determined from the knowledge of the plant KP for the given $\mu_y(y)$ (Fig. 6.2). It is worth noting that the deterministic decision algorithms $\Psi_a(z)$ or $\Psi_b(z)$ have no clear practical interpretation.

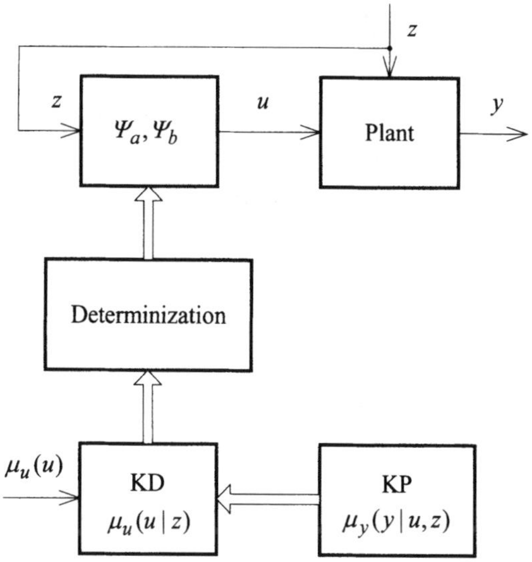

Figure 6.2. Decision system with fuzzy description

From a formal point of view the considerations in this section are the same as in Sect. 5.8 for uncertain variables. Then we can repeat here Theorem 5.3 and the next considerations including Remark 5.1 with $\mu_u(u)$, $\mu_y(y)$, $\mu_z(z)$, $\mu_{uz}(u,z)$, $\mu_u(u|z)$, $\mu_y(y|u,z)$ in place of $h_u(u)$, $h_y(y)$, $h_z(z)$, $h_{uz}(u,z)$, $h_u(u|z)$, $h_y(y|u,z)$, respectively.

Let us note that the condition

$$\bigwedge_{z \in D_z} [\mu_z(z) = 1]$$

corresponding to the condition (5.69) means that $\varphi_z(z)$ is reduced to a crisp property "$z \in D_z$".

The considerations may be extended to the multidimensional case with vectors u, y, z. To formulate the knowledge of the plant one introduces soft properties of the following form: $\varphi_{ui}(j)$ = "$u^{(i)}$ is d_j", $\varphi_{zi}(j)$ = "$z^{(i)}$ is d_j", $\varphi_{yi}(j)$ = "$y^{(i)}$ is d_j" where $u^{(i)}$, $z^{(i)}$, $y^{(i)}$ denote the i-th components of u, z, y, respectively. The determinization of the fuzzy algorithm may be made according to versions I and II presented for the one-dimensional case. In particular, version II consists in the determination of $\mathrm{M}(\hat{u}^{(i)})$ for the fixed z and each component of the vector u, using the membership functions $\mu_{ui}(u^{(i)}, z)$ or $\mu_{ui}(u^{(i)} \mid z)$ where

$$\mu_{ui}(u^{(i)}, z) = \max\{\mu_{ui}(1, z), \mu_{ui}(2, z), ..., \mu_{ui}(m, z)\}$$

and $\mu_{ui}(j,z)$ corresponds to $\varphi_{ui}(j) =$ "$u^{(i)}$ is d_j ".

Example 6.2.
Consider a plant with $u, y, z \in R^1$, described by the following knowledge representation given by an expert:
"If $|u|$ is small and $1 < z < 2$ then y is medium positive". Then
$\varphi_u(u) =$ "$|u|$ is small",
$\varphi_z(z) =$ "z is medium positive",
$\varphi_y(y) =$ "y is medium positive".

The membership function $w[\varphi_u \wedge \varphi_z \to \varphi_y]$ is as follows:

$$\mu_y(y|u,z) = \begin{cases} -z^2(y-u-c)^2+1 & \text{for } u+c-\frac{1}{z} < y < u+c+\frac{1}{z} \\ & \text{and } 1 < z < 2 \\ 0 & \text{otherwise}, \end{cases}$$

$c > 1$.
For the membership function required by a user

$$\mu_y(y) = \begin{cases} -(y-c)^2+1 & \text{for } c-1 \le y \le c+1 \\ 0 & \text{otherwise}, \end{cases}$$

one should determine the fuzzy decision in the form of the membership function $\mu_u(u|z)$.

Substituting $\bar{u} = u + c$, we obtain the same data as in Example 5.9 with $\bar{u}$ in place of u and $\mu_y(y|\bar{u},z)$, $\mu_y(y)$ in place of $h_y(y|u,z)$, $h_y(y)$, respectively. The result is then the same as in Example 5.9:

$$\mu_{uz}(u,z) = \begin{cases} -\frac{z^2}{(z-1)^2}u^2+1 & \text{for } -1+\frac{1}{z} \le u \le 1-\frac{1}{z} \\ 0 & \text{otherwise} \end{cases}$$

and $\mu_u(u|z) = \mu_{uz}(u,z)$. □

Example 6.3.
Consider a plant with $u, y, z \in R^1$ described by the following KP:
" If u is small non-negative and z is large but not greater than b (i.e. $b - z$ is small non-negative) then y is medium". Then
$\varphi_u(u) =$ "u is small non-negative",
$\varphi_z(z) =$ "z is large, not greater than b",
$\varphi_y(y) =$ "y is medium".

The membership function $w[\varphi_u \wedge \varphi_z \to \varphi_y]$ is as follows:

$$\mu_y(y \mid u, z) = -(y-d)^2 + 1 - u - (b-z)$$

for

$$0 \le u \le \frac{1}{2}, \qquad b - \frac{1}{2} \le z \le b,$$

$$-\sqrt{1-u-(b-z)} + d \le y \le \sqrt{1-u-(b-z)} + d$$

and $\mu_y(y \mid u, z) = 0$ otherwise.

For the membership function required by a user

$$\mu_y(y) = \begin{cases} -(y-c)^2 + 1 & \text{for } \; c-1 \le y \le c+1 \\ 0 & \text{otherwise}, \end{cases}$$

one should determine the fuzzy decision algorithm in the form $\mu_u(u \mid z) = \mu_{uz}(u, z)$.

Let us assume that $b > 0, c > 1$ and

$$c + 1 \le d \le c + 2 .$$

Then Equation (6.11) has a unique solution which is reduced to the solution of the equation

$$-(y-c)^2 + 1 = -(y-d)^2 + 1 - u - (b-z) .$$

Further considerations are the same as in Example 5.10, which is identical from the formal point of view. Consequently, we obtain the following result:

$$\mu_{uz}(u,z) = \mu_u(u \mid z)$$

$$= \begin{cases} -[\dfrac{(d-c)^2 + u + b - z}{2(d-c)}]^2 + 1 & \text{for } \; u \le 1 - [(d-c)-1]^2 - (b-z) \\ & 0 \le u \le \dfrac{1}{2}, \; b - \dfrac{1}{2} \le z \le b \\ 0 & \text{otherwise} . \end{cases}$$

By applying the determinization (defuzzification) we can determine the deterministic decision algorithm in an open-loop decision system:

$$u_a = \arg\max_{u \in U} \mu_u(u \mid z) \stackrel{\Delta}{=} \Psi_a(z)$$

or

$$u_b = \mathrm{M}_u(\hat{u} \mid z) \overset{\Delta}{=} \Psi_b(z) . \qquad \square$$

Remark 6.2. The description $\mu_y(y \mid u,z)$ given by an expert and the solution $\mu_u(u \mid z) = \mu_{uz}(u,z)$ do not satisfy the condition $\max \mu = 1$. The normalization in the form analogous to (5.70) is not necessary if we are interested in the deterministic decisions u_a or u_b, which are the same for $\mu_u(u \mid z)$ and the normalized form $\overline{\mu}_u(u \mid z)$. □

6.4 Comparison of Uncertain Variables with Random and Fuzzy Variables

The formal part of the definitions of a random variable, a fuzzy number and an uncertain variable is the same: $< X, \mu(x) >$, that is a set X and a function $\mu : X \to R^1$ where $0 \le \mu(x)$ for every $x \in X$. For the fuzzy number, the uncertain variable and for the random variable in the discrete case, $\mu(x) \le 1$. For the random variable the property of additivity is required, which in the discrete case $X = \{x_1, x_2, ..., x_m\}$ is reduced to the equality $\mu(x_1) + \mu(x_2) + ... + \mu(x_m) = 1$. Without any additional description, one can say that each variable is defined by a fuzzy set $< X, \mu(x) >$. In fact, each definition contains an additional description of semantics which discriminates the respective variables. To compare the uncertain variables with probabilistic and fuzzy approaches, take into account the definitions for $X \subseteq R^1$, using Ω, ω and $g(\omega) = \overline{x}(\omega)$ introduced in Sect. 4.1. *The random variable* $\widetilde{x}$ is defined by X and probability distribution $\mu(x) = F(x)$ (or probability density $f(x) = F'(x)$ if this exists) where $F(x)$ is the probability that $\widetilde{x} \le x$. In the discrete case $\mu(x_i) = p(x_i) = P(\widetilde{x} = x_i)$ (probability that $\widetilde{x} = x_i$). For example, if Ω is a set of 100 persons and 20 of them have the age $\overline{x}(\omega) = 30$, then the probability that a person chosen randomly from Ω has $\overline{x} = 30$ is equal to 0.2. In general, the function $p(x)$ (or $f(x)$ in a continuous case) is an *objective* characteristic of Ω as a whole and $h_\omega(x)$ is a subjective characteristic given by an expert and describes his or her individual opinion of the fixed particular ω.

To compare uncertain variables with fuzzy numbers, let us recall three basic definitions of the fuzzy number in a wide sense of the word, that is the definitions of the fuzzy set based on the number set $X = R^1$.

1. The fuzzy number $\hat{x}(d)$ for the given fixed value $d \in X$ is defined by X and the membership function $\mu(x, d)$, which may be considered as a logic value (*degree of truth*) of the soft property "if $\hat{x} = x$ then $\hat{x} \cong d$".

2. The linguistic fuzzy variable $\hat{x}$ is defined by X and a set of membership functions $\mu_i(x)$ corresponding to different descriptions of the size of $\hat{x}$ (small, medium, large, etc.). For example, $\mu_1(x)$ is a logic value of the soft property "if $\hat{x} = x$ then $\hat{x}$ is small".
3. The fuzzy number $\hat{x}(\omega)$ (where $\omega \in \Omega$ was introduced at the beginning of Sect. 4.1) is defined by X and the membership function $\mu_\omega(x)$, which is a logic value (*degree of possibility*) of the soft property "it is possible that the value x is assigned to ω".

In the first two definitions the membership function does not depend on ω; in the third case there is a family of membership functions (a family of fuzzy sets) for $\omega \in \Omega$. The difference between $\hat{x}(d)$ or the linguistic fuzzy variable $\hat{x}$ and the uncertain variable $\bar{x}(\omega)$ is quite evident. The variables $\hat{x}(\omega)$ and $\bar{x}(\omega)$ are formally defined in the same way by the fuzzy sets $< X, \mu_\omega(x) >$ and $< X, h_\omega(x) >$, respectively, but the interpretations of $\mu_\omega(x)$ and $h_\omega(x)$ are different. In the case of the uncertain variable there exists a function $\bar{x} = g(\omega)$, the value $\bar{x}$ is determined for the fixed ω but is unknown to an expert who formulates the degree of certainty that $\bar{x}(\omega) \cong x$ for the different values $x \in X$. In the case of $\hat{x}(\omega)$ the function g may not exist. Instead we have a property of the type "it is possible that $P(\omega, x)$" (or, briefly, "it is possible that the value x is assigned to ω") where $P(\omega, x)$ is such a property concerning ω and x for which it makes sense to use the words "it is possible". Then $\mu_\omega(x)$ for fixed ω means the degree of possibility for the different values $x \in X$ given by an expert. The example with persons and age is not adequate for this interpretation. In the popular example of the possibilistic approach $P(\omega, x) =$ "John (ω) ate x eggs at his breakfast".

From the point of view presented above, $\bar{x}(\omega)$ may be considered as a special case of $\hat{x}(\omega)$ (when the relation $P(\omega, x)$ is reduced to the function g), with a specific interpretation of $\mu_\omega(x) = h_\omega(x)$. A further difference is connected with the definitions of $w(\bar{x} \,\tilde{\in}\, D_x)$, $w(\bar{x} \,\tilde{\notin}\, D_x)$, $w(\bar{x} \,\tilde{\in}\, D_1 \vee \bar{x} \,\tilde{\in}\, D_2)$ and $w(\bar{x} \,\tilde{\in}\, D_1 \wedge \bar{x} \,\tilde{\in}\, D_2)$. The function $w(\bar{x} \,\tilde{\in}\, D_x) \triangleq m(D_x)$ may be considered as a *measure* defined for the family of sets $D_x \subseteq X$. Two measures have been defined in the definitions of the uncertain variables: $v(\bar{x} \,\tilde{\in}\, D_x) \triangleq \overline{m}(D_x)$ and $v_c(\bar{x} \,\tilde{\in}\, D_x) \triangleq m_c(D_x)$. Let us recall the following special cases of fuzzy measures (see for example [67, 74]) and their properties for every D_1, D_2.

1. If $m(D_x)$ is a *belief measure*, then

$$m(D_1 \cup D_2) \geq m(D_1) + m(D_2) - m(D_1 \cap D_2).$$

2. If $m(D_x)$ is a *plausibility measure*, then

$m(D_1 \cap D_2) \le m(D_1) + m(D_2) - m(D_1 \cup D_2)$.

3. A *necessity measure* is a belief measure for which $m(D_1 \cap D_2) = \min\{m(D_1), m(D_2)\}$.
4. A *possibility measure* is a plausibility measure for which $m(D_1 \cup D_2) = \max\{m(D_1), m(D_2)\}$.

Taking into account the properties of $\overline{m}$ and m_c presented in Definitions 4.5 and 4.8 and in Theorems 4.1, 4.2 and 4.3, 4.4, it is easy to see that $\overline{m}$ is a possibility measure, that $m_n \triangleq 1 - v(\overline{x} \mathrel{\tilde{\in}} \overline{D}_x)$ is a necessity measure and that m_c is neither a belief nor a plausibility measure. To prove this for the plausibility measure, it is enough to take Example 4.3 as a counter-example:

$$m_c(D_1 \cap D_2) = 0.3 > m_c(D_1) + m_c(D_2) - m_c(D_1 \cup D_2) = 0.4 + 0.6 - 0.9\,.$$

For the belief measure, it follows from (4.66) when D_1 and D_2 correspond to the upper case, and from the inequality $\overline{m}(D_1 \cup D_2) = \max\{\overline{m}(D_1), \overline{m}(D_2)\} < \overline{m}(D_1) + \overline{m}(D_2)$ for $D_1 \cap D_2 = \varnothing$.

The interpretation of the membership function $\mu(x)$ as a logic value w of a given soft property $P(x)$, that is $\mu(x) = w[P(x)]$, is especially important and necessary if we consider two fuzzy numbers (x, y) and a relation $R(x, y)$ or a function $y = f(x)$. Consequently, it is necessary if we formulate analysis and decision problems. The formal relationships (see for example [71, 84])

$$\mu_y(y) = \max_x [\mu_x(x) : f(x) = y]$$

for the function and

$$\mu_y(y) = \max_x [\mu_x(x) : (x, y) \in R]$$

for the relation do not determine evidently $P_y(y)$ for the given $P_x(x)$. If $\mu_x(x) = w[P_x(x)]$ where $P_x(x)$ = "if $\hat{x} = x$ then $\hat{x} \cong d$", then we can accept that $\mu_y(y) = w[P_y(y)]$ where $P_y(y)$ = "if $\hat{y} = y$ then $\hat{y} \cong f(\hat{x})$" in the case of the function, but in the case of the relation $P_y(y)$ is not determined. If $P_x(x)$ = "if $\hat{x} = x$ then $\hat{x}$ is small", then $P_y(y)$ may not be evident even in the case of the function, for example $y = \sin x$. For the uncertain variable $\mu_x(x) = h_x(x) = v(\overline{x} \cong x)$ with the definitions (4.53)–(4.56), the property $P_y(y)$ such that $\mu_y(y) = v[P_y(y)]$ is determined precisely: in the case of the function, $\mu_y(y) = h_y(y) = v(\overline{y} \cong y)$ and, in the case of the relation, $\mu_y(y)$ is the certainty index of the property $P_y(y)$ = "there exist $\overline{x}$ such that

$(\bar{x}, \bar{y}) \,\tilde{\in}\, R(x, y)$".

Consequently, using uncertain variables it is possible not only to formulate the analysis and decision problems in the form considered in Chapter 5, but also to define precisely the meaning of these formulations and solutions. This corresponds to the two parts of the definition of the uncertain logic mentioned in Sect. 4.1 after Theorem 4.2: a formal description and its interpretation. The remark concerning ω in this definition is also very important because it makes it possible to interpret precisely the source of the information about the unknown parameter $\bar{x}$ and the term "certainty index".

In the theory of fuzzy sets and systems there exist other formulations of analysis and decision problems (see for example [69]), different from those presented in this chapter. The decision problem with a fuzzy goal is usually based on the given $\mu_y(y)$ as the logic value of the property "$\hat{y}$ is satisfactory" or related properties.

The statements of analysis and decision problems in Chapter 5 for the system with the *known relation* R *and unknown parameter* x considered as an uncertain variable are similar to analogous approaches for the probabilistic model and together with the deterministic case form a unified set of problems. For $y = \Phi(u, x)$ and given y the decision problem is as follows:

1. If x is known (the deterministic case), find u such that $\Phi(u, x) = y$.
2. If x is a value of random variable $\tilde{x}$ with given certainty distribution, find u, maximizing the probability that $\tilde{y} = y$ (for the discrete variable), or find u such that $\mathrm{E}(\tilde{y}, u) = y$ where E denotes the expected value of $\tilde{y}$.
3. If x is a value of uncertain variable $\bar{x}$ with given certainty distribution, find u, maximizing the certainty index of the property $\bar{y} \cong y$, or find u such that $\mathrm{M}_y(u) = y$ where M denotes the mean value of $\bar{y}$.

The definition of the uncertain variable has been used to introduce a *C*-uncertain variable, especially recommended for analysis and decision problems with unknown parameters because of its advantages mentioned in Sect. 4.3. Not only the interpretation but also a formal description of the *C*-uncertain variable differ in an obvious way from the known definitions of fuzzy numbers (see Definition 4.8 and the remark concerning the measure m_c in this section).

6.5 Comparisons and Analogies for Non-parametric Problems

To indicate analogies and differences between the descriptions based on random, uncertain and fuzzy variables let us present together basic non-parametric problems (i.e. the problems based on the non-parametric descriptions), discussed in Sects. 2.3, 3.7, 5.8 and 6.3. The general approach to the decision problem is illustrated in Fig. 6.3, for a static plant with input vector $u \in U$, output vector $y \in Y$ and vector of external disturbances $z \in Z$. The knowledge of the decision making KD is determined from the knowledge of the plant KP and the requirement concerning y, given by a user. The deterministic decision algorithm $u_d = \Psi(z)$ is obtained as a result of the determinization of KD. For simplicity, we shall recall only the mean value as a result of the determinization.

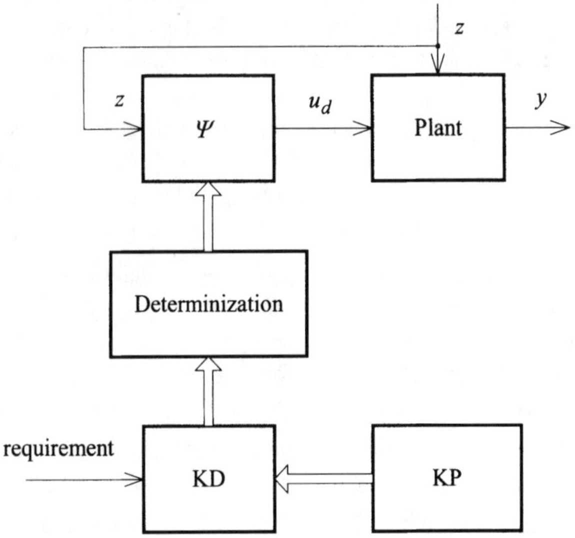

Figure 6.3. General idea of the decision system under consideration

A. Relational system

The knowledge of the plant KP has the form of a relation

$$R(u, y, z) \subset U \times Y \times Z,$$

which determines the set of possible outputs for the given u and z:

$$D_y(u,z) = \{y \in Y : (u, y, z) \in R\}. \tag{6.27}$$

Analysis problem: For the given $D_y(u,z)$, $D_u \subset U$ and $D_z \subset Z$ one should determine the smallest set $D_y \subset Y$ for which the implication

$$(u \in D_u) \wedge (z \in D_z) \rightarrow y \in D_y$$

is satisfied.

According to (2.5) and (6.27)

$$D_y = \bigcup_{u \in D_u} \bigcup_{z \in D_z} D_y(u,z). \tag{6.28}$$

Decision problem: For the given $D_y(u,z)$ and D_y required by a user one should determine the largest set $D_u(z)$ such that for the given z the implication

$$u \in D_u(z) \rightarrow y \in D_y$$

is satisfied.
According to (2.19)

$$D_u(z) = \{u \in U : D_y(u,z) \subseteq D_y\} \stackrel{\Delta}{=} \overline{R}(z,u) \tag{6.29}$$

The knowledge of the decision making $\text{KD} = < \overline{R}(z,u) >$ has been called a *relational decision algorithm* (the description of a *relational controller* in the open-loop control system). The determinization in the form of a mean value gives the deterministic decision algorithm

$$u_d = \int\limits_{D_u(z)} u du \cdot [\int\limits_{D_u(z)} du]^{-1} \stackrel{\Delta}{=} \Psi_d(z).$$

The deterministic decision algorithm $\Psi_d(z)$ is based on the knowledge of the decision making KD, which is determined from the knowledge of the plant KP (reduced to $D_y(u,z)$), for the given D_y (Fig. 6.4).

B. Description based on random variables
The knowledge of the plant has the form of a conditional probability density

$$\text{KP} = < f_y(y|u,z) > . \tag{6.30}$$

Analysis problem: For the given $\text{KP} = < f_y(y|u,z) >$, $f_u(u|z)$ and $f_z(z)$ find the probability density $f_y(y)$:

$$f_y(y) = \int\limits_U \int\limits_Z f_z(z) f_u(u|z) f_y(y|u,z) du\, dz . \tag{6.31}$$

Decision problem: For the given $\text{KP} = < f_y(y|u,z) >$ and $f_y(y)$ required by a user one should determine $f_u(u|z)$.

The determination of $f_u(u|z)$ may be decomposed into two steps. In the first step one, should find the function $f_{uz}(u,z)$ satisfying the equation

$$f_y(y) = \int\limits_U \int\limits_Z f_{uz}(u,z) f_y(y|u,z) du\, dz \tag{6.32}$$

and the conditions for a probability density:

$$\bigwedge_{u \in U} \bigwedge_{z \in Z} f_{uz}(u,z) \geq 0 , \qquad \int\limits_U \int\limits_Z f_{uz}(u,z) du\, dz = 1 .$$

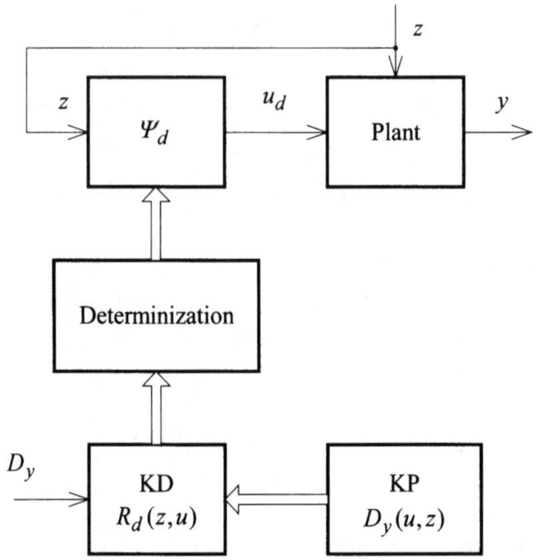

Figure 6.4. Decision system with relational description

In the second step, one should determine the function $f_u(u|z)$:

$$f_u(u|z) = \frac{f_{uz}(u,z)}{\int_U f_{uz}(u,z)\,du} . \tag{6.33}$$

The knowledge of the decision making $\mathrm{KD} = < f_u(u|z) >$ has been called a *random decision algorithm* (the description of a *random controller* in the open-loop control system). The deterministic decision algorithm

$$u_d = \int_U u\, f_u(u|z)\,du \triangleq \Psi_d(z)$$

is based on KD determined from KP, for the given $f_y(y)$ (Fig. 6.5).

C. Description based on uncertain variables

The knowledge of the plant has the form of a conditional certainty distribution given by an expert:

$$\mathrm{KP} = < h_y(y|u,z) > . \tag{6.34}$$

Analysis problem: For the given $\mathrm{KP} = < h_y(y|u,z) >$, $h_u(u|z)$ and $h_z(z)$ find the certainty distribution $h_y(y)$.

According to (5.60)

$$h_y(y) = \max_{u \in U,\, z \in Z} \min\{h_z(z), h_u(u \mid z), h_y(y \mid u, z)\}. \tag{6.35}$$

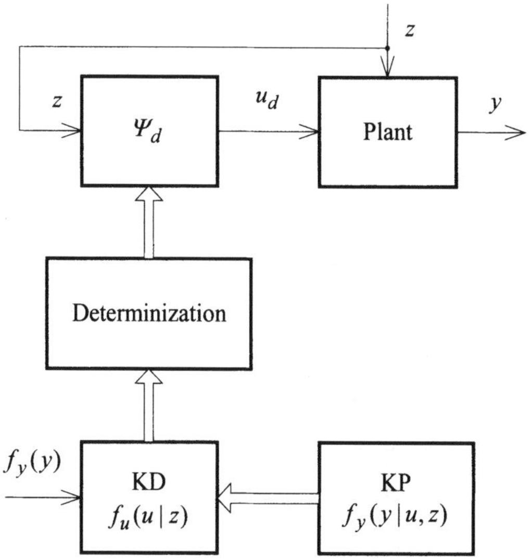

Figure 6.5. Decision system with description based on random variables

Decision problem: For the given $\text{KP} = < h_y(y \mid u, z) >$ and $h_y(y)$ required by a user one should determine $h_u(u \mid z)$.

According to (5.61) and (5.62), the determination of $h_u(u \mid z)$ may be decomposed into two steps. First, one should find the function $h_{uz}(u, z)$ satisfying the equation

$$h_y(y) = \max_{u \in U,\, z \in Z} \min\{h_{uz}(u, z),\ h_y(y \mid u, z)\} \tag{6.36}$$

and the conditions for a certainty distribution

$$\bigwedge_{u \in U} \bigwedge_{z \in Z} h_{uz}(u, z) \geq 0, \qquad \max_{u \in U,\, z \in Z} h_{uz}(u, z) = 1.$$

Then, one should determine the function $h_u(u \mid z)$ satisfying the equation

$$h_{uz}(u, z) = \min\{\max_{u \in U} h_{uz}(u, z), h_u(u \mid z)\} \tag{6.37}$$

and the conditions for a certainty distribution. The knowledge of the decision making $\text{KD} = < h_u(u \mid z) >$ has been called an *uncertain decision algorithm* (the description of an *uncertain controller* in the open-loop control system). The deterministic decision algorithm

$$u_d = \int_U u\, h_u(u \mid z)\, du \cdot [\int_U h_u(u \mid z)\, du\,]^{-1} \stackrel{\Delta}{=} \Psi_d(z)$$

is based on KD determined from KP, for the given $h_y(y)$ (Fig. 6.6).

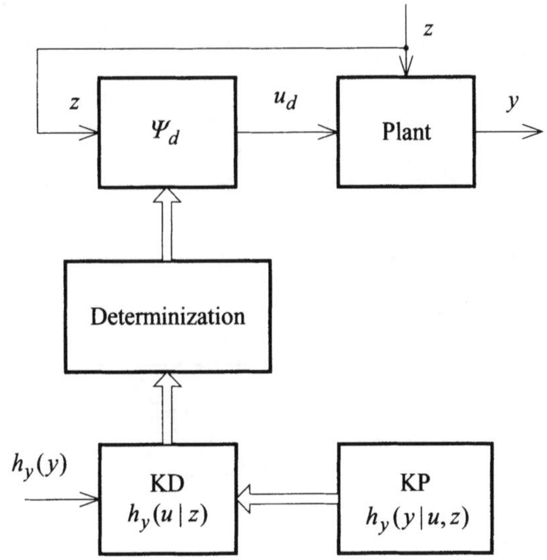

Figure 6.6. Decision system with description based on uncertain variables

D. Description based on fuzzy variables

For the determined soft properties $\varphi_u(u)$, $\varphi_z(z)$ and $\varphi_y(y)$, the knowledge of the plant has the form of a membership function

$$\text{KP} = < \mu_y(y \mid u, z) > . \tag{6.38}$$

Analysis problem: For the given $\text{KP} = < \mu_y(y \mid u, z) >$, $\mu_u(u \mid z)$ and $\mu_z(z)$ find the membership function $\mu_y(y)$.

The solution is given by the formula (6.23).

Decision problem: For the given $\text{KP} = < \mu_y(y \mid u, z) >$ and $\mu_y(y)$ required by a user one should determine $\mu_u(u \mid z)$.

Two steps of the solution are described by the formulas (6.24) and (6.25). The deterministic decision algorithm

$$u_d = \int_U u\, \mu_u(u \mid z)\, du \cdot [\int_U \mu_u(u \mid z)\, du\,]^{-1} \stackrel{\Delta}{=} \Psi_d(z)$$

is based on the *fuzzy decision algorithm* (the description of a *fuzzy controller* in the open-loop control system) KD = $< \mu_u(u \mid z) >$, and is determined from KP for the given $\mu_y(y)$ (Fig. 6.2) with Ψ_d in place of Ψ_a, Ψ_b.

Remark 6.3. In special cases of the decision problem considered in Sects. 5.8 and 6.3, when the solution in the first step in the form of $h_{uz}(u,z)$ or $\mu_{uz}(u,z)$ is not unique, the distribution $h_z(z)$ or $\mu_z(z)$ may be given a priori. □

The different cases of KP are described by (6.27), (6.30), (6.34), (6.38) and the respective results of the analysis problem are given by (6.28), (6.31), (6.35), (6.23). The solution of the decision problem (6.29) corresponds to the solution in two steps described by (6.32) and (6.33) for the random variables, by (6.36) and (6.37) for the uncertain variables, and by (6.24) and (6.25) for the fuzzy variables. The essential differences are the following:

1. Cases A, B are based on the objective descriptions of KP, and cases C, D are based on the subjective descriptions given by an expert.
2. The descriptions in cases B, C are concerned directly with values of (u, y, z), and the description in case D is concerned with determined properties of (u, y, z).

6.6 Introduction to Soft Variables

The uncertain, random and fuzzy variables may be considered as special cases of a more general description of the uncertainty in the form of *soft variables* and *evaluating functions* [54, 59], which may be introduced as a tool for a unification and generalization of non-parametric analysis and decision problems based on the uncertain knowledge representation. The definition of a soft variable should be completed with the determination of relationships for the pair of soft variables.

Definition 6.1 (*soft variable and the pair of soft variables*). A soft variable $\check{x} = < X, g(x) >$ is defined by the set of values X (a real number vector space) and a bounded evaluating function $g : X \to R^+$, satisfying the following condition:

$$\int_X x g(x) < \infty$$

for the continuous case and

$$\sum_{i=1}^{\infty} x_i g(x_i) < \infty$$

for the discrete case.

Let us consider two soft variables $\check{x} = < X, g_x(x) >$, $\check{y} = < Y, g_y(y) >$ and the variable $(\check{x}, \check{y})$ described by $g_{xy}(x, y): X \times Y \to R^{+}$. Denote by $g_y(y \mid x)$ the evaluating function of $\check{y}$ for the given value x (the conditional evaluating function). The pair $(\check{x}, \check{y})$ is defined by $g_{xy}(x, y)$ and two operations:

$$g_{xy}(x, y) = O_1[g_x(x), g_y(y \mid x)], \tag{6.39}$$

$$g_x(x) = O_2[g_{xy}(x, y)], \tag{6.40}$$

i.e.

$$O_1 : D_{gx} \times D_{gy} \to D_{g,xy}, \qquad O_2 : D_{g,xy} \to D_{g,x}$$

where D_{gx}, $D_{gy}(x)$ and $D_{g,xy}$ are sets of the functions $g_x(x)$, $g_y(y \mid x)$ and $g_{xy}(x, y)$, respectively. The mean value $\mathrm{M}(\check{x})$ is defined in the same way as for an uncertain variable, i.e. by (4.70) with $g_x(x)$ in place of $h_x(x)$ □

The evaluating function may have different practical interpretations. In the random case, a soft variable is a random variable described by the probability density $g(x) = f(x)$ or by probabilities $g(x_i) = P(\widetilde{x} = x_i)$. In the case of an uncertain variable, $g(x) = h(x)$ is the certainty distribution. In the case of the fuzzy description, a soft variable is a fuzzy variable described by the membership function $g(x) = \mu(x) = w[\varphi(x)]$ where w denotes a logic value of a given soft property $\varphi(x)$. In general, we can say that $g(x)$ describes an evaluation of the set of possible values X, characterizing for every value x its significance (importance or weight). This description presents a knowledge concerning the variable $\check{x}$, which may be given by an expert describing his / her subjective opinion, or may have an objective character such as in the case of a random variable.

The presented definition of a soft variable is sufficient for non-parametric problems considered in the next section. To formalize a parametric uncertainty and to consider parametric problems it would be necessary to extend our definition by introducing an "evaluating index" that the value of $\check{x}$ belongs to a given set $D_x \subset X$ and to define operations, i.e. the evaluating index for $X - D_x$, $D_1 \cup D_2$ and $D_1 \cap D_2$ where $D_{1,2} \subset X$ (in a similar way as in the definitions of uncertain variables in Chapter 4). These definitions should correspond to the definitions of O_1 and O_2. In other words, it would be necessary to introduce a *measure* in the family of sets D_x, and to define properties of this measure. Then it would be possible to introduce a function of a soft variable and operations in the set of soft variables. The generalization and unification based on such a form of soft variables may be comparable with the unified description based on monotonic measures called fuzzy measures [67, 96].

The further considerations will concern the non-parametric problems, using Definition 6.1 and generalizing the considerations for random, uncertain and fuzzy variables. It is worth noting that the parametric problems for fuzzy variables have no clear interpretation (see remarks in Sect. 6.4) and that is why in the case of the fuzzy description we have presented non-parametric problems only.

6.7 Application of Soft Variables to Non-parametric Problems

The non-parametric problems considered for random, uncertain and fuzzy variables may be generalized by using soft variables. For the plant with input $u \in U$ and output $y \in Y$ we assume that (u, y) are values of soft variables $(\check{u}, \check{y})$ and the knowledge of the plant has the form of a conditional evaluating function

$$\mathrm{KP} = < g_y(y|u) > .$$

According to (6.39) and (6.40)

$$g_y(y) = O_2\{O_1[g_u(u), g_y(y|u)]\} . \tag{6.41}$$

In the analysis problem one should determine $g_y(y)$ for the given KP and $g_u(u)$, and the decision problem consists in finding $g_u(u)$ for the required evaluating function $g_y(y)$. To find the solution one should solve Equation (6.41) with respect to the function $g_u(u)$.

For the plant with $u \in U$, $y \in Y$ and the external disturbance $z \in Z$, the knowledge of the plant has the form of a conditional evaluating function

$$\mathrm{KP} = < g_y(y|u,z) > .$$

Analysis problem: For the given $\mathrm{KP} = < g_y(y|u,z) >$, $g_u(u|z)$ and $g_z(z)$ find the evaluating function $g_y(y)$.

According to (6.39) and (6.40)

$$g_y(y) = O_2\{O_1[\,O_1(g_z(z), g_u(u|z)), g_y(y|u,z)]\} . \tag{6.42}$$

Decision problem: For the given $\mathrm{KP} = < g_y(y|u,z) >$ and $g_y(y)$ required by a user one should determine $g_u(u|z)$.

The determination of $g_u(u|z)$ may be decomposed into two steps. In the first step, one should find the evaluating function $g_{uz}(u,z)$ satisfying the equation

$$g_y(y) = O_2\{O_1[\, g_{uz}(u,z), g_y(y|u,z)]\}.$$

In the second step, one should determine the function $g_u(u|z)$ satisfying the equation

$$g_{uz}(u,z) = O_1[\, g_z(z), g_u(u|z)]$$

where

$$g_z(z) = O_2[\, g_{uz}(u,z)\,] \ .$$

The function $g_u(u|z)$ may be called a knowledge of the decision making $\mathrm{KD} = < g_u(u|z) >$ or a *soft decision algorithm* (the description of a *soft controller* in the open-loop control system). Having $g_u(u|z)$ one can obtain the deterministic decision algorithm as a result of the determinization of the soft decision algorithm. Two versions of the determinization are the following:

Version I.

$$u_a = \arg\max_{u \in U} \, g_u(u|z) \triangleq \Psi_a(z)$$

Version II.

$$u_b = \mathrm{M}(\check{u}|z) = \int_U u\, g_u(u|z)\, du \cdot [\, \int_U g_u(u|z)\, du\,]^{-1} \triangleq \Psi_b(z).$$

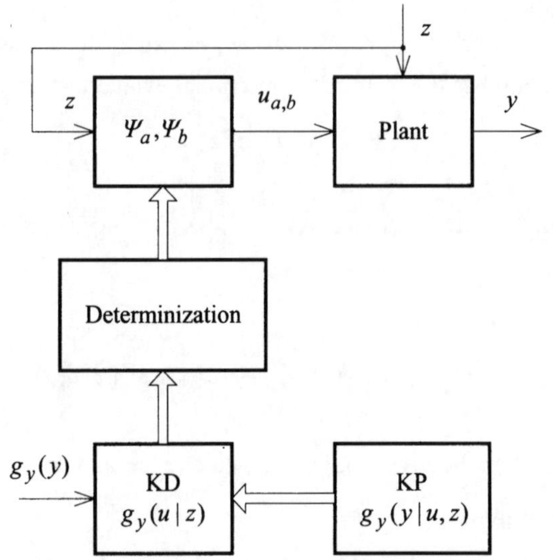

Figure 6.7. Decision system with description based on soft variables

The deterministic decision algorithms $\Psi_a(z)$ or $\Psi_b(z)$ are based on the knowledge of the decision making KD = $< g_u(u|z) >$ determined from the knowledge of the plant KP for the given $g_y(y)$ (Fig. 6.7).

6.8. Generalized Non-parametric Problems

A generalization of the description used in the previous section may be presented as a relation (a relational knowledge representation of the plant):

$$R_g[\,g_z(z), g_u(u|z), g_y(y)] \subset D_{gz} \times D_{gu}(z) \times D_{gy}\,.$$

In the case (6.42) the relation R_g is reduced to an operation (a function) determined by $g_y(y|u,z)$ and the definitions of the operations O_1, O_2. In general, the pair $g_z(z)$, $g_u(u|z)$ determines a set of possible functions $g_y(y)$:

$$\overline{D}_{gy} = \{g_y(y) \in D_{gy} : [\,g_z(z), g_u(u|z), g_y(y)\,] \in R_g\}. \qquad (6.43)$$

For the given functions $g_z(z)$ and $g_u(u|z)$ the analysis problem consists in finding the set $\overline{D}_{gy}$ and the decision problem consists in finding the largest set $\overline{D}_{gu}(z) \subset D_{gu}(z)$ such that the implication

$$g_u(u|z) \in \overline{D}_{gu}(z) \to g_y(y) \in \overline{D}_y$$

is satisfied, where $\overline{D}_y \subset D_{gy}$ is given by a user. It is easy to see that

$$\overline{D}_{gu}(z) = \{g_u(z) \in D_{gu}(z) : \overline{D}_{gy} \subseteq \overline{D}_y\}$$

where $\overline{D}_{gy}$ is given by the formula (6.43).

Now we assume that $g_z(z)$ is not given, but it is known that $g_z(z) \in \overline{D}_z \subset D_{gz}$ (a given set).

If it is known that $g_u(u|z) \in \overline{D}_{gu}(z)$ (where $\overline{D}_{gu}(z) \subset D_{gu}(z)$ is a given set), then the following generalized analysis problem may be formulated: For the given R_g, $\overline{D}_z$ and $\overline{D}_{gu}(z)$ find the smallest set $\overline{D}_y$ such that the implication

$$[g_z(z) \in \overline{D}_z] \wedge [g_u(u|z) \in \overline{D}_{gu}(z)] \to g_y(y) \in \overline{D}_y \qquad (6.44)$$

is satisfied. It is easy to note that

$$\overline{D}_y = \{g_y(y) \in D_{gy} : \bigvee_{g_z(z) \in \overline{D}_z} \; \bigvee_{g_u(u|z) \in \overline{D}_{gu}(z)} [\,g_z(z), g_u(u|z), g_y(y)\,] \in R_g\}. \qquad (6.45)$$

Generalized decision problem: For the given R_g, $\overline{D}_z$ and $\overline{D}_y$ find the largest set $\overline{D}_{gu}(z)$ such that the implication (6.44) is satisfied.

The general form of the solution is as follows:

$$\overline{D}_{gu}(z) = \{g_u(u \mid z) \in D_{gu}(z) : \bigwedge_{g_z(z) \in \overline{D}_z} \overline{D}_{gy} \subseteq D_y\} \qquad (6.46)$$

where $\overline{D}_{gy}$ is given by the formula (6.43). The formula (6.43) is a special case of (6.46) for the given $g_z(z)$.

It is useful to see the analogy between our generalized analysis and decision problems, and the respective problems presented in Chapter 2 for the relation $R(u, y, z)$. The formula (6.46) corresponds to the formula (2.17) for the plant described by the relation R. The set $\overline{D}_{gu}(z)$ is a knowledge of the decision making and may be called a *soft relational decision algorithm*. Having $\overline{D}_{gu}(z)$, one can obtain a set of algorithms $\Psi_a(z)$ or $\Psi_b(z)$ by using the determinization of $g_u(u \mid z)$. It is worth noting that in the case under consideration we have assumed a complex (or two-level) uncertainty, described by a combination of soft variables and relations. The different cases of the complex uncertainty will be presented in Chapter 12.

7 Systems with Logical Knowledge Representation

The relations introduced in Chapter 2 may have a specific form of logical formulas concerning input, output and additional variables. In this case the so-called *logic-algebraic method* may be used to formulate and solve the analysis and decision problems [17, 18, 19, 21–24, 26, 29]. The main idea of this method consists in replacing individual reasoning concepts based on inference rules by unified algebraic procedures based on the rules in two-value logic algebra.

7.1 Logical Knowledge Representation

Now we shall consider the knowledge representation in which the relations R_i (2.3) have the form of logic formulas concerning u, y, w. Let us introduce the following notation:

1. $\alpha_{uj}(u)$ – simple formula (i.e. simple property) concerning u, $j = 1, 2, ..., n_1$, e.g. $\alpha_{u1}(u) =$ "$u^{\mathrm{T}} u \le 2$ "
2. $\alpha_{wr}(u, w, y)$ – simple formula concerning u, w and y, $r = 1, 2, ..., n_2$.
3. $\alpha_{ys}(y)$ – simple formula concerning y, $s = 1, 2, ..., n_3$.
4. $\alpha_u = (\alpha_{u1}, \alpha_{u2}, ..., \alpha_{un_1})$ – subsequence of simple formulas concerning u.
5. $\alpha_w = (\alpha_{w1}, \alpha_{w2}, ..., \alpha_{wn_2})$ – subsequence of simple formulas concerning u, w and y.
6. $\alpha_y = (\alpha_{y1}, \alpha_{y2}, ..., \alpha_{yn_3})$ – subsequence of simple formulas concerning y.
7. $\alpha(u, w, y) \triangleq (\alpha_1, \alpha_2, ..., \alpha_n) = (\alpha_u, \alpha_w, \alpha_y)$ – sequence of all simple formulas in the knowledge representation, $n = n_1 + n_2 + n_3$.
8. $F_i(\alpha)$ – the i-th fact given by an expert. It is a logic formula composed of the subsequence of α and the logic operations: $\vee$ – *or*, $\wedge$ – *and*, $\neg$ – *not*, $\rightarrow$.– *if ... then*, $i = 1, 2, ..., k$.

For example $F_1 = \alpha_1 \wedge \alpha_2 \rightarrow \alpha_4$, $F_2 = \alpha_3 \vee \alpha_2$ where $\alpha_1 =$ "$u^{\mathrm{T}} u \le 2$ ", $\alpha_2 =$ "the temperature is low or $y^{\mathrm{T}} y \le 3$", $\alpha_3 =$ "$y^{\mathrm{T}} y > w^{\mathrm{T}} w$",

$\alpha_4 =$ " $y^{\mathrm{T}} y = 4$ ".

9. $F(\alpha) = F_1(\alpha) \wedge F_2(\alpha) \wedge ... \wedge F_k(\alpha)$.
10. $F_u(\alpha_u)$ – input property, i.e. the logic formula using α_u.
11. $F_y(\alpha_y)$ – output property.
12. $a_m \in \{0,1\}$ – logic value of the simple property α_m, $m = 1,2,...,n$.
13. $a = (a_1, a_2, ..., a_n)$ – zero-one sequence of the logic values.
14. $a_u(u)$, $a_w(u,w,y)$, $a_y(y)$ – zero-one subsequences of the logic values corresponding to $\alpha_u(u)$, $\alpha_w(u,w,y)$, $\alpha_y(y)$.
15. $F(a)$ – the logic value of $F(\alpha)$.

All facts given by an expert are assumed to be true, i.e. $F(a) = 1$.

The description

$$< \alpha, F(\alpha) > \stackrel{\Delta}{=} \mathrm{KP}$$

may be called a *logical knowledge representation* of the plant. For illustration purposes let us consider a very simple example:

$u = (u^{(1)}, u^{(2)})$, $y = (y^{(1)}, y^{(2)})$, $w \in \mathrm{R}^1$,

$\alpha_{u1} =$ " $u^{(1)} + u^{(2)} > 0$ ", $\alpha_{u2} =$ " $u^{(2)} > 2$ ", $\alpha_{y1} =$ " $y^{(2)} < y^{(1)}$ ",

$\alpha_{y2} =$ " $y^{(1)} + y^{(2)} = 4$ ", $\alpha_{w1} =$ " $u^{(1)} - 2w + y^{(2)} < 0$ ",

$\alpha_{w2} =$ " $u^{(2)} > y^{(1)}$ ",

$F_1 = \alpha_{u1} \wedge \alpha_{w1} \to \alpha_{y1} \vee \neg\alpha_{w2}$, $F_2 = (\alpha_{u2} \wedge \alpha_{w2}) \vee (\alpha_{y2} \wedge \neg\alpha_{u1})$,

$F_u = \alpha_{u1} \vee \alpha_{u2}$, $F_y = \neg\alpha_{y2}$.

The expressions $F(a)$ have the same form as the formulas $F(\alpha)$, e.g.

$$F_1(a_{u1}, a_{w1}, a_{w2}, a_{y1}) = a_{u1} \wedge a_{w1} \to a_{y1} \vee \neg a_{w2}.$$

The logic formulas $F_i(\alpha)$, $F_u(\alpha_u)$ and $F_y(\alpha_y)$ are special forms of the relations introduced in Sects 2.1 and 2.2. Now the relation (2.3) has the form

$$R_i(u,w,y) = \{(u,w,y) \in U \times W \times Y : \; F_i[a(u,w,y)] = 1\}, \quad i \in \overline{1,k}. \tag{7.1}$$

The input and output properties may be expressed as follows:

$$u \in D_u, \qquad y \in D_y$$

where

$$D_u = \{u \in U : \; F_u[a_u(u)] = 1\}, \tag{7.2}$$

$$D_y = \{y \in Y : \; F_y[a_y(y)] = 1\}. \tag{7.3}$$

The description with $F(a)$, $F_u(a_u)$, $F_y(a_y)$ may be called a *description on the logical level*. The expressions $F(a)$, $F_u(a_u)$ and $F_y(a_y)$ describe *logical structures* of the plant, the input property and the output property, respectively. The description at the logical level is independent of the particular meaning of the simple formulas. In other words, it is common for the different plants with different practical descriptions but the same logical structures. At the logical level our plant may be considered as a relational plant with input a_u (a vector with n_1 zero-one components) and output a_y (a vector with n_3 zero-one components), described by the relation

$$F(a_u, a_w, a_y) = 1 \tag{7.4}$$

(Fig. 7.1). The input and output properties for this plant corresponding to the properties $u \in D_u$ and $y \in D_y$ for the plant with input u and output y are as follows:

$$a_u \in \overline{S}_u \subset S_u, \qquad a_y \in \overline{S}_y \subset S_y$$

where S_u, S_y are the sets of all zero-one sequences a_u, a_y, respectively, and

$$\overline{S}_u = \{a_u \in S_u : \; F_u(a_u) = 1\}, \quad \overline{S}_y = \{a_y \in S_y : \; F_y(a_y) = 1\}. \tag{7.5}$$

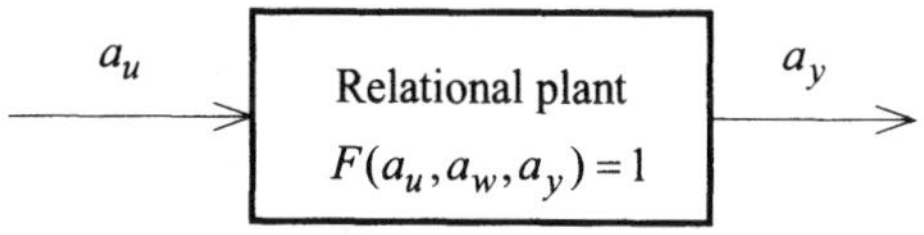

Figure 7.1. Plant at logical level

7.2 Analysis and Decision Making Problems

The analysis and decision making problems for the relational plant described by the logical knowledge representation are analogous to those for the relational plant in Sect. 2.2. The analysis problem consists in finding the output property for the given input property and the decision problem is an inverse problem consisting in finding the input property (the decision) for the required output property.

Analysis problem: For the given $F(\alpha)$ and $F_u(\alpha_u)$ find the best property $F_y(\alpha_y)$ such that the implication

$$F_u(\alpha_u) \rightarrow F_y(\alpha_y) \tag{7.6}$$

is satisfied.

If it is satisfied for F_{y1} and F_{y2}, and $F_{y1} \to F_{y2}$, then F_{y1} is better than F_{y2}. The property F_y is then the best if it implies any other property for which the implication (7.6) is satisfied. The best property F_y corresponds to the smallest set D_y in the formulation presented in Sect. 2.2.

Decision problem: For the given $F(\alpha)$ and $F_y(\alpha_y)$ (the property required by a user) find the best property $F_u(\alpha_u)$ such that the implication (7.6) is satisfied.

If it is satisfied for F_{u1} and F_{u2}, and $F_{u2} \to F_{u1}$, then F_{u1} is better than F_{u2}. The property F_u is then the best if it is implied by any other property for which the implication (7.6) is satisfied. The best property F_u corresponds to the largest set D_u in the formulation presented in Sect. 2.2.

Remark 7.1. The solution of our problem may not exist. In the case of the analysis it means that there is a contradiction between the property $F_u(\alpha_u)$ and the facts $F(\alpha_u, \alpha_w, \alpha_y)$, i.e. the sequence a_u such that $F_u(a_u) \wedge F(a_u, a_w, a_y) = 1$ does not exist. In the case of the decision making it means that the requirement F_y is too strong. The existence of the solution will be explained in the next section. □

Remark 7.2. Our problems are formulated and will be solved on the logic level. Consequently they depend on the logical structures (the forms of F and F_y or F_u) but do not depend on the meaning of the simple formulas. The knowledge representation KP and the problem formulations may be extended for different variables, objects and sets (not particularly the sets of real number vectors) used in the description of the knowledge. For instance, in the example in the previous section we may have the following simple formulas in the text given by an expert:

α_{u1} = "operation O_1 is executed after operation O_2",

α_{u2} = "temperature is low",

α_{w1} = "pressure is high",

α_{w2} = "humidity is low",

α_{y1} = "state S occurs",

α_{y2} = "quality of product is sufficient".

Then the facts F_1 and F_2 in this example mean:

F_1 = "If operation O_1 is executed after operation O_2 and pressure is high then state S occurs or humidity is not low".

F_2 = "Temperature is low and humidity is low or quality is sufficient and operation O_1 is not executed after operation O_2". □

Remark 7.3. The possibilities of forming the input and output properties are restricted. Now the sets D_u and D_y may be determined by the logic formulas $F_u(\alpha_u)$ and $F_y(\alpha_y)$ using the simple formulas α_u and α_y from the sequence of the simple formulas α used in the knowledge representation. □

7.3 Logic-algebraic Method

The solutions of the analysis and decision problems formulated in Sect. 7.2 may be obtained by using the so-called *logic-algebraic method* [17, 18, 23, 24, 29, 54]. It is easy to show that the analysis problem is reduced to solving the following algebraic equation:

$$\widetilde{F}(a_u, a_w, a_y) = 1 \tag{7.7}$$

with respect to a_y, where

$$\widetilde{F}(a_u, a_w, a_y) = F_u(a_u) \wedge F(a_u, a_w, a_y).$$

Now $F(a_u, a_w, a_y)$, $F_u(a_u)$ and $F_y(a_y)$ are algebraic expressions in two-value logic algebra. If S_y is the set of all solutions then F_y is determined by S_y, i.e. $a_y \in S_y \leftrightarrow F_y(a_y) = 1$. For example, if $a_y = (a_{y1}, a_{y2}, a_{y3})$ and $S_y = \{(1,1,0),(0,1,0)\}$ then $F_y(\alpha_y) = (\alpha_{y1} \wedge \alpha_{y2} \wedge \neg\alpha_{y3}) \vee (\neg\alpha_{y1} \wedge \alpha_{y2} \wedge \neg\alpha_{y3})$.

In the decision making problem two sets of the algebraic equations should be solved with respect to a_u:

$$\left.\begin{aligned} F(a_u, a_w, a_y) &= 1 \\ F_y(a_y) &= 1 \end{aligned}\right\}, \qquad \left.\begin{aligned} F(a_u, a_w, a_y) &= 1 \\ F_y(a_y) &= 0 \end{aligned}\right\} \tag{7.8}$$

If S_{u1}, S_{u2} are the sets of the solutions of the first and the second equation, respectively – then $F_u(\alpha_u)$ is determined by $S_u = S_{u1} - S_{u2}$ [23] in the same way as F_y by S_y in the former problem.

The generation of the set S_y requires the testing of all sequences $a = (a_u, a_w, a_y)$ and the execution time may be very long for the large size of the problem. The similar computational difficulties may be connected with the solution of the decision problem. The generation of S_y (and consequently, the solution F_y) may be much easier when the following decomposition is applied:

$$F_u \wedge F = \overline{F}_1(\overline{a}_0, \overline{a}_1) \wedge \overline{F}_2(\overline{a}_1, \overline{a}_2) \wedge \ldots \wedge \overline{F}_N(\overline{a}_{N-1}, \overline{a}_N) \tag{7.9}$$

where $\overline{a}_0 = a_y$, $\overline{F}_1$ is the conjunction of all facts from $\widetilde{F}$ containing the variables from $\overline{a}_0$, $\overline{a}_1$ is the sequence of all other variables in $\overline{F}_1$, $\overline{F}_2$ is the conjunction of all facts containing the variables from $\overline{a}_1$, $\overline{a}_2$ is the sequence of all other variables in $\overline{F}_2$ etc. As a result of the decomposition the following *recursive procedure* may be applied to obtain $\overline{S}_0 = S_y$:

$$\overline{S}_{m-1} = \{\overline{a}_{m-1} \in S_{m-1} : \bigvee_{\overline{a}_m \in \overline{S}_m} [\overline{F}_m(\overline{a}_{m-1}, \overline{a}_m) = 1]\}, \tag{7.10}$$

where S_m is the set of all $\overline{a}_m$, $m = N, N-1, ..., 1$, $\overline{S}_N = S_N$.

The recursive procedure (7.10) has two interesting interpretations.

A. System analysis interpretation

Let us consider the cascade of relation elements (Fig. 7.2) with input $\overline{a}_m$, output $\overline{a}_{m-1}$ (zero-one sequences), described by the relations $\overline{F}_m(\overline{a}_{m-1}, \overline{a}_m) = 1$ ($m = N, N-1, ..., 1$). Then $\overline{S}_{m-1}$ is the set of all possible outputs from the element $\overline{F}_m$ and $\overline{S}_0$ is the set of all possible outputs from the whole cascade.

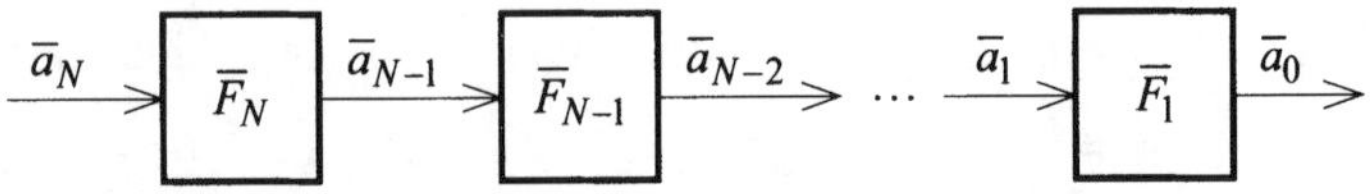

Figure 7.2. Relational system

B. Deductive reasoning interpretation

The set $\overline{S}_{m-1}$ may be considered as the set of all elementary conclusions from $\overline{F}_N \wedge ... \wedge \overline{F}_m$, and $\overline{S}_0$ is the set of all elementary conclusions from the facts $F_u \wedge F$.

A similar approach may be applied to the decision problem. To determine S_{y1} and S_{y2} we may use the recursive procedure (7.10) with F in (7.9) instead of $F_u \wedge F$ and with $\overline{a}_0 = (a_u, a_y)$. After the generation of $\overline{S}_0$ from (7.10) one can determine S_{u1} and S_{u2} in the following way:

$$S_{u1} = \{a_u : \bigvee_{a_y \in \overline{S}_y} [(a_u, a_y) \in \overline{S}_0]\},$$

$$S_{u2} = \{a_u : \bigvee_{a_y \in \hat{S}_y - \overline{S}_y} [(a_u, a_y) \in \overline{S}_0]\}$$

where $\overline{S}_y = \{a_y : F_y(a_y) = 1\}$ and $\hat{S}_y$ is the set of all a_y.

The different versions of the presented procedures have been elaborated and applied in the general purpose expert systems CONTROL-LOG and CLASS-LOG, specially oriented for the applications to a class of knowledge-based control systems and to classification problems.

The main idea of the logic-algebraic method presented here for the generation of the solutions consists in replacing the individual reasoning concepts based on inference rules by unified algebraic procedures based on the rules in two-value logic algebra. The results may be considered as a unification and generalization of the different particular reasoning algorithms for a class of systems with the logical knowledge representation for which the logic-algebraic method has been developed. The logic-algebraic method can be applied to the design of complex knowledge-based computer systems [82, 85, 86, 90].

Example 7.1 (analysis).

The facts $\widetilde{F}$ are the following:

$F_1 = (\alpha_3 \vee \neg\alpha_1) \rightarrow \alpha_4$, $\quad F_2 = (\neg\alpha_1 \wedge \alpha_7) \vee \neg\alpha_3$, $\quad F_3 = (\alpha_9 \wedge \alpha_1) \rightarrow \alpha_2$,

$F_4 = (\alpha_4 \wedge \neg\alpha_7) \vee \alpha_5$, $\quad F_5 = \alpha_6 \rightarrow (\alpha_4 \wedge \alpha_8)$, $\quad F_6 = \alpha_2 \rightarrow (\neg\alpha_4 \wedge \alpha_6)$,

$F_7 = (\alpha_3 \wedge \alpha_2) \vee \alpha_{10}$, $\quad \alpha_y = (\alpha_9, \alpha_{10})$.

It is not important which simple formulas from $\alpha_1 - \alpha_8$ are α_u and which fact from the set $\{F_1, F_2, F_4, F_5, F_6\}$ (not containing α_y) is the input property. It is easy to see that

$\overline{F}_1(\overline{a}_0, \overline{a}_1) = F_3(a_1, a_2, a_9) \wedge F_7(a_2, a_3, a_{10})$, $\overline{a}_1 = (a_1, a_2, a_3)$,

$\overline{F}_2(\overline{a}_1, \overline{a}_2) = F_1(a_1, a_3, a_4) \wedge F_2(a_1, a_3, a_7) \wedge F_6(a_2, a_4, a_6)$, $\overline{a}_2 = (a_4, a_6, a_7)$,

$\overline{F}_3(\overline{a}_2, \overline{a}_3) = F_4(a_4, a_5, a_7) \wedge F_5(a_4, a_6, a_8)$, $\overline{a}_3 = (a_5, a_8)$.

In our case $N = 3$, $S_N = \{(1,1), (1,0), (0,1), (0,0)\}$. According to (7.10) one should put successively the elements of S_N into $\overline{F}_3$ and determine all 0-1 sequences (a_4, a_6, a_7) such that $\overline{F}_3 = 1$. These are the elements of $\overline{S}_2$. In a similar way one determines $\overline{S}_1$ and finally $\overline{S}_0 = \{(0,1), (1,1)\}$. Then $F_y = (\neg\alpha_9 \wedge \alpha_{10}) \vee (\alpha_9 \wedge \alpha_{10}) = \alpha_{10}$. □

Example 7.2 (decision making).

The facts F in the knowledge representation KP are the following:

$F_1 = \alpha_1 \wedge (\alpha_4 \vee \neg\alpha_6)$, $\quad F_2 = (\alpha_2 \wedge \alpha_4) \rightarrow \alpha_6$, $\quad F_3 = \neg\alpha_4 \vee \neg\alpha_3 \vee \alpha_5$,

$F_4 = \alpha_4 \wedge (\alpha_3 \vee \neg\alpha_5)$, $F_5 = (\alpha_4 \wedge \neg\alpha_2) \rightarrow \alpha_7$, $\alpha_u = (\alpha_1, \alpha_2)$, $\alpha_y = (\alpha_6, \alpha_7)$.

Now $\overline{a}_0 = (a_u, a_y) = (a_1, a_2, a_6, a_7)$, $\overline{F}_1 = F_1 \wedge F_2 \wedge F_5$, $\overline{F}_2 = F_3 \wedge F_4$, $\overline{a}_1 = a_4$, $\overline{a}_2 = (a_3, a_5)$.

Using (7.10) (two steps for $m = 2, 1$) we obtain $\overline{S}_0 = \{(1,1,1,1), (1,1,1,0), (1,0,1,1), (1,0,0,1)\}$. We can consider the different cases of $F_y(\alpha_6, \alpha_7)$. It is easy to see that for $F_y = \alpha_6 \vee \alpha_7$ we have S_y = {(1, 1), (1, 0), (0, 1)}, S_{u1} = {(1, 1), (1, 0)}, S_{u2} is an empty set, $S_u = S_{u1}$ and $F_u = (\alpha_1 \wedge \alpha_2) \vee (\alpha_1 \wedge \neg\alpha_2) = \alpha_1$. If $F_y = \alpha_6$ then $F_u = \alpha_1 \wedge \alpha_2$, if $F_y = \alpha_7$ then $F_u = \alpha_1 \wedge \neg\alpha_2$, if $F_y = \alpha_6 \wedge \alpha_7$ then $S_{u1} = S_{u2}$, S_u is an empty set and the solution F_u does not exist.

The formulas α and the facts may have different practical senses. For instance, in the second example $u, y, c \in R^1$ and: α_1 = "$u \le 3c$", α_2 = "$u^2 + c^2 \le 1$", α_3 = "pressure is high", α_4 = "humidity is low", α_5 = "temperature is less than $u + y + c$", α_6 = "$y^2 + (c - 0.5)^2 \le 0.25$", α_7 = "$-c \le y \le c$" for a given parameter c. For example, the fact F_2 means that: "if $u^2 + c^2 \le 1$ and humidity is low then $y^2 + (c - 0.5)^2 \le 0.25$", the fact F_3 means that: "humidity is not low or pressure is not high or temperature is less than $u + y + c$". The required output property $F_y = \alpha_6$ is obtained if $F_u = \alpha_1 \wedge \alpha_2$, i.e. if $u \le 3c$ and $u^2 + c^2 \le 1$. □

7.4 Analysis and Decision Making for a Plant with Random Parameters

Let us consider the plant described by a logical knowledge representation with random parameters in the simple formulas and consequently in the properties F, F_u, F_y [17, 54]. In general, we may have the simple formulas $\alpha_u(u;x)$, $\alpha_w(u, y, w; x)$ and $\alpha_y(y;x)$ where $x \in X$ is an unknown vector parameter which is assumed to be a value of a random variable $\tilde{x}$ described by the probability density $f_x(x)$. For example,

$$\alpha_{u1} = \text{“} u^{\mathrm{T}} u \le 2x^{\mathrm{T}} x \text{”}, \quad \alpha_{w1} = \text{“} y^{\mathrm{T}} y \le x^{\mathrm{T}} x \text{”}, \quad \alpha_{y1} = \text{“} y^{\mathrm{T}} y + x^{\mathrm{T}} x < 4 \text{”}.$$

In particular, some simple formulas may only depend on some components of the vector x.

In the analysis problem the formula $F_u[\alpha_u(u;x)]$ depending on x means that the observed (given) input property is formulated with the help of the unknown parameter (e.g. we may know that u is less than the temperature of a raw material x, but we do not know the exact value of x). By solving the analysis problem described in Sects 7.2 and 7.3 we obtain $F_y[\alpha_y(y;x)]$ and consequently

$$D_y(x) = \{y \in Y: \ F_y[\alpha_y(y;x)] = 1\}. \tag{7.11}$$

Further considerations are the same as in Sect. 3.5. The analysis problem consists in the determination of the probability that the given set D_y belongs to the set of possible outputs (7.11), i.e.

$$P[D_y \subseteq D_y(\widetilde{x})] = \int\limits_{D_x(D_y)} f_x(x)dx$$

where

$$D_x(D_y) = \{x \in X: \ D_y \subseteq D_y(x)\}.$$

In particular, for $D_y = \{y\}$ (a singleton), one can find the probability that y is a possible input (i.e. that y belongs to the set $D_y(x)$):

$$P[y \in D_y(\widetilde{x})] = \int\limits_{D_x(y)} f_x(x)dx$$

where

$$D_x(y) = \{x \in X : y \in D_y(x)\}.$$

In the decision problem the formula $F_y[\alpha_y(y\,;x)]$ depending on x means that the user formulates the required output property with the help of the unknown parameter (e.g. he wants to obtain y less than the temperature of a product x). Solving the decision problem described in Sects 7.2 and 7.3 we obtain $F_u[(u\,;x)]$ and consequently

$$D_u(x) = \{u \in U: \ F_u[a_u(u\,;x)] = 1\}. \tag{7.12}$$

Further considerations are the same as in Sect. 3.5. The decision problem consists in the determination of the optimal decision u^*, maximizing the probability that the set of possible outputs belongs to the set

$$D_y = \{y \in Y : F_y[a_y(y;x)] = 1\},$$

i.e. the probability

$$P[u \in D_u(\widetilde{x})]$$

where $D_u(x)$ is determined by (7.12) and $F_u[(u;x)]$ is obtained as a solution of the decision problem described in Sects 7.2 and 7.3. Then

$$u^* = \arg\max_{u \in D_u} \int\limits_{D_{xd}(u)} f_x(x)dx$$

where

$$D_{xd}(u) = \{x \in X : u \in D_u(x)\}.$$

7.5 Analysis and Decision Making for a Plant with Uncertain Parameters

Now let us consider the plant described by a logical knowledge representation with uncertain parameters in the simple formulas and consequently in the properties F, F_u, F_y [21, 54]. In general, we may have the simple formulas $\alpha_u(u\,;x)$, $\alpha_w(u, y, w\,; x)$ and $\alpha_y(y\,;x)$ where $x \in X$ is an unknown vector parameter which is assumed to be a value of an uncertain variable $\overline{x}$ with the certainty distribution $h_x(x)$ given by an expert. In particular, only some simple formulas may depend on some components of the vector x.

By solving the analysis problem described in Sects 7.2 and 7.3 we obtain $F_y[\alpha_y(y\,;x)]$ and consequently

$$D_y(x) = \{y \in Y : \; F_y[a_y(y\,;x)] = 1\}.$$

Further considerations are the same as in Sect. 5.4 for the given set D_u. In version II (see (5.22) and (5.23)) we have

$$v[D_y(\overline{x}) \subseteq D_y] = \max_{x \in D_x(D_y)} h_x(x)$$

where D_y is given by a user and

$$D_x(D_y) = \{x \in X : \; D_y(x) \subseteq D_y\}.$$

By solving the decision problem described in Sects 7.2 and 7.3 we obtain $F_u[(u\,;x)]$ and consequently

$$D_u(x) = \{u \in U : \; F_u[a_u(u\,;x)] = 1\}. \tag{7.13}$$

Further considerations are the same as in Sect. 5.5 for version II (see (5.36)). The optimal decision, maximizing the certainty index that the requirement $F_y[\alpha_y(y\,;x)]$ is satisfied, may be obtained in the following way:

$$u^* = \arg\max_u \max_{x \in D_{xd}(u)} h_x(x)$$

where

$$D_{xd}(u) = \{x \in X : \; u \in D_u(x)\}.$$

Example 7.3.

The facts are the same as in Example 7.2 where $c \stackrel{\Delta}{=} x$. In Example 7.2 for the required output property $F_y = \alpha_6$ the following result has been obtained:
If

$$u \le 3x \quad \text{and} \quad u^2 + x^2 \le 1 \tag{7.14}$$

then

$$y^2 + (x - 0.5)^2 \le 0.25 .$$

The inequalities (7.14) determine the set (7.13) in our case. Assume that x is a value of an uncertain variable with triangular certainty distribution: $h_x = 2x$ for $0 \le x \le \frac{1}{2}$, $h_x = -2x + 2$ for $\frac{1}{2} \le x \le 2$, $h_x = 0$ otherwise. Then we can use the result in Example 5.5. As the decision u^* we can choose any value from $[-\frac{\sqrt{3}}{2}, \frac{\sqrt{3}}{2}]$ and the requirement will be satisfied with the certainty index equal to 1. The result for a *C*-uncertain variable is $u_c^* = 0$ and $v_c(u_c^*) = 1$. □

7.6 Uncertain and Random Logical Decision Algorithms

Consider the plant with external disturbances $z \in Z$. Then in the logical knowledge representation we have the simple formulas $\alpha_u(u, z\,; x)$, $\alpha_w(u, w, y, z\,; x)$, $\alpha_y(y, z\,; x)$ and $\alpha_z(z\,; x)$ to form the property $F_z(\alpha_z)$ concerning z.

The analysis problem analogous to that described in Sects 7.2 and 7.3 for the fixed x is as follows: for the given $F(\alpha_u, \alpha_w, \alpha_y, \alpha_z)$, $F_z(\alpha_z)$ and $F_u(\alpha_u)$ find the best property $F_y(\alpha_y)$ such that the implication

$$F_z(\alpha_z) \wedge F_u(\alpha_u) \rightarrow F_y(\alpha_y) \tag{7.15}$$

is satisfied. In this formulation $F_z(\alpha_z)$ denotes an observed property concerning z.

The problem solution is the same as in Sect. 7.3 with $F_z \wedge F_u$ in place of F_u. As a result one obtains $F_y[\alpha_y(y, z\,; x)]$ and consequently

$$D_y(z\,; x) = \{y \in Y : \; F_y[a_y(y, z\,; x)] = 1\} . \tag{7.16}$$

Assume that x is a value of an uncertain variable $\bar{x}$ described by the certainty distribution $h_x(x)$. Then further considerations are analogous to those described in Sect. 5.4 for the given z. Now as a result of observation we obtain a set of possible values z, depending on the unknown parameter x:

$$D_z(x) = \{z \in Z : F_z[a_z(z;x)] = 1\}. \tag{7.17}$$

In version II of the analysis problem, for the given set D_y one should determine

$$v\{\bigwedge_{z \in D_z(\bar{x})} [D_y(z;\bar{x}) \tilde{\subseteq} D_y]\} = \max_{x \in D_x(D_y)} h_x(x)$$

where

$$D_x(D_y) = \{x \in X : \bigwedge_{z \in D_z(x)} [D_y(z;x) \subseteq D_y]\}. \tag{7.18}$$

The decision problem analogous to that described in Sects 7.2 and 7.3 for the fixed x is as follows: for the given $F(\alpha_u, \alpha_w, \alpha_y, \alpha_z)$, $F_z(\alpha_z)$ and $F_y(\alpha_y)$ find the best property $F_u(\alpha_u)$ such that the implication (7.15) is satisfied.

The problem solution is the same as in Sect. 7.3 with $F_z \wedge F$ in place of F. As a result one obtains

$$D_u(z;x) = \{u \in U : \ F_u[a_u(u,z;x)] = 1\}. \tag{7.19}$$

Further considerations are analogous to those described in Sect. 5.5 for the given z. In version II of the decision problem, for the given set D_y one should determine the decision u^* maximizing the certainty index

$$v\{\bigwedge_{z \in D_z(\bar{x})} [u \tilde{\in} D_u(z;\bar{x})]\} = \max_{x \in D_{xd}(u)} h_x(x) \stackrel{\Delta}{=} v(u)$$

where

$$D_{xd}(u) = \{x \in X : \bigwedge_{z \in D_z(x)} [u \in D_u(z;x)]\}. \tag{7.20}$$

To obtain the solution of the decision problem another approach may be applied. For the given F and F_y we may state the problem of finding the best input property $F_d(\alpha_u, \alpha_z)$ such that the implication

$$F_d(\alpha_u, \alpha_z) \rightarrow F_y(\alpha_y)$$

is satisfied. The solution may be obtained in the same way as in Sect. 7.3 with (α_u, α_z) and F_d in the place of α_u and F_u, respectively. The formula $F_d(\alpha_u, \alpha_z)$ may be called a logical knowledge representation for the decision making (i.e. the logical form of KD) or a *logical uncertain decision algorithm* corresponding to the relation $\overline{R}$ or the set $D_u(z\,;x)$ in Sect. 5.5. For the given $F_z(\alpha_z)$, the input property may be obtained in the following way. Denote by $\overline{S}_d$ the set of all (a_u, a_z) for which $F_d = 1$ and by $\overline{S}_z$ the set of all a_z for which $F_z = 1$, i.e.

$$\overline{S}_d = \{(a_u, a_z): \; F_d(a_u, a_z) = 1\}.$$

$$\overline{S}_z = \{a_z : \; F_z(a_z) = 1\}.$$

Then $F_u(\alpha_u)$ is determined by the set

$$S_u = \{a_u \in S_u : \bigwedge_{a_z \in \overline{S}_z} (a_u, a_z) \in \overline{S}_d\}. \tag{7.21}$$

The formula (7.19) is analogous to the formula (2.17) for the relational plant. It follows from the fact that at the logical level our plant may be considered as a relational plant with input a_u, disturbance a_z and output a_y (see Fig. 7.1). The decision system is illustrated in Fig. 7.3 where $u^* = \arg\max_{u \in U} v(u)$.

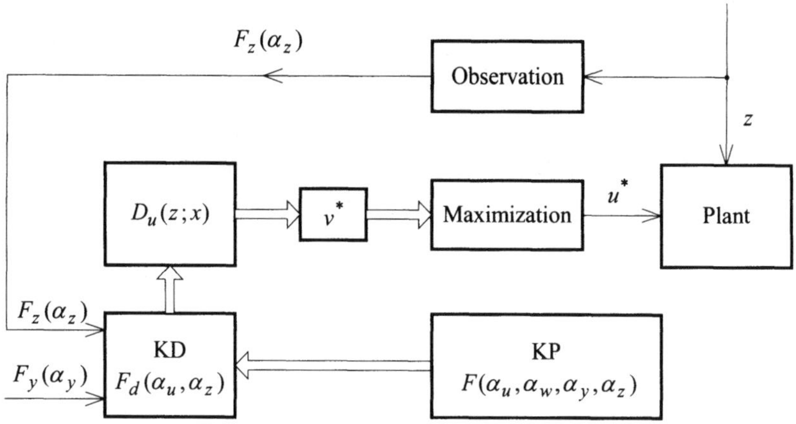

Figure 7.3. Decision system based on uncertain logical knowledge representation

Similar problems may be described for a random parameter x, i.e. under the assumption that x is a value of a random variable $\widetilde{x}$ described by the probability density $f_x(x)$. Then the considerations are analogous to those in Sect. 3.5. The analysis problem (corresponding to version I of the problem described in Sect. 5.4) consists in determining the probability

$$P\{\bigwedge_{z\in D_z(\widetilde{x})}[D_y \subseteq D_y(z;\widetilde{x})]\} = \int_{D_x(D_y)} f_x(x)dx$$

where $D_y \subset Y$ is a given set, $D_z(x)$ is defined by (7.17) and $D_x(D_y)$ is defined by (7.18).

In the decision problem, for the given set $D_y \subset Y$ one should determine the decision u^* maximizing the probability

$$P\{\bigwedge_{z\in D_x(\widetilde{x})}[u \in D_u(z;\widetilde{x})]\} = \int_{D_{xd}(u)} f_x(x)dx$$

where the sets $D_u(z;\widetilde{x})$ and $D_{xd}(u)$ are defined by (7.19) and (7.20), respectively. In another version, the solution (7.21) is based on the knowledge of the decision making $F_d[\alpha_u(u,z;x), \alpha_z(z;x)]$. If x is assumed to be a value of a random variable $\widetilde{x}$ then KD = $<F_d(\alpha_u, \alpha_z)>$ may be called a *logical random decision algorithm*.

8 Dynamical Systems

The aim of this chapter is to show how the approaches and methods presented in the previous chapters may be applied to discrete-time dynamical plants described by traditional functional models or by relational knowledge representations. Special attention is paid to the relational plants and the descriptions based on uncertain variables [22, 54]. The considerations are completed with the optimization of a random and uncertain multistage decision process (dynamic programming under uncertainty) and with applications to a class of assembly systems. Other considerations for dynamical systems are presented in Chapter 9 (uncertain, random and fuzzy controllers in closed-loop systems), in Chapter 10 (stability) and in Chapter 11 (dynamical learning systems).

8.1 Relational Knowledge Representation

The relational knowledge representation for the dynamical plant may have the form analogous to that of the static plant presented in Sect. 2.1. The deterministic dynamical plant is described by the equations

$$\left.\begin{aligned} s_{n+1} &= \chi(s_n, u_n), \\ y_n &= \eta(s_n) \end{aligned}\right\} \tag{8.1}$$

where n denotes the discrete time and $s_n \in S$, $u_n \in U$, $y_n \in Y$ are the state, the input and the output vectors, respectively. In the relational dynamical plants the functions χ and η are replaced by relations

$$\left.\begin{aligned} R_{\text{I}}(u_n, s_n, s_{n+1}) &\subseteq U \times S \times S, \\ R_{\text{II}}(s_n, y_n) &\subseteq S \times Y. \end{aligned}\right\} \tag{8.2}$$

The relations R_{I} and R_{II} form a *relational knowledge representation of the dynamical plant*. For a non-stationary plant the relations R_{I} and R_{II} depend on n. The relations R_{I} and R_{II} may have the form of equalities and/or inequalities concerning the components of the respective vectors. In particular the relations are described by the inequalities

$$\chi_1(u_n, s_n) \le s_{n+1} \le \chi_2(s_n, u_n),$$
$$\eta_1(s_n) \le y_n \le \eta_2(s_n),$$

i.e. by a set of inequalities for the respective components of the vectors. The formulations of the analysis and decision problems may be similar to those in Sect. 2.2. Let us assume that $s_0 \in D_{s0} \subset S$.

Analysis problem: For the given relations (8.2), the set D_{s0} and the given sequence of sets $D_{un} \subset U$ $(n = 0, 1, ...)$ one should find a sequence of the smallest sets $D_{yn} \subset Y$ $(n = 1, 2, ...)$ for which the implication

$$(u_0 \in D_{u0}) \wedge (u_1 \in D_{u1}) \wedge ... \wedge (u_{n-1} \in D_{u,n-1}) \rightarrow y_n \in D_{yn}$$

is satisfied.

It is an extension of the analysis problem for the deterministic plant (8.1), consisting in finding the sequence y_n for the given sequence u_n and the initial state s_0, and for the known functions χ, η. For the fixed moment n, our plant may be considered as a connection of two static relational plants (Fig. 8.1). The analysis problem is then reduced to the analysis for the relational plants $R_{\rm I}$ and $R_{\rm II}$, described in Sect. 2.2. Consequently, according to the formula (2.5) applying to $R_{\rm I}$ and $R_{\rm II}$, we obtain the following *recursive procedure* for $n = 1, 2, ...$:

1. For the given D_{un} and D_{sn} obtained in the former step, determine the set $D_{s,n+1}$ using $R_{\rm I}(u_n, s_n, s_{n+1})$:

$$D_{s,n+1} = \{s_{n+1} \in S : \bigvee_{u_n \in D_{un}} \bigvee_{s_n \in D_{sn}} [(u_n, s_n, s_{n+1}) \in R_{\rm I}(u_n, s_n, s_{n+1})]\}. \quad (8.3)$$

2. Using $D_{s,n+1}$ and $R_{\rm II}(s_{n+1}, y_{n+1})$, determine $D_{y,n+1}$:

$$D_{y,n+1} = \{y_{n+1} \in Y : \bigvee_{s_{n+1} \in D_{s,n+1}} [(s_{n+1}, y_{n+1}) \in R_{\rm II}(s_{n+1}, y_{n+1})]\}. \quad (8.4)$$

For $n = 0$ in the formula (8.3) we use the given set D_{s0}.

Decision problem: For the given relations (8.2), the set D_{s0} and the sequence of sets $D_{yn} \subset Y$ $(n = 1, 2, ..., N)$ one should determine the sequence D_{un} $(n = 0, 1, ..., N-1)$ such that the implication

$$(u_0 \in D_{u0}) \wedge (u_1 \in D_{u1}) \wedge ... \wedge (u_{N-1} \in D_{u,N-1})$$
$$\rightarrow (y_1 \in D_{y1}) \wedge (y_2 \in D_{y2}) \wedge ... \wedge (y_N \in D_{y,N})$$

is satisfied.

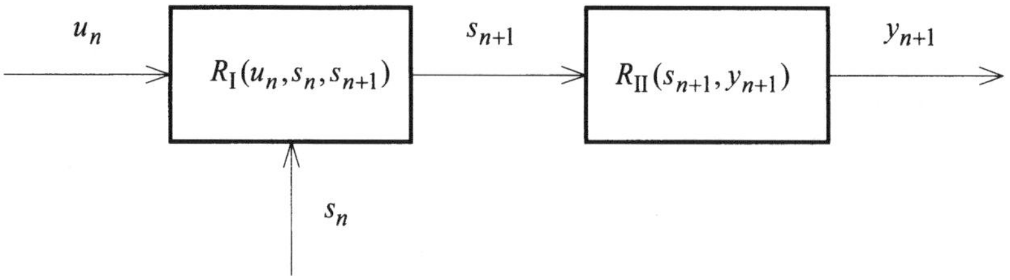

Figure 8.1. Dynamical relational plant

The set D_{yn} is given by a user and the property $y_n \in D_{yn}$ $(n = 1, 2, ..., N)$ denotes the user's requirement. To obtain the solution, one can apply the following *recursive procedure* starting from $n = 0$:

1. For the given $D_{y,n+1}$, using $R_{II}(s_{n+1}, y_{n+1})$ determine the largest set $D_{s,n+1}$ for which the implication

$$s_{n+1} \in D_{s,n+1} \rightarrow y_{n+1} \in D_{y,n+1}$$

is satisfied. This is a decision problem for the part of the plant described by R_{II} (see Fig. 8.1). According to (2.13) with s_{n+1}, y_{n+1} in place of (u, y) we obtain

$$D_{s,n+1} = \{s_{n+1} \in S : D_{y,n+1}(s_{n+1}) \subseteq D_{y,n+1}\} \tag{8.5}$$

where

$$D_{y,n+1}(s_{n+1}) = \{y_{n+1} \in Y : (s_{n+1}, y_{n+1}) \in R_{II}(s_{n+1}, y_{n+1})\}.$$

2. For $D_{s,n+1}$ obtained at point 1 and D_{sn} obtained in the former step, using $R_I(u_n, s_n, s_{n+1})$ determine the largest set D_{un} for which the implication

$$(u_n \in D_{un}) \wedge (s_n \in D_{sn}) \rightarrow s_{n+1} \in D_{s,n+1}$$

is satisfied. This is a decision problem for the part of the plant described by R_I. According to (2.17) with (u_n, s_{n+1}, s_n) in place of (u, y, z) we obtain

$$D_{un} = \{u_n \in U : \bigwedge_{s_n \in D_{sn}} [D_{s,n+1}(u_n, s_n) \subseteq D_{s,n+1}]\} \tag{8.6}$$

where

$$D_{s,n+1}(u_n, s_n) = \{s_{n+1} \in S : (u_n, s_n, s_{n+1}) \in R_I(u_n, s_n, s_{n+1})\}.$$

Remark 8.1. In the formulation of the decision problem we did not use the statement "the largest set D_{un}". Now the set of all possible decisions means the set of all sequences $u_0, u_1, ..., u_{N-1}$ for which the requirements are satisfied. Using the recursive procedure described above we do not obtain the set of all possible decisions. In other words, we determine the set of sequences $u_0, u_1, ..., u_{N-1}$ belonging to the set of all input sequences for which the requirements concerning y_n are satisfied. □

Remark 8.2. The relations R_{I} and R_{II} may be given by the sets of facts in a similar way as described in Sect. 7.1. The formulation and solution of the analysis and decision problems for the plant described by *dynamical logical knowledge representation* are analogous to those presented above and for the fixed n are reduced to the analysis and decision problems considered for the static plant in Sect. 7.2. □

The considerations presented in this section will be used for the plants with random or uncertain parameters in the knowledge representation, described in the next sections.

Assume that the current states s_n may be measured. Then (8.6) becomes

$$\overline{D}_{un}(s_n) = \{u_n \in U : [D_{s,n+1}(u_n, s_n) \subseteq D_{s,n+1}]\}. \tag{8.7}$$

It is easy to note that in this case we can obtain the better result, i.e. the set $\overline{D}_{un}$ may be greater than the set D_{un} obtained according to (8.6). The decisions may be executed in real time, in a closed-loop decision system (control system) with the deterministic decision algorithm

$$u_{dn} = \text{M}(u_n) = \int_{\overline{D}_{un}(s_n)} u\,du \cdot [\int_{\overline{D}_{un}(s_n)} du]^{-1} \stackrel{\Delta}{=} \Psi_n(s_n).$$

The knowledge of the decision making $\text{KD} = \langle \overline{D}_{un}(s_n) \rangle$ may be called a *relational decision algorithm* (the description of a *relational controller* in the closed-loop system) and the deterministic algorithm $\Psi_n(s_n)$ (the description of the *deterministic controller* in the closed-loop system) is based on KD obtained from KP described by (8.2), for the given D_{yn} (Fig. 8.2).

In the decision problem formulated in this section it is not necessary to assume that R_{I} and R_{II} do not depend on n. The considerations are identical for the time-varying plant described by $R_{\text{I}n}$ and $R_{\text{II}n}$, which are used in the formulas (8.5) and (8.6) in place of R_{I} and R_{II}. In a similar way we may consider the plant with external disturbances $z_n \in Z$, described by the relations

$$\left.\begin{aligned} &R_{\text{I}}(u_n, s_n, s_{n+1}, z_n) \subseteq U \times S \times S \times Z, \\ &R_{\text{II}}(s_{n+1}, y_{n+1}) \subseteq S \times Y \ . \end{aligned}\right\} \tag{8.8}$$

Consequently, as a result of the decision problem we obtain $D_{un}(z_n)$ or $\overline{D}_{un}(z_n)$, and after the determinization – the decision algorithm $u_{dn} = \Psi_n(z_n)$ or $\overline{u}_{dn} = \Psi_n(s_n, z_n)$, which may be used for the determination of the current decisions if z_n are measured in the successive moments.

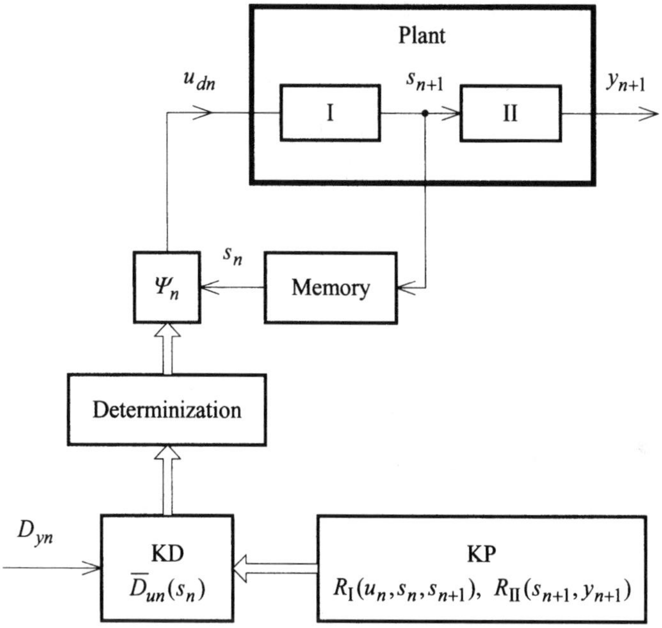

Figure 8.2. Closed-loop dynamical decision system based on relational knowledge representation

Example 8.1.
As a very simple example let us consider a first-order one-dimensional plant described by the inequalities

$$a_1 s_n + b_1 u_n \le s_{n+1} \le a_2 s_n + b_2 u_n ,$$

$$c_1 s_{n+1} \le y_{n+1} \le c_2 s_{n+1} .$$

It is known that $s_{01} \le s_0 \le s_{02}$; b_1, b_2, c_1, $c_2 > 0$. The requirement concerning y_n is as follows:

$$\bigwedge_{n \ge 1} (y_{\min} \le y_n \le y_{\max}),$$

i.e. $D_{yn} = [y_{\min}, y_{\max}]$ for every n. For the given s_{01}, s_{02}, $y_{\min}$, $y_{\max}$ and the coefficients a_1, a_2, b_1, b_2, c_1, c_2 one should determine the sequence D_{un} such that if $u_n \in D_{un}$ for every n, then the requirement is satisfied. For $n = 0$ the set D_{s1} according to (8.5) is determined by the inequalities

$$c_2 s_1 \le y_{\max}, \quad c_1 s_1 \ge y_{\min}.$$

Then

$$D_{s1} = [\frac{y_{\min}}{c_1}, \frac{y_{\max}}{c_2}].$$

Using (8.6) for u_0 we obtain the following inequalities:

$$a_2 s_{02} + b_2 u_0 \le \frac{y_{\max}}{c_2},$$

$$a_1 s_{01} + b_1 u_0 \ge \frac{y_{\min}}{c_1}$$

and

$$D_{u0} = [\frac{y_{\min}}{b_1 c_1} - \frac{a_1 s_{01}}{b_1}, \frac{y_{\max}}{b_2 c_2} - \frac{a_2 s_{02}}{b_2}].$$

For $n \ge 1$ $D_{s,n+1} = D_{s1}$, and according to (8.6), D_{un} is determined by the inequalities

$$a_2 \frac{y_{\max}}{c_2} + b_2 u_n \le \frac{y_{\max}}{c_2},$$

$$a_1 \frac{y_{\min}}{c_1} + b_1 u_n \ge \frac{y_{\min}}{c_1}.$$

Consequently,

$$D_{un} = [\frac{y_{\min}(1-a_1)}{b_1 c_1}, \frac{y_{\max}(1-a_2)}{b_2 c_2}].$$

The final result is then as follows: If

$$\frac{y_{\min}}{b_1 c_1} - \frac{a_1 s_{01}}{b_1} \le u_0 \le \frac{y_{\max}}{b_2 c_2} - \frac{a_2 s_{02}}{b_2} \tag{8.9}$$

and for every $n > 0$

$$\frac{y_{\min}(1-a_1)}{b_1 c_1} \le u_n \le \frac{y_{\max}(1-a_2)}{b_2 c_2} \tag{8.10}$$

then the requirement concerning y_n will be satisfied. The conditions for the existence of the solution are the following:

$$\frac{y_{\min}}{b_1 c_1} - \frac{a_1 s_{01}}{b_1} \le \frac{y_{\max}}{b_2 c_2} - \frac{a_2 s_{02}}{b_2}, \tag{8.11}$$

$$\frac{y_{\min}}{c_1} \le \frac{y_{\max}}{c_2}, \tag{8.12}$$

$$\frac{y_{\min}(1-a_1)}{b_1 c_1} \le \frac{y_{\max}(1-a_2)}{b_2 c_2}. \tag{8.13}$$

If $y_{\min} > 0$ and $a_2 < 1$ then these conditions are reduced to the inequality

$$\frac{y_{\max}}{y_{\min}} \ge \max(\alpha, \beta)$$

where

$$\alpha = \frac{b_2 c_2}{b_1 c_1} \cdot \frac{1-a_1}{1-a_2}, \qquad \beta = \frac{c_2}{c_1}.$$

Then D_{un} are not empty sets if the requirement concerning y_n is not too strong, i.e. the ratio $y_{\max} \cdot y_{\min}^{-1}$ is respectively high. One should note that the inequalities (8.11), (8.12), (8.13) form the sufficient condition for the existence of the sequence u_n, satisfying the requirement.

8.2 Analysis and Decision Making for Dynamical Plants with Uncertain Parameters

The analysis and decision problems for dynamical plants described by a relational knowledge representation with uncertain parameters may be formulated and solved in a similar way as for the static plants in Sects. 5.4 and 5.5. Let us consider the plant described by the relations

$$\left.\begin{aligned} &R_{\mathrm{I}}(u_n, s_n, s_{n+1}; x) \subseteq U \times S \times S, \\ &R_{\mathrm{II}}(s_n, y_n; w) \subseteq S \times Y, \end{aligned}\right\} \tag{8.14}$$

where $x \in X$ and $w \in W$ are unknown vector parameters which are assumed to be values of uncertain variables $(\bar{x}, \bar{w})$ with the joint certainty distribution $h(x, w)$. We shall consider the analysis and decision problems in version II (see Sects 5.4 and 5.5), which has a better practical interpretation. The considerations in version I are analogous.

Analysis problem: For the given relations (8.14), $h(x, w)$, D_{s0} and the sequences D_{un}, D_{yn} one should determine

$$v[D_{yn}(\bar{x},\bar{w}) \tilde{\subseteq} D_{yn}] \triangleq v_n$$

where $D_{yn}(x,w)$ is the result of the analysis problem formulated in the previous section, i.e. the set of all possible outputs y_n for the fixed x and w.

In a similar way as for the static plant considered in Sect. 5.4 (see the formulas (5.25) and (5.26)), we obtain

$$v_n = v[(\bar{x},\bar{w}) \tilde{\in} D(D_{yn}, D_{u,n-1})] = \max_{(x,w)\in D(D_{yn},D_{u,n-1})} h(x,w) \quad (8.15)$$

where

$$D(D_{yn}, D_{u,n-1}) = \{(x,w) \in X \times W : D_{yn}(x,w) \subseteq D_{yn}\}.$$

In the case where $(\bar{x},\bar{w})$ are considered as *C*-uncertain variables, it is necessary to find v_n (8.15) and

$$v[(\bar{x},\bar{w}) \tilde{\in} \overline{D}(D_{yn}, D_{u,n-1})] = \max_{(x,w)\in \overline{D}(D_{yn},D_{u,n-1})} h(x,w)$$

where $\overline{D}(D_{yn}, D_{u,n-1}) = X \times W - D(D_{yn}, D_{u,n-1})$. Then

$$v_c[D_{yn}(\bar{x},\bar{w}) \tilde{\subseteq} D_{yn}] = \frac{1}{2}\{v[(\bar{x},\bar{w}) \tilde{\in} \overline{D}(D_{yn}, D_{u,n-1})] + 1 - v[(\bar{x},\bar{w}) \tilde{\in} \overline{D}(D_{yn}, D_{u,n-1})]\}$$

(see (5.27) and (5.28)).

For the given value u_n, using (8.3) and (8.4) for the fixed (x, w) we obtain

$$D_{s,n+1}(u_n;x) = \{s_{n+1} \in S: \bigvee_{s_n \in D_{sn}(x)} (u_n, s_n, s_{n+1}) \in R_{\mathrm{I}}(u_n, s_n, s_{n+1}; x)\}, \quad (8.16)$$

$$D_{y,n+1}(u_n;x,w) = \{y_{n+1} \in Y: \bigvee_{s_{n+1} \in D_{s,n+1}(u_n;x)} (s_{n+1}, y_{n+1}) \in R_{\mathrm{II}}(s_{n+1}, y_{n+1}; w)\}. \quad (8.17)$$

The formulation and solution of the analysis problem are the same as described above with u_n, $D_{yn}(u_{n-1}; x, w)$ and $D(D_{yn}, u_{n-1})$ instead of D_{un}, $D_{yn}(x, w)$

and $D(D_{yn}, D_{u,n-1})$, respectively.

Decision problem: For the given relations (8.14), $h(x, w)$, D_{s0} and the sequence D_{yn} $(n = 1, 2, ..., N)$ find the sequence of the optimal decisions

$$u_n^* = \arg \max_{u_n \in U} v[D_{y,n+1}(u_n; x, w) \tilde{\subseteq} D_{y,n+1}]$$

for $n = 0, 1, ..., N - 1$, where $D_{y,n+1}(u_n; x, w)$ is the result of the analysis problem (8.17). Then

$$u_{n-1}^* = \arg \max_{u_{n-1} \in U} \max_{(x,w) \in D(D_{yn}, u_{n-1})} h(x, w)$$

where

$$D(D_{yn}, u_{n-1}) = \{(x, w) \in X \times W: D_{yn}(u_{n-1}; x, w) \subseteq D_{yn}\}.$$

In a similar way as in Sect. 5.5 the determination of u_n^* may be replaced by the determination of $u_{dn}^* = u_n^*$ where

$$u_{dn}^* = \arg \max_{u_n \in U} v[u_n \tilde{\in} D_{un}(\overline{x}, \overline{w})]$$

where $D_{un}(x, w)$ is the result of the decision problem considered in the previous section for the fixed x and w. Then

$$u_{dn}^* = u_n^* = \arg \max_{u_n \in U} \max_{(x,w) \in D_d(u_n)} h(x, w)$$

where

$$D_d(u_n) = \{(x, w): u_n \in D_{un}(x, w)\}.$$

Assume that the current states s_n may be measured. Then (8.16) becomes

$$\overline{D}_{s,n+1}(u_n; x) = \{s_{n+1} \in S: (u_n, s_n, s_{n+1}) \in R_{\mathrm{I}}(u_n, s_n, s_{n+1}; x)\} \qquad (8.18)$$

and $\overline{D}_{s,n+1}$ is used in (8.17) in place of $D_{s,n+1}$. Consequently, we obtain the relationship $u_n^* = \Psi_n(s_n)$ as a deterministic decision algorithm in a closed-loop decision system.

In the decision problem formulated in this section it is not necessary to assume that (x, w) are constant parameters and that R_{I}, R_{II} do not depend on n. The considerations are identical for (x_n, w_n) described by $h_n(x, w)$ and for $R_{\mathrm{I}n}$, $R_{\mathrm{II}n}$, which are used in the place of $h(x, w)$, $R_{\mathrm{I}}, R_{\mathrm{II}}$, respectively. In a similar way we may consider the plant with external disturbances $z_n \in Z$, described by the relations (8.8). Now the sets $\overline{D}_{s,n+1}$ in (8.18) and $D_{y,n+1}$ in (8.17) depend on z_n, i.e.

$$\overline{D}_{s,n+1}(u_n, z_n; x) = \{s_{n+1} \in S: (u_n, s_n, s_{n+1}, z_n) \in R_{\rm I}(u_n, s_n, s_{n+1}, z_n)\},$$

$$D_{y,n+1}(u_n, z_n; x, w) = \{y_{n+1} \in Y: \bigvee_{s_{n+1} \in \overline{D}_{s,n+1}(u_n, z_n; x)} (s_{n+1}, y_{n+1}) \in R_{\rm II}(s_{n+1}, y_{n+1}; w)\}.$$

Then

$$u_n^* = \arg\max_{u_n \in U} v_n(u_n, s_n, z_n) \stackrel{\Delta}{=} \Psi_n(s_n, z_n)$$

where

$$v_n(u_n, s_n, z_n) = \max_{(x,w) \in D(D_{y,n+1}, u_n)} h(x, w),$$

and

$$D(D_{y,n+1}, u_n, z_n) = \{(x, w) \in X \times W: D_{y,n+1}(u_n, z_n; x, w) \subseteq D_{y,n+1}\}.$$

Consequently, in the case when s_n and z_n may be measured, the decisions u_n^* may be executed in a real-time combined decision system (control system with open and closed loops), according to the deterministic algorithm $u_n^* = \Psi_n(s_n, z_n)$ (Fig. 8.3).

Example 8.2.

Let us assume that in the plant considered in Example 8.1 the parameters $c_1 \stackrel{\Delta}{=} x_1$ and $c_2 \stackrel{\Delta}{=} x_2$ are unknown and are the values of the independent uncertain variables $\overline{x}_1$ and $\overline{x}_2$, respectively. The certainty distributions $h_{x1}(x_1)$ and $h_{x2}(x_2)$ have the triangular form with the parameters d_1, γ_1 for $\overline{x}_1$ (Fig. 8.4) and d_2, γ_2 for $\overline{x}_2$; $\gamma_1 < d_1$, $\gamma_2 < d_2$. By using the results (8.9) and (8.10) one may determine the optimal decisions u_n^*, maximizing the certainty index $v[u_n \mathrel{\tilde{\in}} D_{un}(\overline{x}_1, \overline{x}_2)] = v(u_n)$. From (8.10) we have

$$v(u_n) = v\{[\overline{x}_1 \mathrel{\tilde{\in}} D_1(u_n)] \wedge [\overline{x}_2 \mathrel{\tilde{\in}} D_2(u_n)]\} = \min\{v[\overline{x}_1 \mathrel{\tilde{\in}} D_1(u_n)], v[\overline{x}_2 \mathrel{\tilde{\in}} D_2(u_n)]\}.$$

Under assumption $a_{1,2} < 1$, the sets $D_1(u_n)$ and $D_2(u_n)$ for $n > 0$ are determined by the inequalities

$$x_1 \geq \frac{\alpha}{u_n}, \qquad x_2 \leq \frac{\beta}{u_n},$$

respectively, where

$$\alpha = \frac{y_{\min}(1-a_1)}{b_1}, \qquad \beta = \frac{y_{\max}(1-a_2)}{b_2}.$$

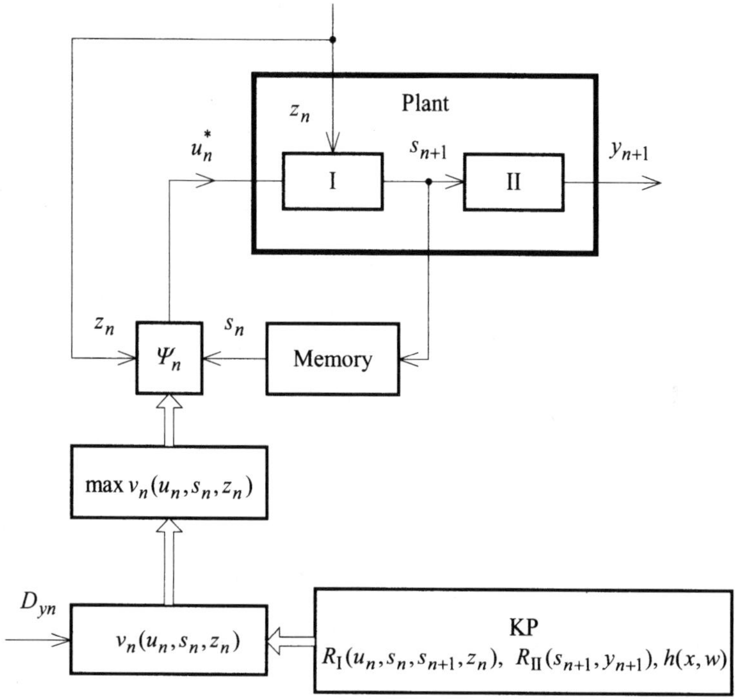

Figure 8.3. Combined decision system with description based on uncertain variables

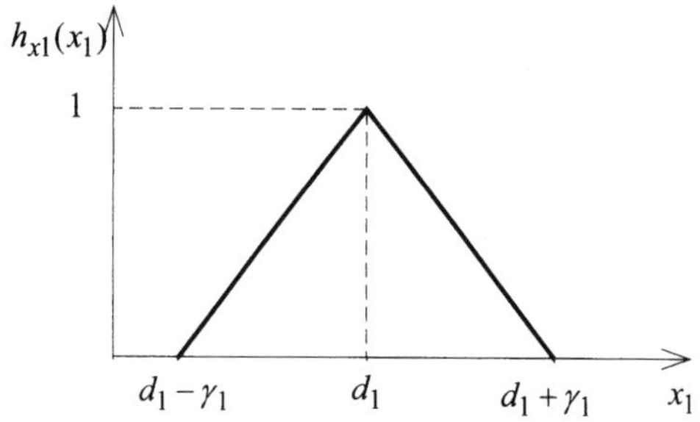

Figure 8.4. Example of certainty distribution

The certainty indexes

$$v[\bar{x}_1 \mathrel{\tilde{\in}} D_1(u_n)] = \max_{x_1 \in D_1(u_n)} h_{x1}(x_1) \stackrel{\Delta}{=} v_1(u_n)$$

and

$$v[\bar{x}_2 \mathrel{\tilde{\in}} D_2(u_n)] = \max_{x_2 \in D_2(u_n)} h_{x2}(x_2) \stackrel{\Delta}{=} v_2(u_n)$$

may be obtained by using the inequalities determining the sets $D_1(u_n)$, $D_2(u_n)$, and the distributions h_{x1}, h_{x2}:

$$v_1(u_n) = \begin{cases} 1 & \text{for } u_n \ge \dfrac{\alpha}{d_1} \\ -\dfrac{\alpha}{u_n \gamma_1} + 1 + \dfrac{d_1}{\gamma_2} & \text{for } \dfrac{\alpha}{d_1 + \gamma_1} \le u_n \le \dfrac{\alpha}{d_1} \\ 0 & \text{for } u_n \le \dfrac{\alpha}{d_1 + \gamma_1}, \end{cases}$$

$$v_2(u_n) = \begin{cases} 1 & \text{for } u_n \le \dfrac{\beta}{d_2} \\ \dfrac{\beta}{\gamma_2 u_n} + 1 - \dfrac{d_2}{\gamma_2} & \text{for } \dfrac{\beta}{d_2} \le u_n \le \dfrac{\beta}{d_2 - \gamma_2} \\ 0 & \text{for } u_n \ge \dfrac{\beta}{d_2 - \gamma_2}. \end{cases}$$

Now we can consider three cases illustrated in Figs. 8.5, 8.6 and 8.7:

1.

$$\frac{\alpha}{d_1} \le \frac{\beta}{d_2}.$$

Then

$$u_n^* = \arg\max_{u_n} \min\{v_1(u_n), v_2(u_n)\}$$

is any value satisfying the inequality

$$\frac{\alpha}{d_1} \le u_n \le \frac{\beta}{d_2}$$

and $v(u_n^*) = 1$.

2.

$$\frac{\alpha}{d_1} > \frac{\beta}{d_2}, \qquad \frac{\alpha}{d_1 + \gamma_1} < \frac{\beta}{d_2 - \gamma_2}.$$

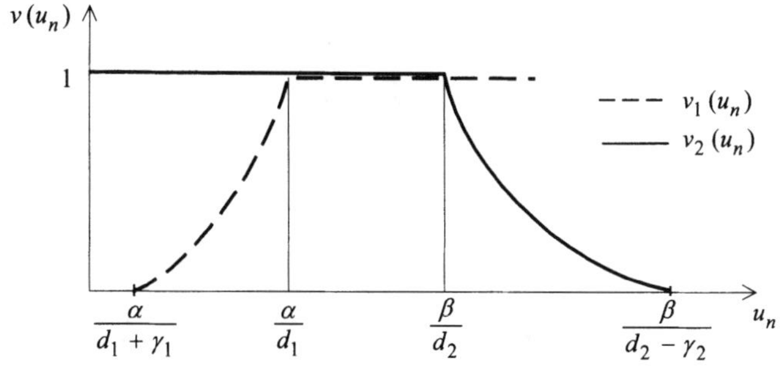

Figure 8.5. Relationship between v and u – the first case

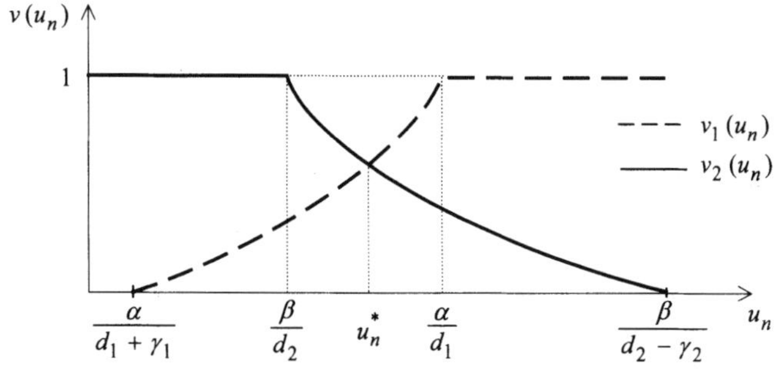

Figure 8.6. Relationship between v and u – the second case

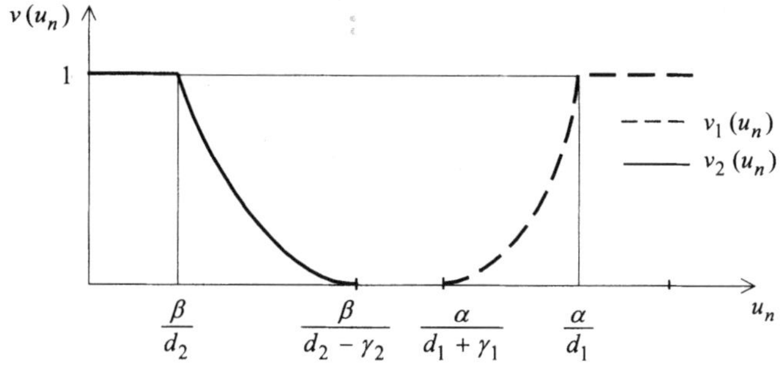

Figure 8.7. Relationship between v and u – the third case

Then

$$u_n^* = \arg\max_{u_n} \min\{v_1(u_n), v_2(u_n)\} = \frac{\gamma_1\beta + \gamma_2\alpha}{\gamma_1 d_2 + \gamma_2 d_1}$$

and

$$v(u_n^*) = \frac{\beta d_1 - \alpha d_2}{\beta \gamma_1 + \alpha \gamma_2} + 1 .$$

3.

$$\frac{\alpha}{d_1 + \gamma_1} \geq \frac{\beta}{d_2 - \gamma_2}$$

Then for every u_n

$$v(u_n) = \min\{v_1(u_n), v_2(u_n)\} = 0$$

which means that the decision for which the requirement is satisfied with the certainty index greater than 0 does not exist.

The results for u_0 based on the inequality (8.9) have a similar form. It is important to note that the results are correct under the assumption

$$\frac{y_{\min}}{d_1 - \gamma_1} \leq \frac{y_{\max}}{d_2 + \gamma_2}$$

which means that the condition (8.12) is satisfied for every x_1 and x_2. Otherwise $v(u_n)$ may be smaller:

$$v(u_n) = \min\{v_1(u_n), v_2(u_n), v_3\}$$

where v_3 is the certainty index that the condition

$$\frac{y_{\min}}{x_1} \leq \frac{y_{\max}}{x_2}$$

is approximately satisfied, i.e.

$$v_3 = v(\frac{x_2}{x_1} \mathrel{\tilde{\in}} D)$$

where $D \subset R^1 \times R^1$ is determined by the inequality

$$\frac{x_2}{x_1} \leq \frac{y_{\max}}{y_{\min}} .$$

□

8.3 Analysis and Decision Making for Dynamical Plants with Random Parameters

The analysis and decision problems for dynamical plants described by a relational knowledge representation with random parameters may be formulated and solved in a similar way as for the static plants in Sect. 3.5. Let us consider the plant described by the relations

$$\left.\begin{aligned} &R_{\rm I}(u_n, s_n, s_{n+1}; x) \subseteq U \times S \times S, \\ &R_{\rm II}(s_n, y_n; w) \subseteq S \times Y \end{aligned}\right\} \tag{8.19}$$

where $x \in X$ and $w \in W$ are unknown vector parameters which are assumed to be values of random variables $(\widetilde{x}, \widetilde{w})$ with the joint probability density $f(x, w)$.

Analysis problem: For the given relations (8.19), $f(x, w)$, D_{s0} and the sequences D_{un}, D_{yn} one should determine the probability

$$P[D_{yn} \subseteq D_{yn}(\widetilde{x}, \widetilde{w})] \triangleq p_n$$

where $D_{yn}(x, w)$ is the result of the analysis problem formulated in Sect. 8.1, that is the set of all possible outputs y_n for the fixed x and w.

In a similar way as for the static plant considered in Sect. 3.5 (see formula (3.28)) we obtain

$$p_n = P[(\widetilde{x}, \widetilde{w}) \in D(D_{yn}, D_{u,n-1}))] = \int\limits_{D(D_{yn}, D_{u,n-1})} f(x, w) dx dw$$

where

$$D(D_{yn}, D_{u,n-1}) = \{(x, w) \in X \times W : D_{yn} \subseteq D_{yn}(x, w)\}.$$

For the given value u_n we have Equations (8.16) and (8.17), and the formulation and solution of the analysis problem are the same as described above with u_n, $D_{yn}(u_{n-1}; x, w)$ and $D(D_{yn}, u_{n-1})$ instead of D_{un}, $D_{yn}(x, w)$ and $D(D_{yn}, D_{u,n-1})$, respectively.

Decision problem: For the given relations (8.19), $f(x, w)$, D_{s0} and the sequence $D_{yn} (n = 1, 2, ..., N)$ find the sequence of the optimal decisions

$$u_n^* = \arg \max_{u_n \in U} P[D_{y,n+1}(u_n; \widetilde{x}, \widetilde{w}) \subseteq D_{y,n+1}]$$

for $n = 0, 1, ..., N-1$, where $D_{y,n+1}(u_n; x, w)$ is the result of the analysis problem (8.17). Then

$$u_n^* = \arg \max_{u_n \in U} \int\limits_{D(D_{y,n+1}, u_n)} f(x, w) dx dw$$

where

$$D(D_{y,n+1},u_n) = \{(x,w) \in X \times W : D_{y,n+1}(u_n;x,w) \subseteq D_{y,n+1}\}.$$

The determination of u_n^* may be replaced by the determination of $u_{dn}^* = u_n^*$ where

$$u_{dn}^* = \arg\max_{u_n \in U} P[u_n \in D_{un}(\widetilde{x},\widetilde{w})]$$

and $D_{un}(x,w)$ is the result of the decision problem considered in Sect. 8.1 for the fixed x and w. Then

$$u_{dn}^* = u_n^* = \arg\max_{u_n \in U} \int_{D_d(u_n)} f(x,w)dxdw$$

where

$$D_d(u_n) = \{(x,w): u_n \in D_{un}(x,w)\}.$$

The considerations for a plant with disturbances z_n and in the case when s_n may be measured are analogous to those presented in the previous section for uncertain variables.

8.4 Optimization of Random and Uncertain Multistage Decision Process

Let us consider a functional dynamical plant described by the equations

$$\left.\begin{aligned} s_{n+1} &= \chi(s_n,u_n;x_n) \\ y_n &= \eta(s_n;w_n) \end{aligned}\right\} \tag{8.20}$$

i.e. by Equations (8.1) with unknown parameters (x,w). In this case we can formulate the optimization problem of the decision process "as a whole" (from $n=1$ to $n=N$), and apply *dynamical programming* as a tool for the problem-solving. This approach is well known for the random description (see e.g. [2, 60]). Assume that (x_{n-1},w_n) are values of the random variables $(\widetilde{x}_{n-1},\widetilde{w}_n)$; the variables $(\widetilde{x}_{n-1},\widetilde{w}_n)$ are stochastically independent for different n and are described by the same joint probability density $f(x,w)$, not depending on n. As a global quality index we introduce

$$Q_N = \mathrm{E}[\sum_{n=1}^{N} \varphi(\widetilde{y}_n, y^*)] \tag{8.21}$$

where y^* is a desirable output, $\varphi(y_n, y^*)$ denotes a local quality index for a particular n and E is the expected value with respect to a sequence

$$(\widetilde{x}_0, \widetilde{w}_1), (\widetilde{x}_1, \widetilde{w}_2), ..., (\widetilde{x}_{n-1}, \widetilde{w}_n).$$

The quality index (8.21) corresponds to the quality index in version III for the static plant presented in Sect. 3.2. Let us assume that the states s_n may be measured.

Decision problem: For the given functions χ, η, φ and the probability density $f(x, w)$, one should determine the decision algorithm

$$u_n^* = \Psi_n(s_n)$$

such that the sequence of the decisions $u_0^*, u_1^*, ..., u_{N-1}^*$ minimizes the quality index (8.21). This problem is usually called a *probabilistic dynamic optimization.*

Let us introduce the following notation:

$$V_{N-n}(s_n) = \min_{u_n, ..., u_{N-1}} [\underset{(x_n, w_{n+1}), ..., (x_{N-1}, w_N)}{\mathrm{E}} \sum_{i=n+1}^{N} \varphi(\widetilde{y}_n, y^*)], \tag{8.22}$$

$n = 0, 1, ..., N-1$,

$$\varphi(y_n, y^*) = \varphi\{\eta[\chi(s_{n-1}, u_{n-1}; x_{n-1}); w_n]\} \triangleq g(s_{n-1}, u_{n-1}; x_{n-1}, w_n). \tag{8.23}$$

Then (8.21) becomes

$$Q_N = \sum_{n=0}^{N-1} g(s_n, u_n; x_n, w_{n+1}). \tag{8.24}$$

The dynamic programming consists in applying the following recursive procedure:

1. For $n = N-1$

$$V_1(s_{N-1}) = \min_{u_{N-1}} \int_{X \times W} g(s_{N-1}, u_{N-1}; x, w) f(x, w) dx dw$$

and as a result of minimization we obtain

$$u^*_{N-1} = \Psi_{N-1}(s_{N-1}). \tag{8.25}$$

2. For $n = N-2$

$$V_2(s_{N-2}) = \min_{u_{N-2}, u_{N-1}} \int_{X \times W} [g(s_{N-2}, u_{N-2}; x, w) + g(s_{N-1}, u_{N-1}; x, w)] f(x, w) dx dw$$

$$= \min_{u_{N-2}} \int_{X \times W} [g(s_{N-2}, u_{N-2}; x, w) + V_1(s_{N-1})] f(x, w) dx dw.$$

Substituting the first equation (8.20) for $n = N-2$ yields

$$V_2(s_{N-2}) = \min_{u_{N-2}} \int_{X \times W} \{g(s_{N-2}, u_{N-2}; x, w) + V_1[\chi(s_{N-2}, u_{N-2}; x)]\} f(x, w) dx dw$$

and as a result of minimization we obtain

$$u^*_{N-2} = \Psi_{N-2}(s_{N-2}). \tag{8.26}$$

.
.
.

N. For $n = 0$

$$V_N(s_0) = \min_{u_0} \int_{X \times W} \{g(s_0, u_0; x, w) + V_{N-1}[\chi(s_0, u_0; x)]\} f(x, w) dx dw$$

and

$$u^*_0 = \Psi_0(s_0). \tag{8.27}$$

In this way we obtain the decision algorithm $u^*_n = \Psi_n(s_n)$, which for $n = N-1, N-2, ..., 0$ is presented by (8.25), (8.26), (8.27), respectively.

A similar approach may be applied for the description using uncertain variables. Assume that (x_{n-1}, w_n) are values of uncertain variables $(\bar{x}_{n-1}, \bar{w}_n)$; the variables $(\bar{x}_{n-1}, \bar{w}_n)$ are independent for different n and are described by the same joint certainty distribution $h(x, w)$, not depending on n. As a global quality index we introduce

$$Q_N = \mathrm{M}[\sum_{n=1}^{N} \varphi(\bar{y}_n, y^*)] \tag{8.28}$$

where M denotes a mean value with respect to a sequence

$$(\bar{x}_0, \bar{w}_1), (\bar{x}_1, \bar{w}_2), ..., (\bar{x}_{n-1}, \bar{w}_n).$$

The quality index (8.28) corresponds to the quality index in version III for the static plant presented in Sect. 5.2. Now the decision problem may consist in the determination of the decisions $u_0^*, u_1^*, ..., u_{N-1}^*$ minimizing the quality index (8.28).

Using the notation (8.23) and (8.22) with M, $\bar{y}_n$ in the place of E, $\tilde{y}_n$, one can apply the following recursive procedure where u_n^* denotes the result of minimization:

1. For $n = N-1$

$$V_1(s_{N-1}) = \min_{u_{N-1}} \{ \int_0^\infty \varepsilon_{N-1} h_N(\varepsilon_{N-1}) d\varepsilon_{N-1} \cdot [\int_0^\infty h_N(\varepsilon_{N-1}) d\varepsilon_{N-1}]^{-1} \}, \quad (8.29)$$

$$u_{N-1}^* = \Psi_{N-1}(s_{N-1}),$$

where

$$\varepsilon_{N-1} = g(s_{N-1}, u_{N-1}; x, w). \quad (8.30)$$

2. For $n \le N-2$

$$V_{N-n}(s_n) = \min_{u_n} \{ \int_0^\infty \varepsilon_n h_n(\varepsilon_n) d\varepsilon_n \cdot [\int_0^\infty h_n(\varepsilon_n) d\varepsilon_n]^{-1} \}, \quad (8.31)$$

$$u_n^* = \Psi_n(s_n),$$

where

$$\varepsilon_n = g(s_n, u_n; x, w) + V_{N-n-1}[\chi(s_n, u_n; x)], \quad (8.32)$$

$n = N-2, N-1, ..., 0$. In this notation $h_n(\varepsilon_n)$ denotes the certainty distribution of $\bar{\varepsilon}_n$ which should be determined for the function (8.30) or (8.32), and the known certainty distribution $h(x, w)$. The procedure is simpler in the discrete case when

$$(x, w) \in \{(x, w)_1, (x, w)_2, ..., (x, w)_m\}.$$

Then, according to the formula for the mean value M_y in Sect. 4.5, the formulas (8.29) and (8.32) become

$$V_1(s_{N-1}) = \min_{u_{N-1}} \sum_{j=1}^{m} g[s_{N-2}, u_{N-2}; (x, w)_j] h[(x, w)_j] \{ \sum_{j=1}^{m} h[(x, w)_j] \}^{-1},$$

$$V_{N-n}(s_n) = \min_{u_n} \sum_{j=1}^{m} \{ g[s_n, u_n; (x, w)_j] + V_{N-n-1}[\chi(s_n, u_n; x_j)] \} h[(x, w)_j \; \{ \sum_{j=1}^{m} h[(x, w)_j] \}^{-1}$$

where x_j denotes the first part in the pair $(x,w)_j$.

Consequently, the optimal decision u_n are determined in real time in the closed-loop decision systems. For the given y^* and φ, the deterministic decision algorithm Ψ_n is based on the knowledge of the plant KP, containing the functions χ, η in (8.20) and the certainty distribution $h(x,w)$ given by an expert (Fig. 8.8).

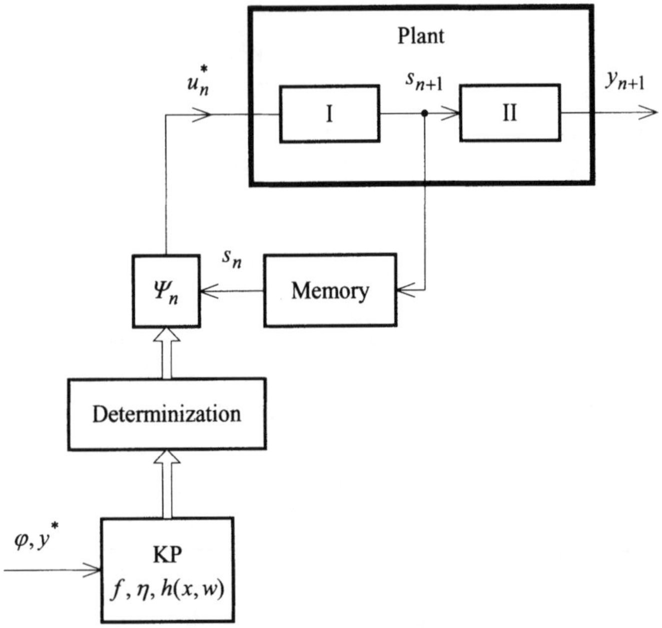

Figure 8.8. Knowledge-based closed-loop decision system with description using uncertain variables

Remark 8.3. The direct application of the dynamic programming, analogous to that presented for the functional plant, is not possible for the relational plant described by (8.14) with uncertain or random parameters due to the following reasons:

1. The certainty index (or probability) that the set of possible sequences $\{y_1, y_2, ..., y_N\}$ belongs to the given set $D_{y_1} \times D_{y_2} \times ... \times D_{y_N}$ does not have an additive form as in (8.21) or (8.28). Then, the optimization of the decision process "as a whole" may not be decomposed into particular moments as was done in the recursive procedures presented above. That is why in Sects. 8.2 and 8.3 a "point to point" optimization has been applied (see also Remark 8.1 in Sect. 8.1).
2. Quite formally (without a clear interpretation), it is possible to apply the additional form of the global quality index, as a sum of certainty indexes

$$v[D_{y,n+1}(u_n;\bar{x},\bar{w}) \subseteq D_{y,n+1}]$$

or probabilities

$$P[D_{y,n+1}(u_n;\tilde{x},\tilde{w}) \subseteq D_{y,n+1}]$$

considered in Sects. 8.2 and 8.3, respectively. Then, by applying the recursive procedure it is not possible to obtain the relationship $u_n = \Psi(s_n)$ because, for the fixed u_n and x, the state s_{n+1} is not determined by s_n. The set of possible states $D_{s,n+1}(u_n,x)$ is determined by D_{sn}, and consequently we may obtain the relationship $u_n = \Psi(D_{sn})$, but it is not possible to "measure" D_{sn}, i.e. to know the state D_{sn} in the successive moments n. □

8.5 Applications of Uncertain Variables for a Class of Knowledge-based Assembly Systems

The problems and methods presented in Sects. 8.1, 8.2 and 8.3 may be illustrated by an example concerning a class of assembly processes [48]. In this case, for each n the states s_n and the decisions u_n belong to finite sets.

In many practical situations there exist uncertainties in the description of an assembly process: the sequence of assembly operations is not a priori determined and the relationships between the successive operations, states and features describing the assembly plant are non-deterministic. This is a frequent situation in small-batch production processes with changes in the relationships describing the assembly plant in different cycles. In such cases it is reasonable to apply special tools and methods of decision making in uncertain systems to the planning and control of the assembly process.

This section deals with a class of knowledge-based assembly systems described by a relational knowledge representation consisting of relations between the operations, states and features, i.e. variables characterizing the current effect of the assembly process (e.g. dimensions or sizes evaluating the precision, accuracy or tolerance in the placement and fastening of elements). The problem of choosing assembly operations from the given sets of operations at each stage is considered as a specific multistage decision process for a relational plant with unknown parameters. The unknown parameters in the description of the assembly process are assumed to be values of uncertain variables described by certainty distributions given by an expert. The certainty distributions express the expert's knowledge concerning different approximate values of the unknown parameters.

8.5.1 Knowledge Representation and Decision Problem

Let us consider an assembly process as a sequence of assembly operations $O_n \in \{O_{n1}, O_{n2}, ..., O_{nl_n}\}$ executed at successive stages n. At each stage the assembly plant is characterized by a state $s_n \in \{S_{n1}, S_{n2}, ..., S_{nm_n}\}$. The state s_{n+1} depends on the state s_n and the operation O_n (Fig. 8.9). To formulate the description of the assembly process, let us introduce the following notation:
i_n – index of the operation at the n-th stage, e.g. if $i_n = 4$ then $O_n = O_{n4}$,

$i_n \in \{1, 2, ..., l_n\} \triangleq L_n$ – set of the operations, $n \in \overline{1,N}$,
j_n – index of the state, e.g. if $j_n = 2$ then $s_n = S_{n2}$,

$j_n \in \{1, 2, ..., m_n\} \triangleq M_n$ – set of the states,
$y_n \in Y_n$ – real number vector of features describing the assembly plant at the n-th stage. The evaluation of the result of the n-th operation and the quality of the process concern the components of y_n, e.g. $a \le y_n^{(v)} \le b$ may denote the requirement concerning the size $y_n^{(v)}$ (a component of y_n).

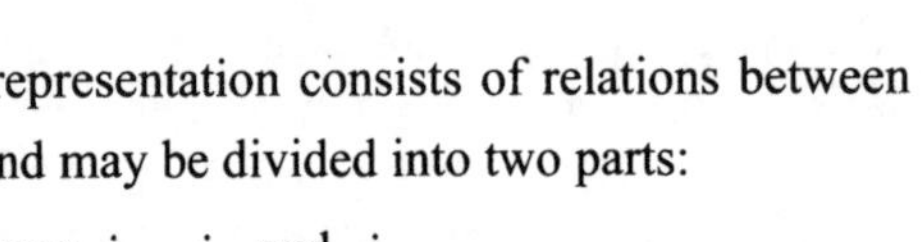

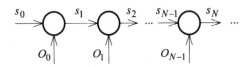

Figure 8.9. Illustration of the assembly process

The knowledge representation consists of relations between the variables i_n, j_n, j_{n+1} and y_n, and may be divided into two parts:
I. A relation between i_n, j_n and j_{n+1}

$$R_{In}(i_n, j_n, j_{n+1}) \subset L_n \times M_n \times M_{n+1}.$$

II. A relation between j_{n+1} and y_{n+1}

$$R_{IIn}(j_{n+1}, y_{n+1}) \subset M_{n+1} \times Y_{n+1}.$$

The relations R_I and R_{II} may be presented in the form of families of sets:

I. $$D_{j,n+1}(i_n, j_n) \subset M_{n+1} \tag{8.33}$$

for all pairs $(i_n, j_n) \in L_n \times M_n$,

II. $$D_{y,n+1}(j_{n+1}) \subset Y_{n+1} \tag{8.34}$$

for all $j_{n+1} \in M_{n+1}$.

Consequently, the relational knowledge representation consists of $l_n \cdot m_n$ sets (8.33) and m_{n+1} sets (8.34). The decision making (control) in the assembly process should be based on the knowledge representation (8.33), (8.34) given by an expert.

At each stage the decision consists in the proper choosing of the assembly operation O_n (i.e. the index i_n) from the given set of operations (i.e. from the set L_n), satisfying the requirement concerning the features y_{n+1} presented in the form $y_{n+1} \in D_{y,n+1}$ where the set $D_{y,n+1} \subset Y_{n+1}$ is given by a user. For making the decision, the knowledge of the state j_n in the form $j_n \in D_{j,n} \subset M_n$ is used.

The decision problem at the n-th stage may be formulated as follows: for the given knowledge representation, the set $D_{j,n}$ (the knowledge of the state j_n, determined at the stage $n-1$) and the set $D_{y,n+1}$ (a user's requirement), one should find the set of all operations i_n satisfying the requirement, i.e. the largest set $D_{i,n} \subset L_n$ such that the implication

$$i_n \in D_{i,n} \rightarrow y_{n+1} \in D_{y,n+1}$$

is satisfied. Consequently, the assembly operation should be chosen from the set $D_{i,n}$.

The decision problem may be decomposed into two parts:

A. For the given sets (8.34) and $D_{y,n+1}$ find the largest set $D_{j,n+1} \subset M_{n+1}$ such that the implication

$$j_{n+1} \in D_{j,n+1} \rightarrow y_{n+1} \in D_{y,n+1}$$

is satisfied.

B. For the given sets (8.33) and the sets $D_{j,n}$, $D_{j,n+1}$ find the largest set $D_{i,n}$ such that the implication

$$i_n \in D_{i,n} \rightarrow j_{n+1} \in D_{j,n+1}$$

is satisfied for each $j_n \in D_{j,n}$.

For the problem solving a general solution of the decision problem based on relational knowledge representation may be used. It is easy to note that:

A.

$$D_{j,n+1} = \{ j_{n+1} \in M_{n+1} : D_{y,n+1}(j_{n+1}) \subseteq D_{y,n+1} \}, \tag{8.35}$$

B.

$$D_{i,n} = \{ i_n \in L_n : \bigwedge_{j_n \in D_{j,n}} D_{j,n+1}(i_n, j_n) \subseteq D_{j,n+1} \}. \qquad (8.36)$$

Using (8.35) and (8.36) for $n = 0, 1, ..., N-1$, we can determine an *assembly plan* in the form of a sequence $D_{i,1}$, $D_{i,2}$, ..., $D_{i,N-1}$. The set of initial states $D_{j,0}$ must be known, and at the n-th stage we use $D_{j,n}$ determined at the former stage. The decision process is then performed in an open-loop control system: at each stage the assembly operation i_n may be chosen randomly from the set $D_{i,n}$ according to the assembly plan.

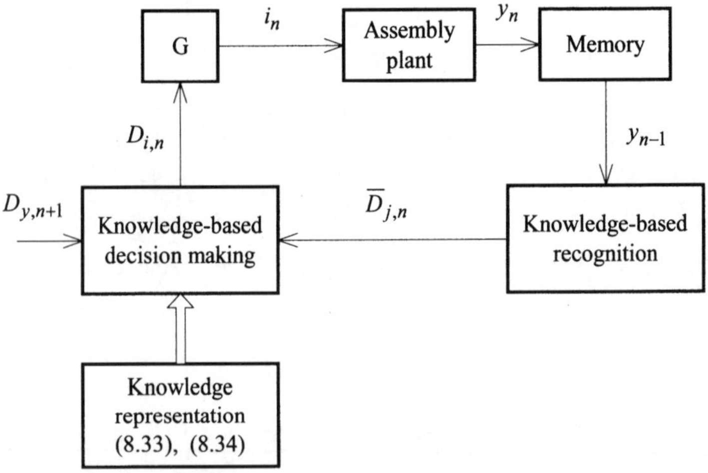

Figure 8.10. Knowledge-based control of an assembly process in a closed-loop system

The decision process may be performed in a closed-loop control system if it is possible to measure y_n. Then we can introduce the *knowledge-based recognition* of the state j_n, which consists in the determination of the set $\overline{D}_{j,n}$ of the possible indexes j_n:

$$\overline{D}_{j,n}(y_n) = \{ j_n \in M_n : \; y_n \in D_{y,n}(j_n) \} \qquad (8.37)$$

where the sets $D_{y,n}(j_n)$ are given in the knowledge representation. Consequently, in (8.37) we use $\overline{D}_{j,n}$ instead of $D_{j,n}$, which may give a less restricted set of the possible decisions $D_{i,n}$. The block scheme of the closed-loop control system is presented in Fig. 8.10 where G denotes the generator of random numbers for the random choosing of the assembly operation i_n from $D_{i,n}$.

8.5.2 Assembly Process with Uncertain Parameters

Now let us consider the assembly process described by the knowledge representation with constant unknown parameters $x \in X$ and $w \in W$ in the first and the second part, respectively. Now the sets $D_{j,n+1}(i_n, j_n; x)$ in (8.33) depend on *x*, and the sets $D_{y,n+1}(j_{n+1}; w)$ in (8.34) depend on *w*. Consequently, the set of the decisions $D_{i,n}(x, w)$ in (8.36) depends on (*x*, *w*). The values (*x*, *w*) are assumed to be values of uncertain variables $(\bar{x}, \bar{w})$ with the certainty distribution

$$h(x, w) = v\,[(\bar{x} \cong x) \wedge (\bar{w} \cong w)]$$

given by an expert. The decision problem consists in the determination of the optimal decisions i_n^* maximizing the certainty index that i_n approximately belongs to $D_{i,n}(\bar{x}, \bar{w})$ (precisely speaking: that i_n belongs to $D_{i,n}(x, w)$ with the approximate values *x* and *w*), i.e.

$$i_n^* = \arg \max_{i_n \in L_n} v(i_n)$$

where

$$v(i_n) = v[i_n \,\tilde{\in}\, D_{i,n}(\bar{x}, \bar{w})]\,.$$

It is easy to note that

$$v[i_n \,\tilde{\in}\, D_{i,n}(\bar{x}, \bar{w})] = v[(\bar{x}, \bar{w}) \,\tilde{\in}\, D_{xw}(i_n)]$$

where

$$D_{xw}(i_n) = \{(x, w) : i_n \in D_{i,n}(x, w)\}\,.$$

Then

$$v(i_n) = \max_{(x,w) \in D_{xw}(i_n)} h(x, w)\,.$$

As a result we obtain an *assembly plan* in the form of a sequence of the assembly operations i_n^*. If $\bar{x}$ is considered as a *C*-uncertain variable then

$$v_c(i_n) = \frac{1}{2}\{v[i_n \,\tilde{\in}\, D_{i,n}(\bar{x}, \bar{w})] + 1 - v[i_n \,\tilde{\in}\, \overline{D}_{i,n}(\bar{x}, \bar{w})]\}$$

and $\overline{D}_{i,n} = I_n - D_{i,n}$.

If it is possible to recognize the current states j_n (i.e. to determine the set $\overline{D}_{j,n}$ such that $j_n \in \overline{D}_{j,n}$), then the decision process may be performed in a closed-loop control system (Fig. 8.11) and $\overline{D}_{j,n}$ may be used in (8.36) instead of $D_{j,n}$.

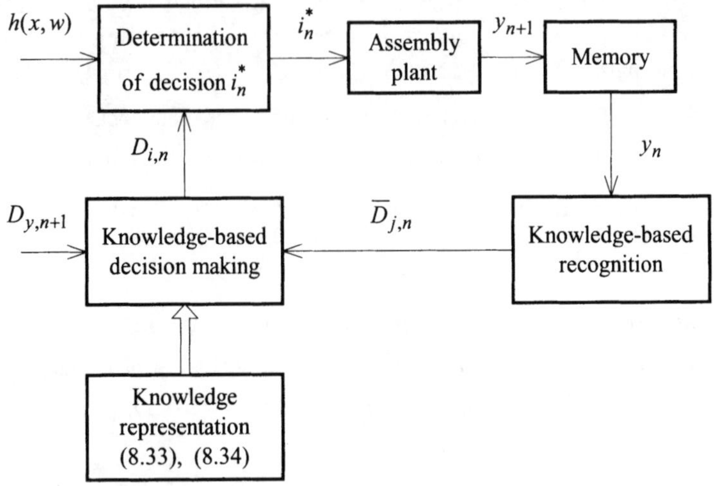

Figure 8.11. Closed-loop decision system with state recognition

In particular, the relation R_{I} describing the first part of the knowledge representation, may be a function $j_{n+1} = F_n(j_n, i_n)$, i.e. a matrix with the entries j_{n+1} in the j_n-th line and i_n-th column. Then we can determine $D_{j,n+1}(w)$ (see part A in the procedure described in Sect. 8.5.1) and

$$j_{n+1}^* = \arg\max_{j_{n+1}} v[j_{n+1} \mathrel{\tilde{\in}} D_{j,n+1}(\overline{w})]. \tag{8.38}$$

Consequently, the set of the decisions $D_{i,n}$ for the given j_{n+1}^* and j_n^* obtained at the former stage ($j_0^* = j_0$ is given) may be obtained from the matrix F_n:

$$D_{i,n} = \{i_n \in L_n : F_n(j_n^*, i_n) = j_{n+1}^*\}. \tag{8.39}$$

The procedure of the determination of the assembly operation i_n^* is then the following:

1. Determination of the set $D_{j,n+1}(w)$ according to (8.35)

$$D_{j,n+1}(w) = \{j_{n+1} : D_{y,n+1}(j_{n+1}; w) \subseteq D_{y,n+1}\}. \tag{8.40}$$

2. Determination of the set

$$D_{w,n+1}(j_{n+1}) = \{w \in W : j_{n+1} \in D_{j,n+1}(w)\}. \tag{8.41}$$

3. Choosing the best index j_{n+1}^* from the set M_{n+1} according to (8.38)

$$j_{n+1}^{*} = \arg\max_{j_n} \max_{w \in D_{w,n+1}(j_{n+1})} h_w(w) . \tag{8.42}$$

4. Determination of $D_{i,n}$ according to (8.39).

5. Random choosing of i_n from $D_{i,n}$.

The decisions i_n are executed in the open-loop control system (Fig. 8.12) where G denotes the generator of random numbers for the random choosing of the assembly operation i_n from $D_{i,n}$.

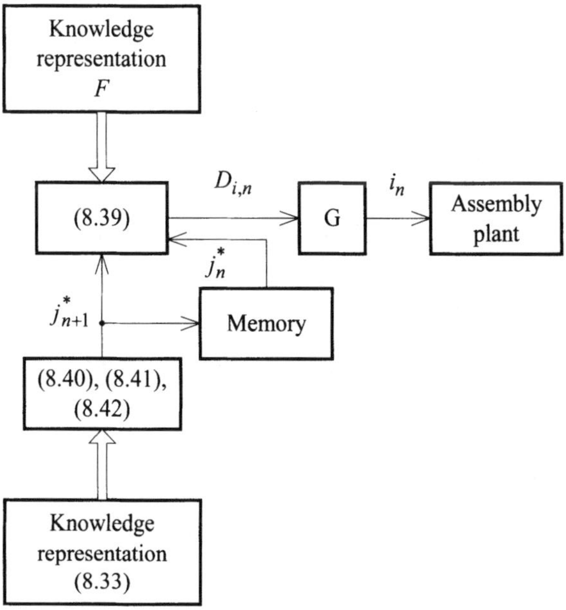

Figure 8.12. Another version of decision system for assembly plant

Example 8.3.
Let the sets $D_{y,n}(j_n;w)$ be described by the inequalities

$$wa(j_n) \le y_n^{\mathrm{T}} y_n \le 2wa(j_n), \quad w > 0, \quad a(j_n) > 0, \tag{8.43}$$

given by an expert. It is then known that the value $y_n^{\mathrm{T}} y_n$ (where y_n is a vector of the features characterizing the assembly plant) satisfies the inequality (8.43), and the bounds are the coefficients a depending on the state, i.e. for the different assembly operations i_n and consequently – the different states j_n, we have the different bounds for the value $y_n^{\mathrm{T}} y_n$ which denotes a quality index. The requirement concerning the quality (i.e. the set $D_{y,n}$) is the following:

$$\beta_n \le y_n^{\mathrm{T}} y_n \le \alpha_n, \quad \beta_n > 0, \quad \alpha_n \ge 2\beta_n.$$

Then, according to (8.35)

$$D_{j,n}(w) = \{ j_n \in M_n : \beta_n \le wa(j_n) \le \frac{\alpha_n}{2} \}.$$

Consequently,

$$j_n^* = \arg\max_{j_n} \max_{w \in D_{w,n}(j_n)} h_w(w) \stackrel{\Delta}{=} \arg\max_{j_n} v(j_n) \tag{8.44}$$

where $h_w(w)$ is the certainty distribution for $\overline{w}$ and

$$D_{w,n}(j_n) = \{ w \in W : j_n \in D_{j,n}(w) \}.$$

Having j_n^* and j_{n-1}^*, one may obtain the set of possible assembly operations (in particular, one operation i_n^*) from the matrix F. For example, assume that $h_w(w)$ has the parabolic form: $h_w(w) = -(w-d)^2 + 1$ for $d - 1 \le w \le d + 1$ and $h_w(w) = 0$ otherwise. Assume also that $j_n \in \{1, 2, 3\}$, $a(1) = 1$, $a(2) = 0.6$, $a(3) = 0.9$ and $\alpha_n = 6.4$, $\beta_n = 2.8$.

Then for $j = 1$

$$\frac{\beta_n}{a(1)} \triangleq g_1 = 2.8, \quad \frac{\alpha_n}{2a(1)} \triangleq g_2 = 3.2$$

and according to (8.44), $v(1) = h_w(3.2) = 0.36$.

For $j = 2$ we obtain $g_1 \approx 4.7$, $g_2 \approx 5.3$, $v(2) = h_w(4.7) = 0.51$.

For $j = 3$ we obtain $g_1 \approx 3.1$, $g_2 \approx 3.5$, $v(3) = h_w(3.5) = 0.75$. Consequently, $j_n^* = 3$. □

8.6 Non-parametric Problems

For the dynamical plants, non-parametric problems may be formulated in different ways and the considerations may be much more complicated than those for the static plant. The direct extension of the non-parametric problems discussed in Sects 3.7, 5.7, 5.8, 6.2, 6.3 and presented together in Sect. 6.5, is possible for the case when the requirement concerns the final output for $n = N$, and may be considered as a *goal* of the decision process. Let us present together such formulations of the decision problem for the descriptions based on random, uncertain, fuzzy and soft variables.

A. Description based on random variables

The knowledge of the plant has a form of two conditional probability densities (exactly speaking – a sequence of these densities for a time-varying plant):

$$\text{KP} = < f_{sn}(s_n \mid s_{n-1}, u_{n-1}), f_{yn}(y_n \mid s_n) >, \qquad n = 1,2,...,$$

what corresponds to the functions χ and η in (8.1) for the deterministic plant.

Decision problem: For the given KP ($n = 1, 2, ..., N$) and $f_{yN}(y_N)$ required by a user one should determine the sequence

$$f_{un}(u_n \mid s_n) \qquad \text{for} \qquad n = 0, 1, ..., N-1.$$

The solution may be obtained by using a recursive procedure starting from $n = N$. For each $n = N, N{-}1,...,1$, the determination of $f_{u,n-1}(u_{n-1} \mid s_{n-1})$ may be decomposed into two parts corresponding to the parts described in Sect. 8.1. In the first part one should solve the non-parametric problem presented in Sect. 3.7 for the part of the plant with input s_n and output y_n, described by $f_{yn}(y_n \mid s_n)$. As a result one obtains $f_{sn}(s_n)$. In the second part one should solve the non-parametric problem for the part of the plant with input (s_{n-1}, u_{n-1}) and output s_n, described by $f_{sn}(s_n \mid s_{n-1}, u_{n-1})$. To obtain the solution, one should apply the approach presented in Sect. 3.7 with (u_{n-1}, s_{n-1}, s_n) in place of (u, z, y), and with the probability density $f_{sn}(s_n)$ in place of the required probability density $f_y(y)$. As a result, one obtains $f_{us,n-1}(u_{n-1}, s_{n-1})$ in the first step and $f_{u,n-1}(u_{n-1} \mid s_{n-1})$ in the second step (see also (6.32) and (6.33) in Sect. 6.5).

The *recursive procedure* is then as follows:

For $n = N, N{-}1,..., 1$

1. Find $f_{us,n-1}(u_{n-1}, s_{n-1})$ satisfying the equation

$$f_{sn}(s_n) = \int\limits_{U}\int\limits_{S} f_{us,n-1}(u_{n-1}, s_{n-1}) f_{sn}(s_n \mid s_{n-1}, u_{n-1})\, du_{n-1} ds_{n-1}. \qquad (8.45)$$

2. Find

$$f_{s,n-1}(s_{n-1}) = \int\limits_{U} f_{us,n-1}(u_{n-1}, s_{n-1})\, du_{n-1}. \qquad (8.46)$$

3. Find

$$f_{u,n-1}(u_{n-1} \mid s_{n-1}) = \frac{f_{us,n-1}(u_{n-1}, s_{n-1})}{f_{s,n-1}(s_{n-1})}. \qquad (8.47)$$

4. For $n=N$ one should find $f_{sN}(s_N)$ satisfying the equation

$$f_{yN}(y_N) = \int_S f_{sN}(s_N) f_{yN}(y_N \mid s_N) ds_N \tag{8.48}$$

where $f_{yN}(y_N)$ is given.

For $n<N$, in (8.45) one should use $f_{sn}(s_n)$ obtained at the former stage, i.e. for $n+1$.

B. Description based on uncertain variables

The knowledge of the plant has a form of two conditional certainty distributions

$$\text{KP} = < h_{sn}(s_n \mid s_{n-1}, u_{n-1}), h_{yn}(y_n \mid s_n) >, \qquad n = 1,2,...,$$

given by an expert.

Decision problem: For the given KP ($n = 1, 2, ..., N$) and $h_{yN}(y_N)$ required by a user one should determine the sequence $h_{un}(u_n \mid s_n)$ for $n = 0, 1, ..., N-1$.

According to the approach described in Sects. 5.7, 5.8 and 6.5 for the static plant, the recursive procedure for finding the solution is analogous to that for random variables, described by using (8.45)–(8.48):

For $n = N, N-1, ..., 1$

1. Find $h_{us,n-1}(u_{n-1}, s_{n-1})$ satisfying the equation

$$h_{sn}(s_n) = \max_{u\in U,\ s\in S} \min\{h_{us,n-1}(u_{n-1}, s_{n-1}), h_{sn}(s_n \mid s_{n-1}, u_{n-1})\}.$$

2. Find

$$h_{s,n-1}(s_{n-1}) = \max_{u\in U} h_{us,n-1}(u_{n-1}, s_{n-1}).$$

3. Find $h_{u,n-1}(u_{n-1} \mid s_{n-1})$ satisfying the equation

$$h_{us,n-1}(u_{n-1}, s_{n-1}) = \min\{h_{s,n-1}(s_{n-1}), h_{u,n-1}(u_{n-1} \mid s_{n-1})\}.$$

4. For $n=N$ one should find $h_{sN}(s_N)$ satisfying the equation

$$h_{yN}(y_N) = \max_{s\in S} \min\{h_{sN}(s_N), h_{yN}(y_N \mid s_N)\}$$

where $h_{yN}(y_N)$ is given.

C. Description based on fuzzy variables

For the determined soft properties $\varphi_{un}(u_n)$, $\varphi_{sn}(s_n)$ and $\varphi_{yn}(y_n)$, the knowledge of the plant has a form of two membership functions

$$\text{KP} = < \mu_{sn}(s_n \mid s_{n-1}, u_{n-1}), \mu_{yn}(y_n \mid s_n) >, \quad n = 1,2,...$$

given by an expert.

Decision problem: For the given KP ($n=1,2,\ ...,\ N$) and $\mu_{yN}(y_N)$ required by a user one should determine the sequence $\mu_{un}(u_n \mid s_n)$ for $n=0,1,\ldots,N-1$.

According to the approach described in Sects. 6.2 and 6.3 for the static plant, the recursive procedure is the same as for uncertain variables with μ instead of h.

The functions $f_{un}(u_n \mid s_n)$, $h_{un}(u_n \mid s_n)$ and $\mu_{un}(u_n \mid s_n)$ describe the knowledge of the decision making KD and may be called a *random, uncertain and fuzzy decision algorithm,* respectively. The deterministic decision algorithms $u_n = \Psi_n(s_n)$, obtained via the determinization, may be executed in a closed-loop decision system where s_n is measured. It is worth noting that this is one of the possible formulations of a decision problem for multistage processes. Different and much more complicated cases are described in [69].

Remark 8.4. It is easy to see that, for the given sequence $f_{sn}(s_n \mid y_n)$, it is not possible to determine the sequence $f_{un}(u_n \mid s_n)$ such that the probability densities $f_{yn}(y_n)$ have a required form for $n=1,2,\ ...,\ N-1$. It follows from the fact that $f_{y,n-1}(y_{n-1})$ is determined by $f_{us,n-1}(u_{n-1},s_{n-1})$ obtained from (8.45), i.e.

$$f_{y,n-1}(y_{n-1}) = \int_S f_{us,n-1}(u_{n-1},s_{n-1})ds_{n-1} \ ,$$

which means that $f_{y,n-1}(y_{n-1})$ cannot be given independently. The same remark concerns the descriptions based on uncertain and fuzzy variables. □

The considerations based on random, uncertain and fuzzy variables may be generalized by using soft variables (see Sects. 6.6 and 6.7).

D. Description based on soft variables

The knowledge of the plant has a form of two conditional evaluating functions

$$\text{KP} =< g_{sn}(s_n \mid s_{n-1},u_{n-1}), g_{yn}(y_n \mid s_n) >, \quad n=1,2,\ldots \ .$$

Decision problem: For the given KP ($n=1,2,\ ...,\ N$) and $g_{yN}(y_N)$ required by a user one should determine the sequence $g_{un}(u_n \mid s_n)$ for $n=0,1,\ ...,\ N-1$.

According to the approach described in Sect. 6.7, the recursive procedure for finding the solution is as follows:

For $n = N, N-1,..., 1$

1. Find $g_{us,n-1}(u_{n-1},s_{n-1})$ satisfying the equation

$$g_{sn}(s_n) = O_2\{O_1[g_{us,n-1}(u_{n-1},s_{n-1}), g_{sn}(s_n \mid s_{n-1},u_{n-1})]\}.$$

2. Find

$$g_{s,n-1}(s_{n-1}) = O_2[g_{us,n-1}(u_{n-1},s_{n-1})].$$

3. Find $g_{u,n-1}(u_{n-1} \mid s_{n-1})$ satisfying the equation

$$g_{us,n-1}(u_{n-1}, s_{n-1}) = O_1[g_{s,n-1}(s_{n-1}), g_{u,n-1}(u_{n-1} \mid s_{n-1})].$$

4. For $n = N$ one should find $g_{sN}(s_N)$ satisfying the equation

$$g_{yN}(y_N) = O_2\{O_1[g_{sN}(s_N), g_{yN}(y_N \mid s_N)]\}.$$

where $g_{yN}(y_N)$ is given.

The function $g_{un}(u_n \mid s_n)$ may be called a knowledge of the decision making KD $= < g_{un}(u_n \mid s_n) >$, $n = 0, 1, ..., N-1$ or a *soft decision algorithm* (the description of a *soft controller* in the closed-loop control system, under the assumption that s_n can be measured). Having $g_{un}(u_n \mid s_n)$ and applying the determinization presented in Sect. 6.7, one can obtain the deterministic decision algorithm $u_n = \Psi_n(s_n)$ and use it in the closed-loop system (Fig. 8.13). The system has an analogous structure for the non-parametric description of uncertainties based on random, uncertain and fuzzy variables.

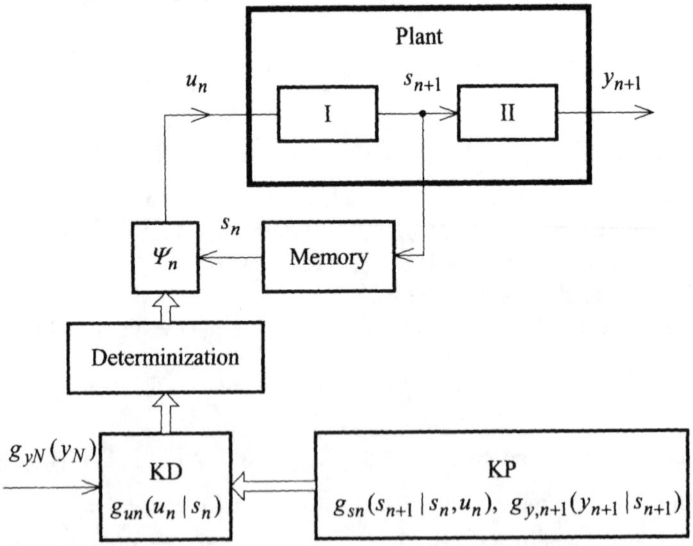

Figure 8.13. Knowledge-based closed-loop decision system with a non-parametric description of uncertainties

9 Parametric Optimization of Decision Systems

9.1 General Idea of Parametric Optimization and Adaptation

In the previous chapters the decision problems for non-deterministic plants with different descriptions of the uncertainty (different forms of the knowledge representation) have been considered. The typical procedure of finding the deterministic decision algorithm has been as follows:

1. The determination of the non-deterministic decision algorithm (the knowledge of the decision making KD) using the knowledge of the plant KP and the requirement concerning the output of the plant.
2. The determination of the deterministic decision algorithm by applying a determinization of KD.

Denote by $a \in A$ a vector of parameters in KP. They may be the parameters in probability distributions, certainty distributions, membership functions or other parameters in the description of the non-deterministic plant. Consequently, these parameters appear in KD and in the deterministic algorithm (the deterministic controller)

$$u = \Psi(z;a) \qquad \text{or} \qquad u_n = \Psi(s_n;a)$$

in an open-loop system with the disturbance z or in a closed-loop system with the state s_n, respectively, under the assumption that z or s_n may be measured and put at the input of the controller. Let $Q(a)$ denote a performance index (a quality index) of the decision system. For example, in the case of the dynamical system

$$Q(a) = \sum_{n=1}^{N} (y_n - y^*)^{\mathrm{T}} (y_n - y^*) \tag{9.1}$$

where N is sufficiently large to estimate the quality of the control. The performance index $Q(a)$ for the decision system with the real plant may be used to evaluate the quality of the algorithm Ψ based on the non-deterministic description KP, and to compare the different forms of the knowledge and different methods. In the case where KP is given by an expert, it may be used to compare different experts by estimating the effects of their knowledge in the form of Q. The performance index $Q(a)$ may also be used as a basis of the *parametric optimization* consisting in the determination of the optimal value a^* of the parameter a in the decision algorithm (the best value of the parameter in the controller), minimizing $Q(a)$, i.e.

$$a^* = \arg\max_{a \in A} Q(a) .$$

Assume that the real plant is deterministic and the non-deterministic description KP is used because we have no full knowledge of the plant. If the exact model of the deterministic plant is given, then it may be possible to determine the function $Q(a)$ in an analytical form or the approximate values $Q(a)$ for the given values a, using numerical methods. Consequently, a^* may be determined in an analytical or numerical way. If the controller with the deterministic algorithm Ψ is used in the decision system with the real plant or its simulator, then $Q(a)$ may be obtained (measured or calculated) in the decision system. In this case the parameter optimization of the controller Ψ may consist in a *step by step* adjusting process and the basic decision system with the additional adjusting algorithm forms an *adaptive decision system* or a system with an *adaptive controller*. Such an approach may be especially recommended when the subjective knowledge given by an expert has a form of certainty distributions or membership functions with rather arbitrary shapes and parameters. In the adaptive system these parameters may adapt to the real plant indirectly, by minimizing the effect $Q(a)$.

If $Q(a)$ is a differentiable function, a^* is a unique point in which $\operatorname{grad} Q(a) = \overline{0}$ and $\operatorname{grad} Q(a)$ may be estimated in the system, then the adaptation algorithm may have the form

$$a_{m+1} = a_m - K \operatorname{grad}_a Q(a)\big|_{a=a_m} \tag{9.2}$$

where a_m is the m-th approximation of a^* and K is a matrix of coefficients. Usually in practice the algorithm (9.2) is replaced by the algorithm with trial steps:

$$a_{m+1} = a_m - K w_m$$

where the i-th component of the vector w is the following:

$$w_m^{(i)} = \frac{Q(a_m + \delta_i) - Q(a_m - \delta_i)}{2\sigma_i} \tag{9.3}$$

and δ_i is a vector with zero components except the i-th component equal to σ_i (the value of a trial step for the i-th component of a). The values of K should be chosen in a way assuring the convergence of a_m to a^* for $m \to \infty$. The period of the adaptation (the time between the moments when a_m and a_{m+1} are determined and put into the controller in place of a) should contain N moments of the decision making, according to (9.1). The adaptive decision system in the case of a closed-loop basic decision system is illustrated in Fig. 9.1.

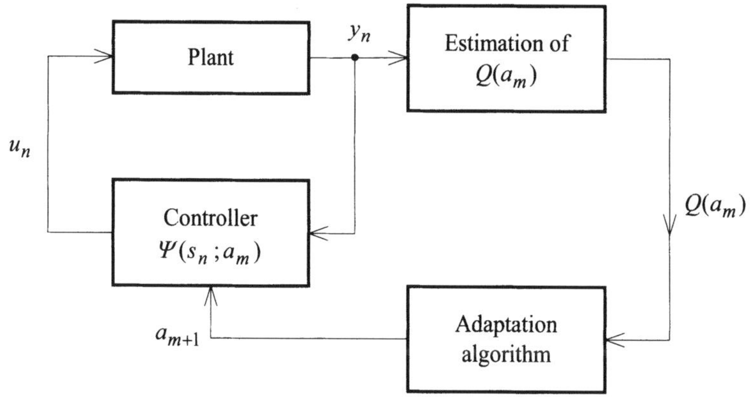

Figure 9.1. Adaptive decision system under consideration

In the above considerations the deterministic algorithms have been obtained from KD determined from KP given by an expert. We can also take into account another approach (especially popular in the case of fuzzy controllers) in which KD is given directly by an expert. Then a does not denote the parameter in KP but the parameter in the non-deterministic description of the decision algorithm (the controller) given by an expert. For example, the components of a may denote the parameters in membership functions used by an expert expressing his/her knowledge of the decision making. Both versions of the expert's knowledge (on the plant or directly on the decision making) will be described more precisely in Sect. 9.4. The general idea of the determination and optimization of Ψ in both versions is illustrated in Fig. 9.2.

A special situation arises for the functional (static or dynamical) plant with an unknown vector parameter x. Denote by $u = \Psi(z,x)$ or $u_n = \Psi(s_n,x)$ the optimal decision algorithm minimizing $Q(x)$. The optimal decision algorithm is then the result of non-parametric optimization for the given model of the plant with the fixed x, and for the given form of the performance index. Special cases of such algorithms denoted by Φ_d have been considered for the static functional plants in Sects. 3.2 and 5.3, where minimization of Q has been reduced to the equality $y = y^*$. By applying the determinization of the non-deterministic (random or uncertain) decision algorithms, one can obtain the deterministic algorithms $u_d = \Psi_d(z)$ or $u_{dn} = \Psi_d(s_n)$. The determinization may consist in the determination of u_d as a mean value of u or as a value maximizing a distribution of u. In both cases, it is necessary to determine the distribution of u for the known function Ψ and the distribution of x (see Sect. 3.2 for random variables and Sect. 5.3 for uncertain variables). If the determination of the distribution of u is too complicated, one may put the mean value of x in the place of x in the non-deterministic algorithm. It should be noted that a result of such an approach has no clear interpretation.

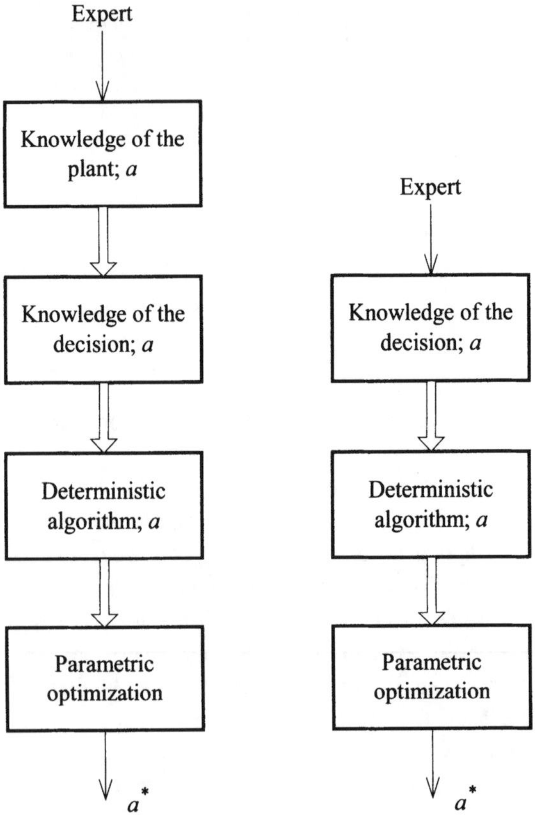

Figure 9.2. Determination and optimization of Ψ in two versions

To complete the list of cases, it is necessary to consider the case when KD is given directly by an expert and the plant is described by a functional model with x. Assume that KD has a functional form $u=\Psi(z;b)$ or $u_n=\Psi(s_n;b)$ where the function Ψ is given and b is a vector parameter whose value may be determined by a designer. Then the decision system as a whole may be considered as a static plant with input b, output Q and the unknown parameter x, for which we can formulate and solve the decision problem analogous to that described in Sect. 3.2 and 5.2. This is a *parametric* decision (or control) problem consisting in the determination of b in the known form of the control algorithm. A functional algorithm Ψ with the parameter b may be given directly by an expert or may be obtained via a determinization of the initial non-deterministic KD given by an expert in the form of a function or relation with unknown parameters, or in the non-parametric form, together with respective distributions. Some of the cases listed above will be described more precisely for random, uncertain and fuzzy variables in the next sections.

Finally, let us note that the dynamical functional plant considered here may be discrete-time or continuous as well. In the second case the plant may be described by the equation

$$\dot{s}(t) = \chi[s(t), u(t); x]$$

and

$$Q = \int_0^T \varphi[s(t)]dt$$

where T is given and φ is a local quality index for the fixed t, e.g. $\varphi(s) = (s - s^*)^{\mathrm{T}}(s - s^*)$ where s^* is a required value of the state. In the closed-loop control system with a single-input–single-output (one-dimensional) plant we may choose as a state vector $s^{\mathrm{T}}(t) = [\varepsilon(t), \dot{\varepsilon}(t), ..., \varepsilon^{(k-1)}(t)]$ where k is an order of the observable plant and $\varepsilon(t) = y(t) - y^*$ is a control error. In particular, for the stable system with one-dimensional error

$$Q = \int_0^\infty \varepsilon^2(t)dt . \tag{9.4}$$

In traditional control systems the plant and/or the controller may be described by transmittances and the parametric form of the linear controller proposed by an expert may have the form of a transmittance $K_{\mathrm{C}}(p;b)$, e.g. if

$$u(t) = b_1\varepsilon(t) + b_2\dot{\varepsilon}(t) \tag{9.5}$$

or

$$u(t) = b_1\varepsilon(t) + b_2\dot{\varepsilon}(t) + b_3\int_0^t \varepsilon(t)dt$$

then

$$K_{\mathrm{C}}(p;b) = b_1 + b_2 p$$

or

$$K_{\mathrm{C}}(p;b) = b_1 + b_2 p + \frac{b_3}{p},$$

respectively.

9.2 Uncertain Controller in a Closed-loop System

Let us explain the idea of the parametric optimization more precisely for the uncertain controller in the closed-loop system with a dynamical plant [27, 55]. The plant is described by the equations

$$\dot{s}(t) = \chi[s(t), u(t); x],$$

$$y(t) = \eta[s(t)]$$

where s is the state vector, or by the transfer function $K_P(p;x)$ in a linear case; x is an unknown vector parameter which is assumed to be a value of an uncertain variable $\overline{x}$ described by $h_x(x)$. The controller with input y (or control error ε) is described by the analogous model with a vector of parameters b which is to be determined. Consequently, for the given T and φ, the performance index

$$Q = \int_0^T \varphi(y, u)\, dt \triangleq \Phi(b, x)$$

is a function of b and x. In particular, for a one-dimensional plant

$$Q = \int_0^\infty \varepsilon^2(t) dt = \Phi(b, x).$$

The closed-loop control system is then considered as a static plant with input b, the output Q and an unknown parameter x, for which we can formulate and solve the decision problem described in Sects. 5.2. and 5.3. The control problem consisting in the determination of b in the known form of the control algorithm may be formulated as follows.

Control problem: For the given models of the plant and the controller find the value b^* minimizing $\mathrm{M}(\overline{Q})$, i.e. the mean value of the performance index.

The procedure for solving the problem is then the following:

1. To determine the function $Q = \Phi(b, x)$.
2. To determine the certainty distribution $h_q(q;b)$ for $\overline{Q}$ using the function Φ and the distribution $h_x(x)$ in the same way as in the formula (5.1) for $\overline{y}$.
3. To determine the mean value $\mathrm{M}(\overline{Q};b)$.
4. To find b^* minimizing $\mathrm{M}(\overline{Q};b)$.

In order to apply the second case of the determinization, corresponding to the determination of Ψ_d for the static plant (see Sect. 5.3), it is necessary to find the value $b(x)$ minimizing $Q = \Phi(b, x)$ for the fixed x. The control algorithm with the uncertain parameter $b(x)$ may be considered as a knowledge of the control in our case, and the controller with this parameter is an *uncertain controller* in the closed-loop system. To obtain the deterministic control algorithm, one should substitute $\mathrm{M}(\overline{b})$ in place of $b(x)$ in the uncertain control algorithm, where the mean value $\mathrm{M}(\overline{b})$ should be determined by using the function $b(x)$ and the certainty distribution $h_x(x)$.

Assume that the state $s(t)$ is put at the input of the controller. Then the uncertain controller has a form

$$u = \Psi(s, x)$$

which may be obtained as a result of non-parametric optimization, i.e. Ψ is the optimal control algorithm for the given model of the plant with the fixed x and for the given form of a performance index. Then

$$u_d = \mathrm{M}(\overline{u}; s) \stackrel{\Delta}{=} \Psi_d(s)$$

where $\mathrm{M}(\overline{u}; s)$ is determined by using the distribution

$$h_u(u; s) = v[\overline{x} \,\tilde{\in}\, D_x(u; s)] = \max_{x \in D_x(u;s)} h_x(x)$$

and

$$D_x(u; s) = \{x \in X : u = \Psi(s, x)\}.$$

A similar approach may be applied in the discrete-time case with the plant described by the equations

$$s_{n+1} = \chi(s_n, u_n; x)$$
$$y_n = \eta(s_n)$$

or by the respective form of a transmittance.

Example 9.1.

The data for the linear control system under consideration (Fig. 9.3) are the following:

$$K_{\mathrm{P}}(p; x) = \frac{x}{(pT_1 + 1)(pT_2 + 1)}, \quad K_{\mathrm{C}}(p; b) = \frac{b}{p}$$

$z(t) = 0$ for $t < 0$, $z(t) = 1$ for $t \geq 0$, $h_x(x)$ has a triangular form presented in Fig. 8.4 with $d_1 \stackrel{\Delta}{=} a$ and $\gamma_1 \stackrel{\Delta}{=} d$.

It is easy to determine

$$Q = \int_0^\infty \varepsilon^2(t)\, dt = \frac{x^2(T_1 + T_2)}{2xb(T_1 + T_2 - xbT_1T_2)} = \Phi(b, x). \tag{9.6}$$

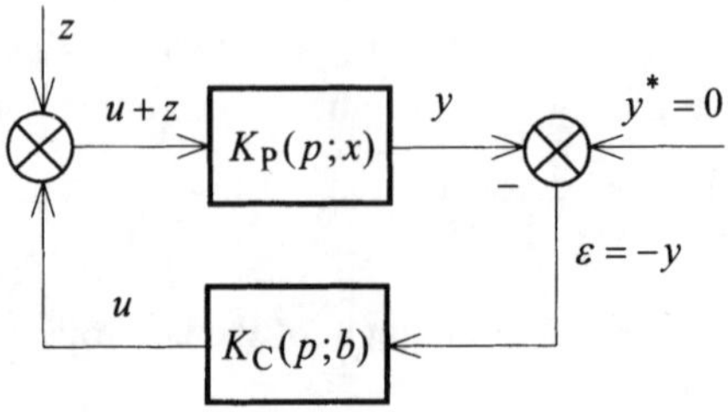

Figure 9.3. Closed-loop control system

The minimization of Q with respect to b gives

$$b(x) = \frac{\alpha}{x}, \qquad \alpha = \frac{T_1 + T_2}{T_1 T_2},$$

i.e. the uncertain controller is described by

$$K_C(p) = \frac{b(x)}{p} = \frac{\alpha}{xp}. \tag{9.7}$$

The certainty distribution $h_b(b)$ is as follows:

$$h_b(b) = \begin{cases} 0 & \text{for} \quad 0 < b \le \dfrac{\alpha}{a+d} \\ \dfrac{ab - \alpha}{db} + 1 & \text{for} \quad \dfrac{\alpha}{a+d} \le b \le \dfrac{\alpha}{a} \\ \dfrac{-ab + \alpha}{db} + 1 & \text{for} \quad \dfrac{\alpha}{a} \le b \le \dfrac{\alpha}{a-d} \\ 0 & \text{for} \quad \dfrac{\alpha}{a-d} \le b < \infty . \end{cases}$$

From the definition of a mean value we obtain

$$\mathrm{M}(\bar{b}) = \frac{\alpha d(d^2 + 2a^2)}{2a^2 \ln \dfrac{a^2}{a^2 - d^2}} . \tag{9.8}$$

Finally, the deterministic controller is described by

$$K_{C,d}(p) = \frac{\mathrm{M}(\bar{b})}{p}. \qquad \square$$

To apply the first approach described in the previous section, it is necessary to find the certainty distribution for Q using formula (9.6) and the distribution $h_x(x)$, then to determine $\mathrm{M}(\bar{Q};b)$ and to find the value b^* minimizing $\mathrm{M}(\bar{Q};b)$. It may be shown that $b^* \neq \mathrm{M}(\overline{b})$ given by the formula (9.8).

Example 9.2.

Let us consider the time-optimal control of the plant with $K_P(p;x) = xp^{-2}$ (Fig. 9.4), subject to constraint $|\, u(t)\,| \leq M$.

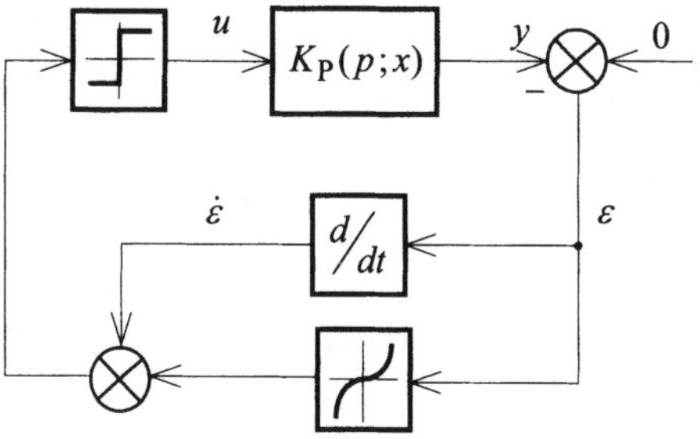

Figure 9.4. Example of control system

It is well known that the optimal control algorithm $u = \Psi(s, x)$ is the following:

$$u(t) = M\,\mathrm{sgn}(\varepsilon + \frac{|\dot{\varepsilon}|\dot{\varepsilon}}{2xM})$$

where $s = [\varepsilon, \dot{\varepsilon}]$. For the given $h_x(x)$ we can determine $h_u(u;\varepsilon,\dot{\varepsilon})$, which is reduced to three values $v_1 = v(\overline{u} \cong M)$, $v_2 = v(\overline{u} \cong -M)$, $v_3 = v(\overline{u} \cong 0)$. Then

$$u_d(t) = \mathrm{M}(\overline{u}) = M(v_1 - v_2)(v_1 + v_2 + v_3)^{-1}.$$

It is easy to see that

$$v_1 = \max_{x \in D_{x1}} h_x(x), \qquad v_2 = \max_{x \in D_{x2}} h_x(x)$$

where

$$D_{x1} = \{x : x\,\mathrm{sgn}\,\varepsilon > -|\dot{\varepsilon}|\dot{\varepsilon}(2M|\varepsilon|)^{-1}\},$$

$$D_{x2} = \{x : x\,\mathrm{sgn}\,\varepsilon < -|\dot{\varepsilon}|\dot{\varepsilon}(2M|\varepsilon|)^{-1}\}$$

and

$$v_3 = h_x(\frac{-|\dot{\varepsilon}|\dot{\varepsilon}}{2M\varepsilon}).$$

Assume that the certainty distribution of $\overline{x}$ is the same as in Example 9.1. For

$\varepsilon > 0$, $\dot{\varepsilon} < 0$ and $x_g < a$ it is easy to obtain the following control algorithm

$$u_d = \mathrm{M}(\bar{u}) = \begin{cases} M & \text{for } d \le a - x_g \\ M \dfrac{a - x_g}{3d - 2(a - x_g)} & \text{for } d \ge a - x_g \end{cases}$$

where $x_g = (\dot{\varepsilon})^2 (2M\varepsilon)^{-1}$. For example, for $M = 0.5$, $\dot{\varepsilon} = -3$, $\varepsilon = 1$, $a = 16$ and $d = 10$ we obtain $u_d = 0.2$. □

9.3 Random Controller in a Closed-loop System

The considerations for the random controller are analogous to those for the uncertain controller in the previous section. Now we assume that the unknown parameter x is a value of a random variable $\widetilde{x}$ described by a probability density $f(x)$.

Control problem: For the given models of the plant and the controller find the value b^* that minimizes $\mathrm{E}(\widetilde{Q}; b)$, i.e. the expected value of the performance index.

The procedure for solving the problem is as follows:

1. To determine the function $Q = \Phi(b, x)$.
2. To determine the expected value

$$\mathrm{E}(\widetilde{Q}; b) = \int_X \Phi(b, x) f_x(x) dx .$$

3. To find b^* minimizing $\mathrm{E}(\widetilde{Q}; b)$.

In the second case of the determinization one should determine

$$b(x) = \arg\min_b \Phi(b, x) .$$

The control algorithm with the random parameter $b(x)$ may be called a knowledge of the control in our case, and the controller with this parameter is a *random controller* in the closed-loop system. To obtain the deterministic control algorithm, one should substitute $\mathrm{E}(\widetilde{b})$:

$$\mathrm{E}(\widetilde{b}) = \mathrm{E}[b(\widetilde{x})] = \int_X b(x) f_x(x) dx$$

in place of $b(x)$ in the random algorithm. In the case when the random algorithm has the form $u = \Psi(s, x)$, one should determine

$$u_d = \mathrm{E}(\tilde{u};s) = \int_X \Psi(s,x) f_x(x)dx \stackrel{\Delta}{=} \Psi_d(s).$$

Example 9.3.
In the system considered in Example 9.1 x is a value of a random variable with a triangular probability density presented in Fig. 9.5. Now (9.7) is a transmittance of a random controller. It is easy to obtain

$$\mathrm{E}[b(\tilde{x})] = \int b(x) f_x(x)dx = \frac{\alpha \ln 4}{a}.$$

Then the deterministic controller is described by

$$K_{\mathrm{C},d}(p) = \frac{\alpha \ln 4}{ap} = \frac{(T_1 + T_2)\ln 4}{aT_1T_2 p}.$$ □

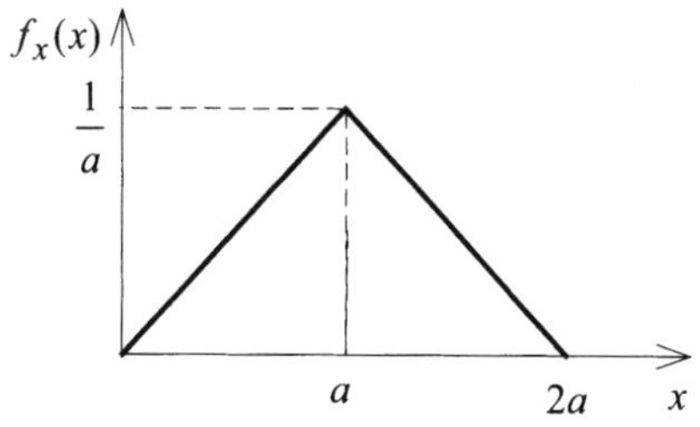

Figure 9.5. Example of probability density

Example 9.4.
In the system considered in Example 9.2 x is a value of a random variable $\tilde{x}$ with the probability density $f_x(x)$. Then

$$P(\tilde{u} = M) \stackrel{\Delta}{=} p_1 = \int_{D_{x1}} f_x(x)dx,$$

$$P(\tilde{u} = -M) \stackrel{\Delta}{=} p_2 = \int_{D_{x2}} f_x(x)dx = 1 - p_1, \qquad P(\tilde{u} = 0) \stackrel{\Delta}{=} p_3 = 0,$$

and

$$u_d(t) = \mathrm{E}(\tilde{u}) = M(p_1 - p_2) = M(2p_1 - 1). \tag{9.9}$$

Assume that $\tilde{x}$ has a probability density presented in Fig. 9.5 and $\varepsilon > 0$. Then

$$p_1 = \begin{cases} 0 & \text{for} \quad x_g \geq 2a \\ \dfrac{(2a - x_g)^2}{2a^2} & \text{for} \quad a \leq x_g \leq 2a \\ 1 - \dfrac{x_g^2}{2a^2} & \text{for} \quad x_g \leq a \end{cases}$$

where $x_g = -|\dot{\varepsilon}|\dot{\varepsilon}(2M\varepsilon)^{-1}$. Using (9.9) we obtain the following deterministic control algorithm:

$$u_d = \begin{cases} -M & \text{for} \quad x_g \geq 2a \\ M[\dfrac{(2a - x_g)^2}{a^2} - 1] & \text{for} \quad a \leq x_g \leq 2a \\ M(1 - \dfrac{x_g^2}{a^2}) & \text{for} \quad x_g \leq a. \end{cases}$$

□

9.4 Descriptive and Prescriptive Approaches

In the analysis and design of knowledge-based uncertain systems it may be important to investigate a relation between two concepts concerning two different subjects of the knowledge given by an expert, which have been mentioned in Sect. 9.1 [38, 39, 47]. In the *descriptive approach* an expert gives the knowledge of the plant KP, and the knowledge of the decision making KD is obtained from KP for the given requirement. This approach is widely used in the traditional decision and control theory. The deterministic decision algorithm may be obtained via the determinization of KP or the determinization of KD based on KP. Such a situation is illustrated in Figs. 3.2 and 3.3 for the description using random variables and in Figs. 5.2 and 5.3 for the formulation based on uncertain variables. In the *prescriptive approach* the knowledge of the decision making $\overline{\text{KD}}$ is given directly by an expert. This approach is used in the design of fuzzy controllers where the deterministic control algorithm is obtained via the defuzzification of the knowledge of the control given by an expert. The descriptive approach to the decision making based on the fuzzy description may be found in [69].

Generally speaking, the descriptive and prescriptive approaches may be called *equivalent* if the deterministic decision algorithms based on KP and $\overline{\text{KD}}$ are the same. Different particular cases considered in the previous chapters may be illustrated in Figs. 9.6 and 9.7 for two different concepts of the determinization. Fig. 9.8 illustrates the prescriptive approach. In the first version (Fig. 9.6) the approaches are equivalent if $\Psi(z) = \overline{\Psi}_d(z)$ for every z. In the second version (Fig.

9.7) the approaches are equivalent if KD = $\overline{\text{KD}}$. Then $\Psi_d(z) = \overline{\Psi}_d(z)$ for every z.

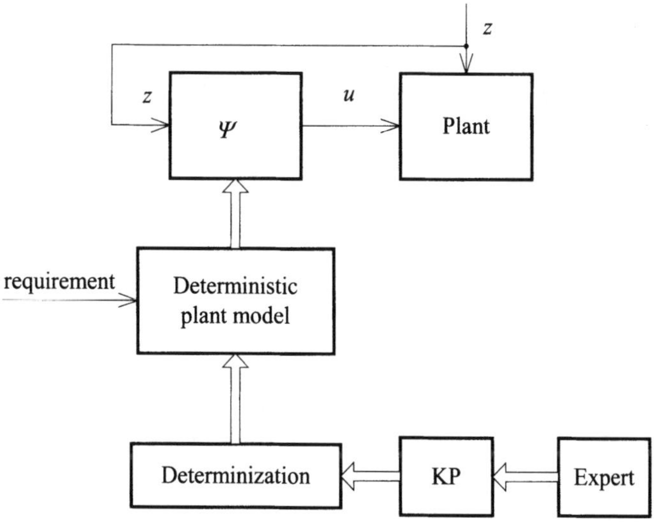

Figure 9.6. Illustration of descriptive approach – the first version

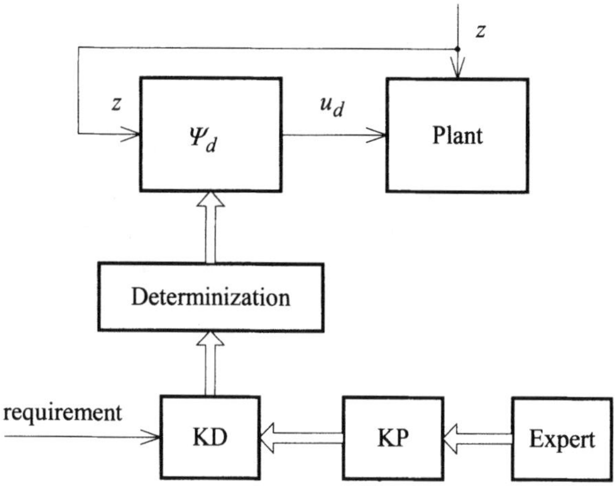

Figure 9.7. Illustration of descriptive approach – the second version

Let us consider more precisely version I of the decision problem described in Sect. 5.3. An expert formulates KP $=<\Phi, h_x>$ (the descriptive approach) or $\overline{\text{KD}} =<\overline{\Phi}_d, h_x>$ (the prescriptive approach). In the first version of the determinization illustrated in Fig. 5.2, the approaches are equivalent if $\Psi_a(z) = \overline{\Psi}_{ad}(z)$ for every z, where $\Psi_a(z)$ is determined by (5.8) and $\overline{\Psi}_{ad}(z)$ is determined by (5.13) with $\overline{\Phi}_d(z,x)$ instead of $\Phi_d(z,x)$ obtained as a solution of the equation

$$\Phi(u,z,x) = y^* . \tag{9.10}$$

In the second version of the determinization illustrated in Fig. 5.3, the approaches are equivalent if the solution of Equation (9.10) with respect to u has the form $\overline{\Phi}_d(z,x)$, i.e.

$$\Phi[\overline{\Phi}_d(z,x), z, x] = y^* .$$

For the non-parametric problem described in Sect. 5.8 only the second version of the determinization illustrated in Fig. 5.10 may be applied.

The similar formulation of the equivalency may be given for the random and fuzzy descriptions presented in Chapters 3 and 6, respectively. The generalization for the soft variables and evaluating functions described in Sects. 6.7 and 6.8 may be formulated as a principle of equivalency.

Principle of equivalency: If the knowledge of the decision making $\overline{KD}$ given by an expert has the form of a set of evaluating functions $\hat{D}_{gu}(z)$ and $\hat{D}_{gu}(z) \subseteq \overline{D}_{gu}(z)$ where $\overline{D}_{gu}(z)$ is determined by (6.46), then $\hat{S}_\Psi \subseteq \overline{S}_\Psi$ where $\hat{S}_\Psi$ and $\overline{S}_\Psi$ are the sets of the decision algorithms Ψ corresponding to $\hat{D}_{gu}(z)$ and $\overline{D}_{gu}(z)$, respectively. In particular, if an expert gives one evaluating function $\hat{g}_u(u \mid z)$, i.e. $\hat{D}_{gu}(z) = \{\hat{g}_u(u \mid z)\}$ and $\hat{g}_u(u \mid z) \in D_{gu}(z)$, then the decision algorithm based on the knowledge of the decision making given by an expert is equivalent to one of the decision algorithms based on the knowledge of the plant. □

For the non-parametric cases described together in Sects. 6.5 and 6.7, descriptive and prescriptive approaches are equivalent if:

1. $\bar{f}_u(u \mid z)$ given by an expert satisfies Equation (6.31).
2. $\bar{h}_u(u \mid z)$ given by an expert satisfies Equation (6.35).
3. $\bar{\mu}_u(u \mid z)$ given by an expert satisfies Equation (6.23).
4. $\bar{g}_u(u \mid z)$ given by an expert satisfies Equation (6.42).

It is worth noting that the determination of the decision algorithm $\overline{\Psi}$ based on $\overline{KD}$ means the solution of the analysis problem for the unit (the plant) described by $\overline{KD}$ and for the given input of this unit: z in the open-loop system and s in the closed-loop system. It may be useful to present together the determinizations of $\overline{KD}$ in non-parametric cases for static plants in an open-loop system (see Fig. 9.8).

A. Description based on random variables

For the given $\bar{f}_u(u \mid z)$ one should determine

$$\bar{u}_d = \mathrm{E}(\tilde{u} \mid z) = \int_U u\bar{f}_u(u \mid z)du \stackrel{\Delta}{=} \bar{\Psi}_d(z).$$

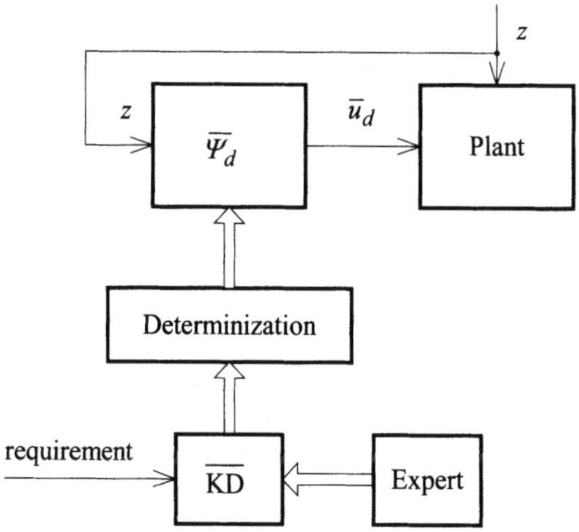

Figure 9.8. Illustration of prescriptive approach

B. Description based on uncertain variables

For the given $\bar{h}_u(u \mid z)$ one should determine

$$\bar{u}_d = \mathrm{M}(\bar{u} \mid z) = \int_U u\bar{h}_u(u \mid z)du \cdot [\int_U \bar{h}_u(u,z)du]^{-1} \stackrel{\Delta}{=} \bar{\Psi}_d(z).$$

C. Description based on fuzzy variables ($U = R^1$)

In this case we can consider two versions (see Sects 6.1–6.3):

1. For the given $\bar{\mu}_u(u \mid z) = w[\varphi_z(z) \to \varphi_u(u)]$ one should determine

$$\bar{u}_d = \mathrm{M}(\hat{u} \mid z) = \int_{-\infty}^{\infty} u\bar{\mu}_u(u \mid z)du \cdot [\int_{-\infty}^{\infty} \bar{\mu}_u(u,z)du]^{-1} \stackrel{\Delta}{=} \bar{\Psi}_d(z). \quad (9.11)$$

2. For the given

$$\bar{\mu}_{uz}(u,z) = w[\hat{z} = z \to \varphi_u(u)] = \min\{\bar{\mu}_z(z), \bar{\mu}_u(u \mid z)\}$$

one should determine

$$\bar{u}_d = \mathrm{M}(\hat{u} \mid z) = \int_{-\infty}^{\infty} u\bar{\mu}_{uz}(u \mid z)du \cdot [\int_{-\infty}^{\infty} \bar{\mu}_{uz}(u \mid z)du]^{-1} \stackrel{\Delta}{=} \bar{\Psi}_d(z). \quad (9.12)$$

Cases 1 and 2 are equivalent if $\bar{\mu}_u(u \mid z) = \bar{\mu}_{uz}(u \mid z)$.

Instead of the mean value E or M we can use the value of u maximizing the distribution, e.g. in case A

$$\bar{u}_d = \arg\max_{u \in U} \bar{f}_u(u \mid z).$$

Denote by a the vector of parameters in the description given by an expert. They may be parameters in $\bar{f}_u(u \mid z)$, $\bar{h}_u(u \mid z)$ or $\bar{\mu}_u(u \mid z)$. Consequently, the deterministic decision algorithm $\bar{u}_d = \bar{\Psi}_d(z, a)$ depends on a. Then the problem of a *parametric optimization* consisting in choosing a^* minimizing the performance index Q, and the problem of *adaptation* consisting in adjusting a to a^*, may be considered (see Sect. 9.1).

The cases corresponding to A, B, C may be listed for the dynamical plant with s instead of z. Note that the description for the fuzzy variables is concerned with a simple one-dimensional case. In the next section we shall present it for a multidimensional case in the closed-loop system.

9.5 Fuzzy Controller in a Closed-loop System

Let us consider the closed-loop control system with a dynamical plant (continuous- or discrete-time) in which the state s is put at the input of the controller. In the simple one-dimensional case the knowledge $\overline{\text{KD}}$ (or the fuzzy controller) given by an expert consists of two parts:

1. The *rule*

$$\varphi_s(s) \to \varphi_u(u) \tag{9.13}$$

with the determined properties $\varphi_s(s)$ and $\varphi_u(u)$: "s is d_s" and "u is d_u", i.e. "if $\hat{s} = s$ then s is d_s" and "if $\hat{u} = u$ then u is d_u" (see Sect. 6.1).

2. In the first version corresponding to (9.11) for the open-loop system – the membership function $\mu_u(u \mid s)$ of the property (9.13). In the second version corresponding to (9.12) – the membership function $\mu_u(u \mid s)$ and the membership function $\mu_s(s)$ of the property $\varphi_s(s)$, or directly the membership function

$$\mu_{us}(u, s) = w[\hat{s} = s \to \varphi_u(u)] = \min\{\mu_s(s), \mu_u(u \mid s)\}. \tag{9.14}$$

Then the deterministic control algorithm (or deterministic controller) is described by the following procedure:

1. Put s at the input of the controller.
2. In the first version, determine the decision

$$u_d = \int_{-\infty}^{\infty} u\mu_u(u \mid s)du \cdot [\int_{-\infty}^{\infty} \mu_u(u \mid s)du]^{-1}.$$

In the second version, for the given $\mu_u(u \mid s)$ and $\mu_s(s)$ determine $\mu_{us}(u,s)$ according to (9.14) and find the decision

$$u_d = \int_{-\infty}^{\infty} u\mu_{us}(u,s)du \cdot [\int_{-\infty}^{\infty} \mu_{us}(u,s)du]^{-1} .$$

Instead of the mean value we can determine and use the decision u_d maximizing the membership function $\mu_u(u \mid s)$ or $\mu_{us}(u,s)$.

Let us present the extension of the second version to the controller with k inputs $s^{(1)}, s^{(2)}, ..., s^{(k)}$ (the components of the state vector s) and one output u.The description of the *fuzzy controller* ($\overline{\text{KD}}$) given by an expert has a form analogous to that for the fuzzy description of a multidimensional static plant (see Sect. 6.3) and contains two parts:

1. The set of *rules*

$$\varphi_{j1}(s^{(1)}) \wedge \varphi_{j2}(s^{(2)}) \wedge ... \wedge \varphi_{jk}(s^{(k)}) \rightarrow \varphi_{ju}(s^{(u)}), \quad j = 1,2,...,N \quad (9.15)$$

where N is a number of rules, $\varphi_{ji}(s^{(i)}) =$ "$s^{(i)}$ is d_{ji}" and $\varphi_{ju}(u) =$ "u is d_j", d_{ji} and d_j denote the size of the numbers as in Sects. 6.1 and 6.3. The meaning of the rules (9.15) is then as follows:

IF ($s^{(1)}$ is d_{j1}) AND ($s^{(2)}$ is d_{j2}) AND ... AND ($s^{(k)}$ is d_{jk}) THEN u is d_j.

For example ($k = 3$)
($s^{(1)}$ is small positive) $\wedge$ ($s^{(2)}$ is large negative) $\wedge$ ($s^{(3)}$ is small negative) $\rightarrow$ u is medium positive.

2. The matrix of the membership functions

$$\mu_{sji}(s^{(i)}) = w[\varphi_{ji}(s^{(i)})], \quad \begin{array}{l} i = 1,2,...,k, \\ j = 1,2,...,N \end{array}$$

and the sequence of the membership functions

$$\mu_{uj}(u \mid s^{(1)}, s^{(2)}, ..., s^{(k)}), \quad j = 1,2,...,N$$

for the properties (9.15).

The *deterministic controller* (i.e. the deterministic control algorithm obtained as a result of the determinization of $\overline{\text{KD}}$) is described by the following procedure:

1. Put s at the input of the controller.
2. Find the sequence of values

$$\mu_{sj}(s) = \min\{\mu_{sj1}(s^{(1)}), \mu_{sj2}(s^{(2)}), ..., \mu_{sjk}(s^{(k)})\}, \; j = 1,2,...,N .$$

3. From each rule determine the membership function

$$\mu_{us,j}(u,s) = w[\hat{s} = s \rightarrow u \text{ is } d_j] = \min\{\mu_{sj}(s), \mu_{uj}(u \mid s)\}\ ,\ j = 1,2,..., N\ .$$

4. Determine the membership function $\mu_u(s)$ of the property

$$(\hat{s} = s) \rightarrow (u \text{ is } d_1) \vee (u \text{ is } d_2) \vee ... \vee (u \text{ is } d_N).$$

Then

$$\mu_u(s) = \max\{\mu_{us,1}(u,s),\ \mu_{us,2}(u,s),\ ...,\ \mu_{us,N}(u,s)\}.$$

5. Determine the decision u_d as a result of the determinization (defuzzification) of $\mu_u(s)$:

$$u_{ad} = \arg\max_u \mu_u(s) \triangleq \overline{\Psi}_{ad}(s)$$

or

$$u_{bd} = \int_{-\infty}^{+\infty} u\mu_u(s)\,du\ [\int_{-\infty}^{+\infty} \mu_u(s)\,du]^{-1} \triangleq \overline{\Psi}_{bd}(s)\ .$$

In a discrete case the integrals are replaced by the sums (see (6.5)). For simplicity, it may be assumed that the membership function of the implication (9.15) does not depend on s. Then

$$\mu_{uj}(u \mid s) \triangleq \overline{\mu}_{uj}(u)$$

and

$$\mu_{us,j}(u,s) = w[\hat{s} = s \rightarrow u \text{ is } d_j] = \min\{\mu_{sj}(s), \overline{\mu}_{uj}(u)\}$$

The relations between $\overline{\mu}_{uj}$, μ_{sj} and μ_{usj} are illustrated in Fig. 9.9.

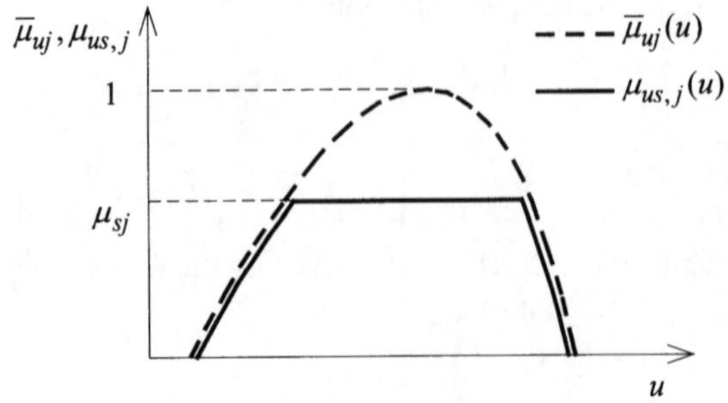

Figure 9.9. Example of $\overline{\mu}_{uj}$ and $\mu_{us,j}$

If for a single-output continuous dynamical plant $s^{\mathrm{T}} = [\varepsilon(t), \dot{\varepsilon}(t), ..., \varepsilon^{(k-1)}(t)]$ (where $\varepsilon(t)$ is the control error put at the input of the controller), then the properties in the rules and the corresponding membership functions concern the control error and its derivatives. If u is a vector (in the case of a multi-input plant) then the knowledge given by an expert and the procedure of finding the decision are for each component of u the same as for one-dimensional u considered above. There exist different versions and modifications of fuzzy controllers described in the literature. To characterize the fuzzy controllers based on the knowledge of the control given by an expert, the following remarks should be taken into account:

1. In fact, the control decisions u_d are determined by the *deterministic controller* Ψ_d in the closed-loop control system.
2. The deterministic control algorithm Ψ_d has a form of the *procedure* presented in this section, based on the description of the *fuzzy controller* (i.e. the knowledge of the control $\overline{\mathrm{KD}}$) given by an expert (Fig. 9.10).

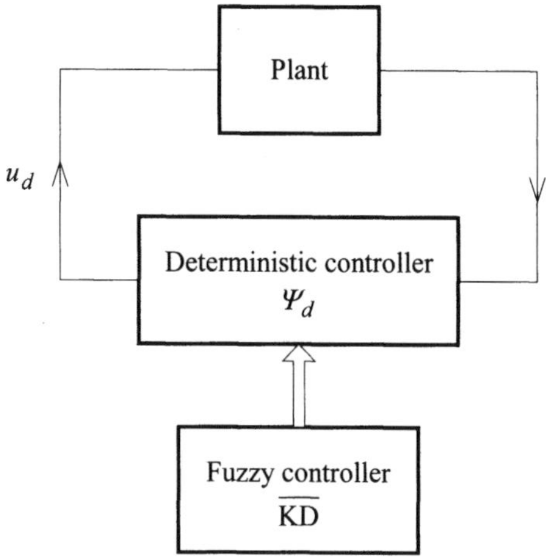

Figure 9.10. Control system based on fuzzy controller

3. The deterministic control algorithm $u_d = \Psi_d(s, a)$ where a is the vector of parameters of the membership functions in $\overline{\mathrm{KD}}$ – may be considered as a parametric form of a deterministic controller. This form is determined by the forms of rules and membership functions in $\overline{\mathrm{KD}}$, i.e. is proposed indirectly by an expert.
4. The parametric form $u_d = \Psi_d(s, a)$ is proposed in a rather arbitrary way, not reasoned by the description of the plant. Besides, it is a rather complicated form (in comparison with traditional and given directly parametric forms of a deterministic controller) and the decisions u_d may be very sensitive to changes of forms and parameters of the membership functions in $\overline{\mathrm{KD}}$.

5. It is reasonable and recommended to apply the *parametric optimization* and *adaptation* described in Sect 9.1, to achieve the value a^* optimal for the accepted form Ψ_d, i.e. for the forms of rules and membership functions in $\overline{\text{KD}}$ given by an expert.

9.6 Quality of Decisions Based on Non-parametric Descriptions

Let us describe more exactly the evaluation of the quality and the parametric optimization mentioned in Sect. 9.1, for a static plant and a non-parametric description of the uncertainties using uncertain and fuzzy variables. Consider a plant described by a function $y = \Phi(u,z)$ and introduce the performance index evaluating the quality of the decision u for the given z

$$Q(u,z) = (y - y^*)^{\mathrm{T}}(y - y^*) = [\Phi(u,z) - y^*]^{\mathrm{T}}[\Phi(u,z) - y^*]$$

where y^* denotes a desirable value of the output. Assume that the function Φ (i.e. the exact deterministic description of the plant) is unknown, (u, y, z) are values of uncertain variables $(\overline{u}, \overline{y}, \overline{z})$ and a user presents the requirement in the form of a certainty distribution $h_y(y)$ in which

$$\arg\max_y h_y(y) = y^* .$$

If $u_d = \Psi(z,a)$ is the deterministic decision algorithm obtained as a result of a determinization of the uncertain decision algorithm $h_u(u \mid z)$ obtained from KP or given directly by a user, then

$$Q(z) = [\Phi(\Psi(z,a),z) - y^*]^{\mathrm{T}}[\Phi(\Psi(z,a),z) - y^*] \triangleq \overline{\Phi}(z) \qquad (9.16)$$

where a is the vector of parameters in the certainty distribution $h_y(y \mid z)$ (in the descriptive approach) or directly in the certainty distribution $h_u(u \mid z)$ (in the prescriptive approach). For the given z, the performance index (9.16) evaluates the quality of the decision u_d based on the uncertain knowledge and applied to the real plant described by Φ. To evaluate the quality of the algorithm Ψ_d for different possible values of z, one can use the mean value

$$\mathrm{M}(\overline{Q}) = \int_0^{+\infty} q\, h_q(q)dq \cdot [\int_0^{+\infty} h_q(q)dq]^{-1} \qquad (9.17)$$

where $h_q(q)$ is the certainty distribution of $\overline{Q}=\overline{\Phi}(\overline{z})$ which should be determined for the given function $\overline{\Phi}$ and the certainty distribution

$$h_z(z)=\max_{u\in U} h_{uz}(u,z),$$

and $h_{uz}(u,z)$ is the distribution obtained in the first step of the decision problem solution (see (5.61) in Sect. 5.8). In the prescriptive approach $h_z(z)$ should be given by an expert. The performance index (9.16) or (9.17) may be used in:

1. Investigation of the influence of the parameter a in the description of the uncertain knowledge on the quality of the decisions based on this knowledge.
2. Comparison of the descriptive and prescriptive approaches in the case when $h_u(u \mid z)$ and $\overline{h}_u(u \mid z)$ have the same form with different values of the parameter a.
3. Parametric optimization and adaptation, when

$$a^* = \arg\min_a Q(a)$$

is obtained by the adaptation process consisting in *step by step* changing of the parameters of the controller in an open-loop decision system with a simulator of the plant described by the model Φ (see Sect. 9.1).

The considerations for fuzzy controllers are analogous, with $\mu_y(y \mid u,z)$, $\mu_z(z)$ and $\mu_u(u \mid z)$ in place of $h_y(y \mid u,z)$, $h_z(z)$ and $h_u(u \mid z)$ (see Sect. 6.3).

Example 9.5.

Consider the decision problem for the plant with $f_u(u \mid z)$ and $f_y(y)$ the same as in Example 5.10. The result (i.e. the uncertain controller) may be rewritten in the form

$$h_{uz}(u,z)=\begin{cases} -[\frac{e^2+\alpha+u}{2e}]^2+1 & \text{for} \quad u\le 1-(e-1)^2-\alpha, \\ & 0\le u\le \frac{1}{2}, \quad 0\le\alpha\le\frac{1}{2} \\ 0 & \text{otherwise} \end{cases}$$

where $e=d-c$ and $\alpha=b-z$. Assuming $h_u(u \mid z)=h_{uz}(u,z)$ we have

$$u_d=\int_0^s u\, h_{uz}(u,z)du\cdot[\int_0^s h_{uz}(u,z)du]^{-1} \triangleq \Psi_d(z,a)$$

where $s=\min\{\frac{1}{2},N\}$, $N=1-(e-1)^2-\alpha$, $a=(e,\alpha)$.

The influence of the parameters α and e on the decision u_d has been investigated by simulations. The results are presented in Fig. 9.11. It is assumed that the value $c = y^*$ is given by a user. Then, the influence of α and e denotes the influence of d and b given by an expert. It is worth noting that the decision u_d may be significantly sensitive to changes of α and e. For example, for $\alpha = 0.4$ we can read the following data from Fig. 9.11:

1. For $e = 1.4$ the decision $u_d \approx 0.15$.

2. For $e = 1.6$ the decision $u_d \approx 0.08$.

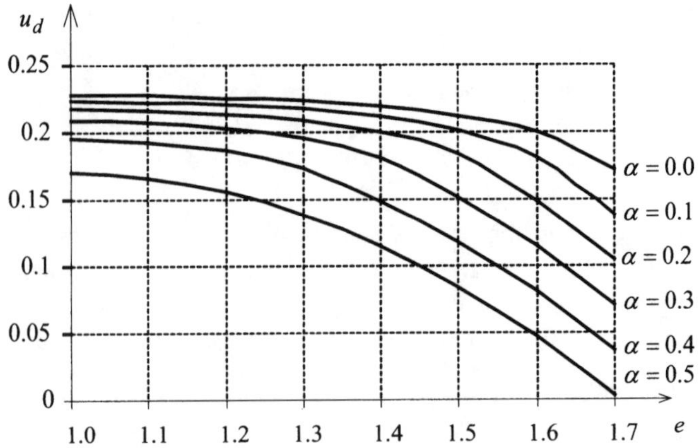

Figure 9.11. The relationship between u_d and e for α = const.

Let us assume that the exact description of the plant is as follows:

$$y = \Phi(u, z) = \frac{u}{(b - z)^2}$$

and the performance index is $Q = (y - y^*)^2$ where $y^* = c$. Then

$$Q = (c - \frac{u_d}{(b - z)^2})^2 = (c - \frac{u_d}{\alpha^2})^2 .$$

The relationships between Q and the parameter of the certainty distribution $d = c + e$ are illustrated in Fig. 9.12. In this case $d^* \approx 2.35$ for $\alpha = 0.4$ and $d^* \approx 2.65$ for $\alpha = 0.3$.

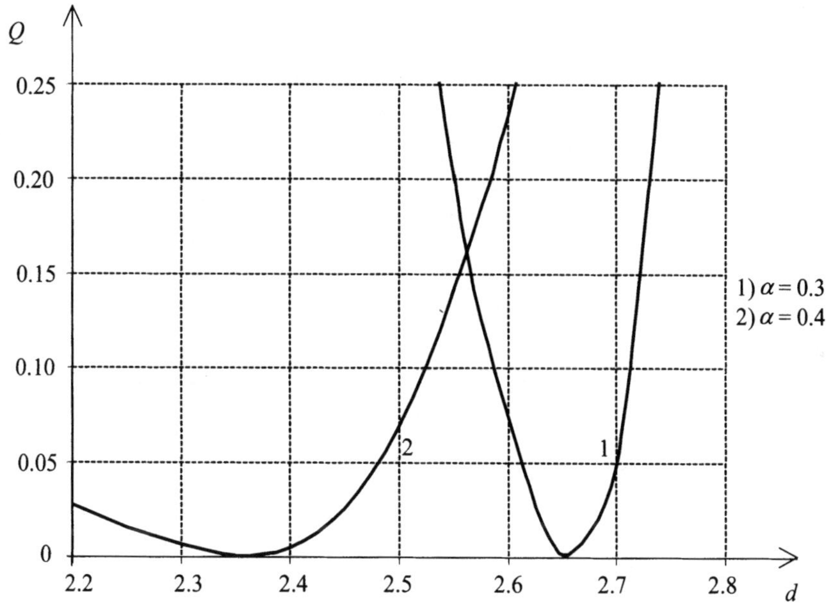

Figure 9.12. The relationship between Q and d for $c = 1$

10 Stability of Uncertain Dynamical Systems

10.1 Introduction

The analysis and decision making problems considered in the previous chapters may be called *quantitative* problems. In many cases there is a need to formulate and solve *quantitative* analysis problems, which consist in an investigation of some *properties* concerning the system under consideration, such as stability, controllability, observability, etc. Let us consider a system with two unknown vector parameters in its description: $c \in C$ and $b \in B$. The uncertainties concerning c and b are formulated as follows:

1. $c \in \Delta_c$ for the constant parameter or

$$\bigwedge_{n \geq 0} (c_n \in \Delta_c)$$

for the time-varying parameter, where Δ_c is a given set in C.

2. b is a value of a random variable $\tilde{b}$ described by the probability density $f_b(b)$ or b is a value of an uncertain variable $\bar{b}$ described by the certainty distribution $h_b(b)$. In particular, we may have only one uncertain parameter c or b. If c is an unknown parameter in a function, e.g. $y = \Phi(u, c)$, then the plant is deterministic but the expert giving Δ_c has no full knowledge of the plant (see Sect. 2.1).

Denote by V the property under consideration. For the system with the parameter c, V depends on c, and the sufficient condition that the property $V(c)$ is satisfied is as follows: $\Delta_c \subseteq D_c$ where

$$D_c = \{c \in C : V(c)\}.$$

In the case with $\tilde{b}$ or $\bar{b}$ we can determine the probability P or the certainty index v that $V(b)$ is satisfied:

$$P[V(\tilde{b})] = \int_{D_b} f_b(b)db,$$

$$v[V(\bar{b})] = \max_{b \in D_b} h_b(b)$$

where

$$D_b = \{b \in B : V(b)\}.$$

Exactly speaking, v denotes the certainty index of the soft property "$\bar{b} \mathrel{\tilde{\in}} D_b$", i.e. "$V$ is satisfied for an approximate value of $\bar{b}$", or "V is approximately satisfied". In the general case with c and b

$$P[\overline{V}(b)] = \int_{\overline{D}_b} f_b(b)db, \qquad v[\overline{V}(b)] = \max_{b \in \overline{D}_b} h_b(b)$$

where

$$\overline{V}(b) = \text{“} \Delta_c \subseteq \{c \in C : V(c,b)\} \text{”}$$

and $\overline{D}_b = \{b \in B : \overline{V}(b)\}$.

The presented approach may be applied for

$$V(c,b) = \text{“} Q(c,b) \le \alpha \text{”}$$

where $Q(c,b)$ denotes a performance (quality) index of a system, in a situation when $Q(c,b)$ may be determined (see the cases described in Chapter 9), and α is a given number.

The further considerations will concern the *stability* of functional time-varying and non-linear discrete-time systems with the parameter c in Sects. 10.2–10.4 and with parameters (c,b) in Sects. 10.5–10.7. The problem of stability and stabilization for uncertain discrete-time systems has been the topic of extensive research. The purpose of this chapter is to present a general and unified approach to the stability and stabilization problem for time-varying and non-linear uncertain discrete-time systems, based on the principle of contraction mapping [31]. The application of the old approach to the convergence problem for approximation processes based on the principle of contraction mapping, in the form presented in [10, 11, 12] gives unified forms of the stability conditions for a wide class of linear time-varying systems, time-invariant non-linear systems, and systems which are both time-varying and non-linear. These conditions are used for the estimation (evaluation) of the probability or the certainty index that the system is stable, described in the second part of this chapter.

It is worth noting that the stability problem may concern a quantitative decision problem for a functional plant with unknown parameters. The proper decision may be achieved by an approximation process performed in a closed-loop system, and the convergence of this process means the stability of the closed-loop decision system. Let us explain this idea for a static plant described by a function $y = \Phi(u,c)$ with an unknown parameter c and the desired output y^*. The decision u^* such that $\Phi(u^*,c) = y^*$ may be achieved as a *limes* of the following *step by step* approximation process (or control algorithm in a closed-loop control system):

$$u_{n+1} = u_n + K(y^* - y_n),\ n = 0,1,...$$

where $y_n = \Phi(u_n,c)$ is the output of the plant in the n-th moment and K is a matrix of coefficients. Knowing that $c \in \Delta_c$ it may be possible to choose K assuring the convergence of u_n to u^*, i.e. the stability of the control system. Such an approach may be applied to a dynamical plant and a more complicated dynamical controller. Consequently, one should investigate the stability of an uncertain dynamical system, which in this case is a closed-loop control system.

10.2 Stability Conditions

Consider a non-linear time-varying system described by

$$x_{n+1} = f(c_n, x_n), \qquad n = 0, 1, 2, ... \tag{10.1}$$

i.e.

$$x_{n+1}^{(i)} = f^{(i)}(c_n, x_n), \qquad i = 1, 2, ..., k$$

where $x_n \in X$ is the state vector, $c_n \in C$ is the vector of time-varying parameters, $X = R^k$, $C = R^p$. Assume that the functions f_i have the form

$$f^{(i)}(c_n, x_n) = \sum_{j=1}^{k} a_{ij}(c_n, x_n) x_n^{(j)}$$

i.e.

$$x_{n+1} = A(c_n, x_n) x_n \tag{10.2}$$

where the matrix $A(c_n, x_n) = [a_{ij}(c_n, x_n)] \in R^{k \times k}$. Assume that for every $c \in C$ the equation

$$x = A(c, x) x$$

has a unique solution $x_e = \overline{0}$ (the vector with zero components).

Definition 10.1 (*Global Asymptotic Stability*). The system (10.2) (or the equilibrium state x_e) is globally asymptotically stable (GAS) iff x_n converges to $\overline{0}$ for any x_0. □

The following theorems are based on [10, 12].

Theorem 10.1. If there exists a norm $\|\cdot\|$ such that

$$\bigwedge_{n\geq 0} \bigwedge_{x\in X} \| A(c_n, x) \| < 1 \tag{10.3}$$

then the system (10.2) is GAS. □

Theorem 10.2. If there exists a norm $\|\cdot\|$ and a non-singular matrix $P \in R^{k\times k}$ such that

$$\bigwedge_{n\geq 0} \bigwedge_{x\in X} \| P^{-1} A(c_n, x) P \| < 1 \tag{10.4}$$

then the system (10.2) is GAS. □

Theorem 10.3. Denote by $\lambda_i(A) = \lambda_i(c_n, x)$ the eigenvalues of the matrix A ($i =$ 1, 2, ..., k). If $A(c_n, x)$ is a symmetric matrix and

$$\bigwedge_{n\geq 0} \bigwedge_{x\in X} \max_i \left|\lambda_i(c_n, x)\right| < 1 \tag{10.5}$$

then the system (10.2) is GAS. □

In particular the norm $\| A \|$ may have the form

$$\| A \|_2 = \sqrt{\lambda_{\max}(A^{\mathrm{T}} A)} \tag{10.6}$$

where $\lambda_{\max}$ is the maximum eigenvalue of the matrix $A^{\mathrm{T}} A$,

$$\| A \|_1 = \max_{1\leq i\leq k} \sum_{j=1}^{k} | a_{ij} |, \tag{10.7}$$

$$\| A \|_\infty = \max_{1\leq j\leq k} \sum_{i=1}^{k} | a_{ij} |. \tag{10.8}$$

Theorem 10.1 follows from the fact that under the assumption (10.3) $A(c_n, x)x$ is a contraction mapping in X. The condition (10.4) is obtained by introducing the new state vector $v_n = P^{-1} x_n$ and using Theorem 10.1 for the equation

$$v_{n+1} = P^{-1} A(c_n, x_n) P v_n .$$

Theorem 10.3 may be easily proved (see [12]) by using the norm (10.6). If A is a symmetric matrix then $\lambda_{\max}(A^{\mathrm{T}}A) = \max_i |\lambda_i(A)|^2$.

Remark 10.1. For the norm $\|\cdot\|_2$ and $(P^{-1})^{\mathrm{T}}P^{-1} \stackrel{\Delta}{=} Q$, the condition (10.4) is reduced to the following statement. If there exists a positive definite matrix Q such that

$$\bigwedge_{n\geq 0}\bigwedge_{x\in X} A^{\mathrm{T}}(c_n, x)QA(c_n, x) - Q < 0$$

then the system (10.2) is GAS. □

Condition (10.3) may be presented in the form

$$\bigwedge_{n\geq 0}\bigwedge_{x\in X} A(c_n, x) \in \mathcal{A} \tag{10.9}$$

where $\mathcal{A}$ is a set of $k \times k$ matrices $\hat{A}$ defined as

$$\mathcal{A} = \{\hat{A} : \|\hat{A}\| < 1\}. \tag{10.10}$$

We shall also use another form of the stability condition (10.3)

$$\bigwedge_{n\geq 0} c_n \in D_c \tag{10.11}$$

where

$$D_c = \{c \in C : \bigwedge_{x\in X} \|A(c,x)\| < 1\}. \tag{10.12}$$

Conditions (10.4) and (10.5) may be presented in an analogous form with $\|P^{-1}\hat{A}P\|$ or $\max_i |\lambda_i(\hat{A})|$ instead of $\|\hat{A}\|$ in (10.10) and with $\|P^{-1}A(c,x)P\|$ or $\max_i |\lambda_i(c,x)|$ instead of $\|A(c,x)\|$ in (10.12).

The condition presented above may be applied to the stability analysis of uncertain systems. The uncertainty may concern the function $A(c,x)$ and the sequence c_n. In general, it may be formulated as

$$\bigwedge_{n\geq 0}\bigwedge_{x\in X} A(c_n, x) \in \mathcal{A}_u \tag{10.13}$$

where $\mathcal{A}_u$ is a given set of the matrices $\hat{A} \in R^{k\times k}$. The typical formulations of $\mathcal{A}_u$ are considered in the next section. Then the general condition of the global asymptotic stability for the uncertain system corresponding to (10.3) is $\mathcal{A}_u \subseteq \mathcal{A}$ and may be expressed in the following way.

Theorem 10.4. If $\| A \|< 1$ for every $A \in \mathcal{A}_u$ then the system is GAS.

When the function $A(c, x)$ is known and the uncertainty concerns only the sequence c_n, it may be formulated as

$$\bigwedge_{n \geq 0} c_n \in \Delta_c \tag{10.14}$$

where $\Delta_c \subset C$ is a given subset of C. □

Theorem 10.5. If $\Delta_c \subseteq D_c$ where D_c is determined by (10.12), i.e. if

$$\bigwedge_{c \in \Delta_c} \bigwedge_{x \in X} \| A(c,x) \| < 1 \tag{10.15}$$

then the system is GAS. □

The theorem follows immediately from (10.11), (10.12) and (10.14). The conditions corresponding to (10.4) and (10.5) have the analogous form.

10.3 Special Cases

Let us now consider two typical cases of the uncertainty: *additive uncertainty* and *multiplicative uncertainty*. For simplicity let us denote $A(c_n, x_n)$ by A_n.

10.3.1 Additive Uncertainty

In this case $A_n = A + \overline{A}_n$, i.e.

$$x_{n+1} = (A + \overline{A}_n) x_n . \tag{10.16}$$

The uncertainty concerns the matrix $\overline{A}_n$ and is formulated by one of the three forms denoted by (10.17), (10.18) and (10.19)

$$\bigwedge_{n \geq 0} \bigwedge_{x \in X} \overline{A}_n^+ \leq A_M \tag{10.17}$$

where A_M is a given non-negative matrix (i.e. all entries of A_M are non-negative) and $\overline{A}_n^+$ is the matrix obtained by replacing the entries of $\overline{A}_n$ by their absolute values

$$\bigwedge_{n \geq 0} \bigwedge_{x \in X} \| \overline{A}_n \| \leq \beta , \tag{10.18}$$

$$\bigwedge_{n\geq 0}\ \bigwedge_{x\in X}\ \max_i\ |\lambda_i(\overline{A}_n)|\leq \overline{\beta} \tag{10.19}$$

where β and $\overline{\beta}$ are given positive numbers. The inequalities (10.17), (10.18) and (10.19) define the set $\mathcal{A}_u$ in (10.13) for the cases under consideration.

Lemma 1. If A and B are quadratic matrices with non-negative entries (some of the entries are positive) and $A \geq B$ (i.e. $a_{ij} \geq b_{ij}$ for each i and j) then $\|A\| \geq \|B\|$ for the norm (10.6), (10.7) and (10.8).

Proof: For (10.7) and (10.8) the lemma follows immediately from the definition of the norm. Denote

$$\overline{x} \stackrel{\Delta}{=} \arg\max_{x\in X}\ (\|Ax\|_2 : \|x\|_2 = 1)$$

i.e.

$$\|A\|_2 = \|A\overline{x}\|_2 \tag{10.20}$$

where

$$\|Ax\|_2^2 = \sum_{i=1}^{k}(a_{i1}x^{(1)} + a_{i2}x^{(2)} + \ldots + a_{ik}x^{(k)})^2 = \sum_{i=1}^{k}\sum_{l=1}^{k}\sum_{j=1}^{k} a_{il}a_{ij}x^{(l)}x^{(j)}\ . \tag{10.21}$$

Suppose that there exist l and j such that $\overline{x}^{(l)} > 0$ and $\overline{x}^{(j)} < 0$. Then from (10.21) under the assumption about the entries of A

$$\|A\hat{x}\|_2 > \|A\overline{x}\|_2 \tag{10.22}$$

where $\hat{x}^{(i)} = \overline{x}^{(i)}$ for $i \neq j$ and $\hat{x}^{(i)} = -\overline{x}^{(i)}$ for $i = j$. From (10.22) we see that $\overline{x}$ does not maximise $\|Ax\|_2$. Hence

$$\bigwedge_{l,j}\ \overline{x}^{(l)}\ \overline{x}^{(j)} \geq 0 \tag{10.23}$$

and from (10.21) it follows that the norm (10.20) is an increasing function of its entries, which proves the lemma. □

Theorem 10.6. Assume that A has distinct eigenvalues. Then the system (10.16) with the uncertainty (10.17) is GAS if

$$\alpha + \|(M^{-1})^{+} A_M M^{+}\| < 1 \tag{10.24}$$

where $\|\cdot\|$ is one of the norms (10.6), (10.7), (10.8), M is the modal matrix of A (i.e. the columns of M are the eigenvectors of A) and

$$\alpha = \max_i\ |\lambda_i(A)|\ .$$

Proof: Let us use Theorem 10.2 with $P = M$ and the equality $M^{-1}AM = \operatorname{diag}[\lambda_1(A), \lambda_2(A), ..., \lambda_k(A)]$. Then

$$\| M^{-1}(A+\overline{A}_n)M \| = \| M^{-1}AM + M^{-1}\overline{A}_n M \| \le \alpha + \| M^{-1}\overline{A}_n M \|. \quad (10.25)$$

It is easy to see that for any matrices A and B

$$(AB)^+ \le A^+B^+. \quad (10.26)$$

It is known (see [76]) that for any matrix A

$$\| A \|_2 \le \| A^+ \|_2. \quad (10.27)$$

For the norms (10.7) and (10.8) the equality $\| A \| = \| A^+ \|$ follows directly from the definitions of the norms. Then, using (10.25), (10.26), (10.27) (or the equality for the norms $\| \cdot \|_1$, $\| \cdot \|_\infty$), Lemma 1 and (10.17), we obtain

$$\| M^{-1}(A+\overline{A}_n)M \| \le \alpha + \| (M^{-1}\overline{A}_n M)^+ \| \le \alpha + \| (M^{-1})^+ A_M M^+ \|.$$

Finally, using Theorem 10.2 yields the desired result. □

The result (10.24) for $\| \cdot \|_2$ and the linear time-varying systems was given in [93] and used in [102]. Now it has been shown that by using Theorem 10.2 based on the general principle of the contraction mapping it is possible to obtain in a much simpler way (much shorter proof of the stability condition) a more general result for non-linear and time-varying systems with different forms of the norm.

Corollary 1. Assume that A has distinct eigenvalues and all eigenvalues of $(M^{-1})^+ A_M M^+$ are real. Then the system (10.16) with the uncertainty (10.17) is GAS if

$$\alpha + \lambda_{\max}[(M^{-1})^+ A_M M^+] < 1 \quad (10.28)$$

where $\lambda_{\max}$ is the maximum eigenvalue of $(M^{-1})^+ A_M M^+$.

Proof: Let N be a diagonal matrix with real positive entries. Then MN is also the modal matrix of A. It is known (see [93]) that

$$\inf_N \| N^{-1}(M^{-1})^+ A_M M^+ N \|_2 = \lambda_{\max}[(M^{-1})^+ A_M M^+]. \quad (10.29)$$

Condition (10.28) follows from (10.24) and (10.29). □

Theorem 10.7. If A is a symmetric matrix and

$$\alpha + \| A_M \|_2 < 1 \quad (10.30)$$

then the system (10.16) with the uncertainty (10.17) is GAS.

Proof: If A is a symmetric matrix then

$$\| A \|_2 = \max_i \ | \lambda_i(A) | = \alpha . \tag{10.31}$$

Using (10.31), (10.27) and Lemma 1, we obtain

$$\| A + \overline{A}_n \|_2 \le \| A \|_2 + \| \overline{A}_n \|_2 \le \alpha + \| \overline{A}_n^+ \|_2 \le \alpha + \| A_M \|_2 .$$

Finally, using Theorem 10.1 yields the desired result. □

Theorem 10.8. If there exists a non-singular matrix $P \in R^{k \times k}$ such that

$$\| (P^{-1})^+ (A^+ + A_M) P^+ \| < 1 \tag{10.32}$$

where $\| \cdot \|$ is one of the norms (10.6), (10.7) and (10.8), then the system (10.16) with the uncertainty (10.17) is GAS.

Proof: Using (10.27) (or the equality for the norms $\| \cdot \|_1$, $\| \cdot \|_\infty$), (10.26) and Lemma 1, we obtain

$$\| P^{-1}(A + \overline{A}_n) P \| \le \| (P^{-1})^+ (A^+ + \overline{A}_n^+) P^+ \| \le \| (P^{-1})^+ (A^+ + A_M) P^+ \| .$$

Consequently, (10.32) implies the inequality $\| P^{-1}(A + \overline{A}_n) P \| < 1$ and according to Theorem 10.2 the system is GAS. □

Remark 10.2. In particular we can apply the diagonal positive matrix P. If $P = I$ (identity matrix) then (10.32) becomes

$$\| A^+ + A_M \| < 1 . \qquad \square \tag{10.33}$$

Condition (10.33) may be easily extended to the following form.

Corollary 2. Let us denote the entries of $\overline{A}_n$ by $\overline{a}_{ij,n}$ and assume that the uncertainty is

$$\bigwedge_{i,j=1,2,\ldots,k} \ \bigwedge_{n \ge 0} \ \bigwedge_{x \in X} \underline{a}_{ij} \le \overline{a}_{ij,n} \le \overline{a}_{ij}$$

where $\underline{a}_{ij}$ and $\overline{a}_{ij}$ are given numbers. Then the system (10.16) is GAS if $\| \widetilde{A} \| < 1$ where $\| \cdot \|$ is one of the norms (10.6), (10.7) and (10.8), the entries of $\widetilde{A}$ are as follows:

$$\widetilde{a}_{ij} \triangleq \max(| a_{ij} + \underline{a}_{ij} |, | a_{ij} + \overline{a}_{ij} |)$$

and a_{ij} are the entries of A.

Proof: Using (10.27) (or the equality for the norms $\| \cdot \|_1$, $\| \cdot \|_\infty$), Lemma 1 and the inequality

$$\bigwedge_{n \ge 0} | a_{ij} + \overline{a}_{ij,n} | \le \widetilde{a}_{ij} ,$$

we obtain

$$\| A+\overline{A}_n \| \le \| (A+\overline{A}_n)^{+} \| \le \| \widetilde{A} \|$$

and the condition $\| \widetilde{A} \| < 1$ follows from Theorem 10.1. □

Remark 10.3. In the case $\overline{a}_{ij} = a_{ij,M}$, $\underline{a}_{ij} = -a_{ij,M}$ where $a_{ij,M}$ are the entries of A_M, it is easy to see that if there exist negative entries of A then the condition in Corollary 2 may be less conservative than (10.33). □

Remark 10.4. Bauer et al. in [6] considered the linear time-varying system with $\overline{A}_n$ belonging to the set of matrices

$$A^{\mathrm{I}} \stackrel{\Delta}{=} \{\overline{A}_n : \ \overline{A}_n = \sum_{l=1}^{p} \lambda_l A_l \,, \quad \lambda_l \in [\underline{\lambda}_l, \overline{\lambda}_l], \ A_l \in R^k \times R^k, \ l = 1,\dots, p\}$$

and used the sufficient stability condition $\| A+\overline{A}_n \| < 1$ for each $\overline{A}_n$ obtained for the sequences λ_l such that $\lambda_l = \underline{\lambda}_l$ or $\lambda_l = \overline{\lambda}_l$, and with the norm $\| \cdot \|_1$ or $\| \cdot \|_\infty$. It is easy to note that the uncertainty formed by A^{I} may be reduced to the uncertainty in Corollary 2. Consequently, the condition in Corollary 2 (obtained in a very simple way) may be considered as an extension of the above condition for a non-linear time-varying system and the norm $\| \cdot \|_2$. □

Remark 10.5. For a linear system with $\overline{A}_n = \overline{A}(c_n)$ and the uncertainty defined by Δ_c in (10.14), if

$$\underline{a}_{ij} = \min_{c \in \Delta_c} \overline{a}_{ij}(c) \,, \qquad \overline{a}_{ij} = \min_{c \in \Delta_c} \overline{a}_{ij}(c)$$

then the condition in Corollary 2 may be applied. A similar condition was obtained in a rather complicated way (see [1]) for a special form of the function $A(c)$ and for

$$\| A \|_P \stackrel{\Delta}{=} \| P^{-1}(A+\overline{A}_n)P \|_2$$

i.e. for $\| x \| = x^{\mathrm{T}} Q x$ where $Q = (P^{-1})^{\mathrm{T}} P^{-1}$ is a positive definite matrix. □

Corollary 3. If A is a symmetric matrix and

$$\alpha + \beta < 1 \tag{10.34}$$

then the system (10.16) with the uncertainty (10.18) for the norm $\| \cdot \|_2$ is GAS. □
The condition (10.34) follows directly from the proof of Theorem 10.7.

Corollary 4. If A and $\overline{A}_n$ are symmetric matrices and

$$\alpha + \overline{\beta} < 1 \tag{10.35}$$

then the system (10.16) with the uncertainty (10.19) is GAS. □

The condition (10.35) follows directly from (10.31) for matrix $\overline{A}_n$ and from the proof of Theorem 10.7, i.e.

$$\| A \|_2 + \| \overline{A}_n \|_2 = \alpha + \max_i \; | \lambda_i(\overline{A}_n) | \; \le \alpha + \overline{\beta} \, . \qquad (10.36)$$

Remark 10.6. It is worth noting that the conditions (10.30), (10.32), (10.34) and (10.35) do not require the assumption about distinct eigenvalues of A. Condition (10.33) does not require any additional assumption about A and from the computational point of view it may be easy to apply (specially for the norms $\| \cdot \|_1$ and $\| \cdot \|_\infty$). The example in the next section shows that the condition (10.33) may be less conservative than condition (10.28). □

10.3.2 Multiplicative Uncertainty

In this case $A_n = A \cdot \overline{A}_n$ where $A, \; \overline{A}_n \in R^{k \times k}$, i.e.

$$x_{n+1} = A\overline{A}_n x_n \, , \qquad (10.37)$$

the uncertainty concerns the matrix $\overline{A}_n$ and is formulated by one of the three forms denoted by (10.17), (10.18) and (10.19).

Theorem 10.9. Assume that A has distinct eigenvalues. Then the system (10.37) with the uncertainty (10.17) is GAS if

$$\alpha \, \| (M^{-1})^+ A_M M^+ \| < 1 \qquad (10.38)$$

where $\| \cdot \|$ is one of the norms (10.6), (10.7), (10.8), M is the modal matrix of A and

$$\alpha = \max_i \; | \lambda_i(A) | \, .$$

If all eigenvalues of $(M^{-1})^+ A_M M^+$ are real then (10.38) may be reduced to

$$\alpha \, \lambda_{\max} [(M^{-1})^+ A_M M^+] < 1 \, . \qquad (10.39)$$

Proof: Using (10.26), (10.27) (or the equality for $\| \cdot \|_1$, $\| \cdot \|_\infty$) and Lemma 1, we obtain

$$\| M^{-1} A\overline{A}_n M \| = \| M^{-1} AMM^{-1} \overline{A}_n M \| \le \| M^{-1} AM \| \cdot \| M^{-1} \overline{A}_n M \|$$

$$\le \alpha \, \| (M^{-1} \overline{A}_n M)^+ \| \le \alpha \, \| (M^{-1})^+ A_M M^+ \| \, .$$

Finally, using Theorem 10.2 for $P = M$ yields condition (10.38). By applying the optimal matrix N as in the proof of Corollary 1 we obtain condition (10.39). □

Theorem 10.10. Assume that A is a symmetric matrix. Then the system (10.37) is GAS if

1. The uncertainty has the form (10.17) and

$$\alpha \| A_M \|_2 < 1 . \tag{10.40}$$

2. The uncertainty has the form (10.18) with the norm $\| \cdot \|_2$ and

$$\alpha\beta < 1 . \tag{10.41}$$

3. The uncertainty has the form (10.19), $\overline{A}_n$ is a symmetric matrix and

$$\alpha\overline{\beta} < 1 . \tag{10.42}$$

Proof: Using (10.31), (10.27) and Lemma 1, we obtain

$$\| A\overline{A}n \|_2 \le \| A \|_2 \cdot \| \overline{A}_n \|_2 \le \alpha \| \overline{A}_n^+ \|_2 \le \alpha \| A_M \|_2 . \tag{10.43}$$

Finally, using Theorem 10.1 yields the condition (10.40). From (10.43)

$$\| A\overline{A}_n \|_2 \le \| A \|_2 \cdot \| \overline{A}_n \|_2 \le \alpha\beta ,$$

which proves the condition (10.41). From (10.43) for the symmetric matrix $\overline{A}_n$ we obtain

$$\| A\overline{A}_n \|_2 \le \alpha \cdot \max_i | \lambda_i(\overline{A}_n) | \le \alpha\overline{\beta}$$

and consequently the condition (10.42). □

Theorem 10.11. If there exists a non-singular matrix $P \in R^{k\times k}$ such that

$$\| (P^{-1})^+ A^+ A_M P^+ \| < 1 \tag{10.44}$$

where $\| \cdot \|$ is one of the norms (10.6), (10.7), (10.8), then the system (10.37) with the uncertainty (10.17) is GAS.

Proof: Using (10.27) (or the equality for the norms $\| \cdot \|_1, \| \cdot \|_\infty$), (10.26) and Lemma 1 we obtain

$$\| P^{-1}(A\overline{A}_n)P \| \le \| (P^{-1})^+ A^+ \overline{A}_n^+ P^+ \| \le \| (P^{-1})^+ A^+ A_M P^+ \| .$$

Consequently (10.32) implies the inequality $\| P^{-1}(A\overline{A}_n)P \| \le 1$ and according to Theorem 10.2 the system is GAS. □

Remark 10.7. In particular we can apply the diagonal positive matrix P. If $P = I$ then (10.44) becomes

$$\| A^+ A_M \| < 1 . \quad \square \tag{10.45}$$

Remark 10.8. It is easy to note that for $A_n = \overline{A}_n A$ we obtain the same conditions

(10.38), (10.39), (10.40), (10.41), and the condition analogous to (10.44) has the form

$$\| (P^{-1})^{+} A_M A^{+} P^{+} \| < 1 . \qquad \square \quad (10.46)$$

Let us recall that all the conditions for additive and multiplicative uncertainty concern time-varying and non-linear systems, i.e. $A_n = A(c_n, x_n)$. For example, take into consideration the system (10.16) and the condition (10.24), and use the notation from the previous section. In this case

$$\mathcal{A} = \{\hat{A} \in R^{k \times k} : \| M^{-1} \hat{A} M \| < 1\} , \qquad (10.47)$$

$$D_c = \{c \in C : \bigwedge_{x \in X} \| M^{-1} [A + \overline{A}(c,x)] M \| < 1\} , \qquad (10.48)$$

$$\mathcal{A}_u = \{\hat{A} \in R^{k \times k} : (\hat{A} - A)^{+} \le A_M\} , \qquad (10.49)$$

$$\Delta_c = \{c \in C : \bigwedge_{x \in X} [\overline{A}(c,x)]^{+} \le A_M\} . \qquad (10.50)$$

If the condition (10.24) is satisfied then $\mathcal{A}_u \subseteq \mathcal{A}$ (or $\Delta_c \subseteq D_c$) and the system is GAS.

Suppose that $A(b)$ depends on the vector b which may be chosen by a designer. Then the vector b should be chosen from the set

$$D_b = \{b \in B : \mathcal{A}_u (б) \subseteq \mathcal{A}\}$$

where the sets $\mathcal{A}(b)$ and $\mathcal{A}_u(b)$ depend on b according to (10.47) and (10.49) ($\mathcal{A}$ depends on b because M depends on b). The same considerations may be taken for the other conditions in this section.

Corollary 5. The non-linear system

$$x_{n+1} = A\, \overline{A}(x_n)\, x_n \qquad (10.51)$$

with the uncertainty

$$\bigwedge_{x \in X} \overline{A}(x)^{+} \le A_M$$

is GAS if there exists a non-singular matrix $P \in R^{k \times k}$ such that

$$\| (P^{-1})^{+} A^{+} A_M P^{+} \| < 1 \qquad (10.52)$$

where $\| \cdot \|$ is one of the norms (10.6), (10.7) and (10.8). For the system

$x_{n+1}=\overline{A}(x_n)A\,x_n$ the analogous condition has the form (10.46). □

The corollary follows directly from Theorem 10.11 and Remark 10.8. Let us note that (10.51) may be presented in the form

$$x_{n+1}=A\,\phi(x_n) \tag{10.53}$$

where $\phi(x_n)=\overline{A}(x_n)\,x_n$ and for $\|\cdot\|_2$ (10.52) is reduced to the following condition: if there exists a positive definite matrix Q such that

$$(A^{+}A_M)^{\mathrm{T}}Q(A^{+}A_M)-Q<0 \tag{10.54}$$

then the system is GAS (see Remark 10.1). For

$$\phi(x)=\begin{bmatrix}\phi_1(x^{(1)})\\ \phi_2(x^{(2)})\\ \cdot\\ \cdot\\ \cdot\\ \phi_k(x^{(k)})\end{bmatrix}$$

i.e. for the diagonal matrix $\overline{A}(x_n)$ in (10.50), sufficient stability conditions in the form (10.54) with A in place of A^{+} have been presented in [70]. The conditions (10.52) and (10.46) may be considered as a generalization of these results for a positive matrix A, obtained in a simple way based on the general principle of the contraction mapping.

It is worth noting that the conditions presented above for time-varying parameters are valid for a particular case when $c_n=\text{const.}=c$, i.e. for the system with a constant unknown parameter belonging to the given set $\Delta_c\subset C$. Now the condition (10.13) takes a form

$$\bigwedge_{c\in\Delta_c}\ \bigwedge_{x\in X} A(c,x)\in\mathcal{A}_u$$

and in the other conditions the expression "for every $n\geq 0$" should be replaced by "for every $c\in\Delta_c$".

10.4 Examples

To illustrate the general conditions and statements presented in the previous section let us consider very simple examples with the notation $A(c_n,x_n)=A_n$ and then special cases for special forms of the matrix $A(c_n,x_n)$.

Example 10.1.
Let in (10.16) and (10.17) $k=2$,

$$A=\begin{bmatrix} a_{11}+b & 0 \\ a_{21}+b & a_{22} \end{bmatrix}, \qquad A_M=\begin{bmatrix} a_{M11} & a_{M12} \\ a_{M21} & a_{M22} \end{bmatrix},$$

$a_{11}, a_{21}, a_{22}, b>0.$
Applying the condition (10.33) with the norm $\|\cdot\|_1$ yields

$$a_{11}+b+a_{M11}+a_{M12}<1,$$

$$a_{21}+b+a_{M21}+a_{22}+a_{M22}<1$$

and finally

$$b<1-\max\{(a_{11}+a_{M11}+a_{M12}), (a_{21}+a_{22}+a_{M21}+a_{M22})\}. \quad (10.55)$$

Let us now apply the condition (10.24) with the norm $\|\cdot\|_1$. We have $\lambda_1(A)=a_{11}+b$, $\lambda_2(A)=a_{22}$, $\lambda_{\max}(A)=\max\,(a_{11}+b, a_{22})$. It is easy to show that

$$M=\begin{bmatrix} 1 & 0 \\ s & 1 \end{bmatrix}$$

with

$$s=\frac{a_{21}+b}{a_{11}+b-a_{22}}$$

is a modal matrix of A and

$$(M^{-1})^{+}A_M M^{+}=\begin{bmatrix} a_{M11}+a_{M12} & a_{M12} \\ a_{M11}|s|+a_{M12}|s|^2+a_{M21}+a_{M22}|s| & a_{M12}|s|+a_{M22} \end{bmatrix}. \quad (10.56)$$

Suppose that $a_{11}\geq a_{22}$, i.e. $\alpha=a_{11}+b$. Applying (10.24), we obtain

$$b\leq 1-a_{11}-\max\{[a_{M11}+2a_{M12}], [a_{M11}|s| + a_{M12}(|s|^2+|s|)+a_{M21}+a_{M22}(|s|+1)]\}. \quad (10.57)$$

Since s depends on b, the final condition for b may be very complicated. To show that the condition (10.57) may be more conservative than (10.55) assume that $a_{21}=a_{11}-a_{22}$, i.e. $s=1$. Then (10.55) and (10.57) become

$$b<1-a_{11}-\max\{a_{M11}+a_{M12}, a_{M21}+a_{M22}\}, \quad (10.58)$$

$$b < 1 - a_{11} - (a_{M11} + 2a_{M12} + a_{M21} + 2a_{M22}). \tag{10.59}$$

Let us now use condition (10.28). The eigenvalues of the matrix (10.56) are

$$\lambda_{1,2} = \frac{a_{M11} + a_{M12}(1+|s|) + a_{M22} \pm \sqrt{[a_{M11} + a_{M12}(1+|s|) + a_{M22}]^2 + 4e}}{2} \tag{10.60}$$

where

$$e = a_{M12}^2(|s|^2 - |s|) + a_{M12}a_{M21} + a_{M12}a_{M22}(|s|-1) - a_{M11}a_{M22}$$

and condition (10.28) becomes

$$b < 1 - a_{11} - \lambda_{\max} \tag{10.61}$$

where $\lambda_{\max}$ is obtained by putting $+$ in the numerator of (10.60). To compare it with (10.58) and (10.59) let us put $s = 1$. Then

$$e = a_{M12}a_{M21} - a_{M11}a_{M22}.$$

If $e \geq 0$ and $a_{M11} + 2a_{M12} > a_{M21}$ then

$$\lambda_{\max} \geq a_{M11} + 2a_{M12} + a_{M22} > a_{M21} + a_{M22}$$

and the condition (10.61) is more conservative than (10.58). The condition (10.59) may be more conservative than (10.61) but it is easier to obtain. When $a_{M21} = a_{M22} = 0$, (10.59) and (10.61) give the same result.
For numerical data $a_{11} = 0.4$, $a_{21} = 0.3$, $a_{22} = 0.1$, $a_{M11} = 0.1$, $a_{M12} = 0.1$, $a_{M21} = 0.1$ and $a_{M22} = 0.05$ we obtain

from condition (10.33), i.e. from (10.55): $b < 0.4$,

from condition (10.24), i.e. from (10.57): $b < 0.1$,

from condition (10.28), i.e. from (10.61): $b < 0.24$.

For $a_{11} = 0.65$ from (10.55) we obtain $b < 0.15$; positive b satisfying condition (10.24) or condition (10.28) does not exist.

The obtained conditions for b may be applied to different forms of the matrix $A(c_n, x_n)$. Let us list the typical cases.

1. Linear time-varying system

$$x_{n+1} = \begin{bmatrix} a_{11} + b + c_n^{(1)} & c_n^{(2)} \\ a_{21} + b + c_n^{(3)} & a_{22} + c_n^{(4)} \end{bmatrix} x_n$$

with the uncertainties

$$\bigwedge_{n\geq 0}[(|c_n^{(1)}|\leq a_{M11})\wedge(|c_n^{(2)}|\leq a_{M12})\wedge(|c_n^{(3)}|\leq a_{M21})\wedge(|c_n^{(4)}|\leq a_{M22})].$$

Now $c_n^{\mathrm{T}}=[c_n^{(1)}\ c_n^{(2)}\ c_n^{(3)}\ c_n^{(4)}]$.

2. Non-linear system

$$x_{n+1}^{(1)}=(a_{11}+b)x_n^{(1)}+f_1^{(1)}(x_n^{(1)})+f_2^{(1)}(x_n^{(2)}),$$

$$x_{n+1}^{(2)}=(a_{21}+b)x_n^{(1)}+a_{22}x_n^{(2)}+f_1^{(2)}(x_n^{(1)})+f_2^{(2)}(x_n^{(2)})$$

with the uncertainties

$$\bigwedge_{-\infty<x^{(1)}<\infty}\left(\left|\frac{f_1^{(1)}(x^{(1)})}{x^{(1)}}\right|\leq a_{M11}\right)\wedge\left(\left|\frac{f_1^{(2)}(x^{(1)})}{x^{(1)}}\right|\leq a_{M21}\right),\quad(10.62)$$

$$\bigwedge_{-\infty<x^{(2)}<\infty}\left(\left|\frac{f_2^{(1)}(x^{(2)})}{x^{(2)}}\right|\leq a_{M12}\right)\wedge\left(\left|\frac{f_2^{(2)}(x^{(2)})}{x^{(2)}}\right|\leq a_{M22}\right).\quad(10.63)$$

For $x=0$ one should put

$$\lim_{x\to 0}\frac{f(x)}{x}$$

under the assumption that the limit exists.

3. Non-linear time-varying system

$$x_{n+1}^{(1)}=(a_{11}+b)x_n^{(1)}+f_1^{(1)}(c_n^{(1)},x_n^{(1)})+f_2^{(1)}(c_n^{(2)},x_n^{(2)}),$$

$$x_{n+1}^{(2)}=(a_{21}+b)x_n^{(1)}+a_{22}x_n^{(2)}+f_1^{(2)}(c_n^{(3)},x_n^{(1)})+f_2^{(2)}(c_n^{(4)},x_n^{(2)}),$$

with the uncertainties analogous to the statements (10.62) and (10.63) which should be satisfied for every $n\geq 0$. For example

$$\bigwedge_{n\geq 0}\ \bigwedge_{-\infty<x^{(1)}<\infty}\left(\left|\frac{f_1^{(1)}(c_n^{(1)},x^{(1)})}{x^{(1)}}\right|\leq a_{M11}\right)\quad(10.64)$$

means that the function $f_1^{(1)}(c^{(1)},x^{(1)})$ and the sequence $c_n^{(1)}$ are such that (10.64) is satisfied, e.g. the function

$$f_1^{(1)}(c^{(1)},x^{(1)}) = \begin{cases} c^{(1)}[1-\exp(-2x^{(1)})] & \text{for } x^{(1)} \geq 0 \\ c^{(1)}[\exp(2x^{(1)})-1] & \text{for } x^{(1)} < 0 \end{cases}$$

and the sequence $c_n^{(1)}$ such that

$$\bigwedge_{n \geq 0} |c_n^{(1)}| \leq \frac{1}{2} a_{M11}$$

satisfy the condition (10.64). For the function $f_1^{(1)}(c^{(1)},x^{(1)}) = c^{(1)} \overline{f}_1(x^{(1)})$, if

$$\bigwedge_{n \geq 0} |c_n^{(1)}| \leq \gamma, \qquad \bigwedge_{-\infty < x^{(1)} < \infty} \left| \frac{\overline{f}_1(x^{(1)})}{x^{(1)}} \right| \leq \delta$$

and $\gamma \cdot \delta = a_{M11}$ then the condition (10.64) is satisfied. □

Example 10.2.

Consider the system (10.37) with the uncertainty (10.17) and the matrices A, A_M the same as in Example 10.1. The application of the condition (10.45) with the norm $\|\cdot\|_1$ gives

$$b < \frac{1}{a_{M11} + a_{M12}} - \max\,(a_{11}, a_{22} \frac{a_{M21} + a_{M22}}{a_{M11} + a_{M12}} + a_{21})\,. \tag{10.65}$$

Let us assume $a_{11} \geq a_{22}$, $a_{21} = a_{11} - a_{22}$ and use the results of Example 10.1. Applying the condition (10.38) with the norm $\|\cdot\|_1$ and the condition (10.39) gives

$$b < \frac{1}{a_{M11} + 2a_{M12} + a_{M21} + 2a_{M22}} - a_{11}\,, \tag{10.66}$$

$$b < \frac{1}{\lambda_{\max}} - a_{11}\,, \tag{10.67}$$

respectively. If $e > 0$ and $a_{M11} + a_{M12} > a_{M21} + a_{M22}$ then

$$\lambda_{\max} \geq a_{M11} + 2a_{M12} + a_{M22}\,,$$

the condition (10.65) takes the form

$$b < \frac{1}{a_{M11} + a_{M12}} - a_{11}$$

and the condition (10.67) is more conservative than (10.65).

For the same numerical data as in Example 10.1 we obtain

from condition (10.45): $b < 4.6$,

from condition (10.38): $b < 1.6$,

from condition (10.39): $b < 2.37$.

The condition obtained for b may be applied to different forms of the matrix $A(c_n, x_n)$ as in Example 10.1. □

10.5 An Approach Based on Random Variables

Consider a non-linear time-varying system described by

$$x_{n+1} = A(c_n, b, x_n)x_n \tag{10.68}$$

where $x_n \in X$ is the state vector, $c_n \in C$ is the vector of time-varying parameters, $b \in B$ is the vector of constant parameters; $X = R^k$, C and B are the real number vector space. The matrix

$$A(c_n, b, x_n) = [a_{ij}(c_n, b, x_n)] \in R^{k \times k}.$$

Assume that for every $c \in C$ and $b \in B$ the equation $x = A(c, b, x)$ has a unique solution $x_e = \overline{0}$ (the vector with zero components). According to Definition 10.1, the system (10.68) (or the equilibrium state x_e) is globally asymptotically stable (GAS) iff x_n converges to $\overline{0}$ for any x_0.

Assume now that the parameters c_n and b are unknown and the uncertainties concerning c_n and b are formulated as follows:

1. $$\bigwedge_{n \geq 0} (c_n \in \Delta_c) \tag{10.69}$$

where Δ_c is a given set in C.

2. b is a value of random variable $\widetilde{b}$ described by the probability density $f_b(b)$, and $f_b(b)$ is known.

Definition 10.2. For the fixed b, the uncertain system (10.68), (10.69) is GAS iff the system (10.68) is GAS for every sequence c_n satisfying (10.69). □

Denote by P_s the probability that the uncertain system (10.68), (10.69) is GAS. The problem considered here consists in the determination of an estimation of P_s [45, 52]. Let $W(b)$ and $V(b)$ denote properties concerning b such that $W(b)$ is a sufficient condition and $V(b)$ is a necessary condition of the global asymptotic stability for the system (10.68), (10.69), i.e.

$W(b) \rightarrow$ the system (10.68), (10.69) is GAS,

the system (10.68), (10.69) is GAS $\rightarrow V(b)$.

Then

$$P_w \le P_s \le P_v \tag{10.70}$$

where

$$P_w = \int\limits_{D_{bw}} f_b(b)db\,, \qquad P_v = \int\limits_{D_{bv}} f_b(b)db\,, \tag{10.71}$$

$$D_{bw} = \{\, b \in B\colon W(b)\}, \qquad D_{bv} = \{\, b \in B\colon V(b)\},$$

P_w is the probability that the sufficient condition is satisfied and P_v is the probability that the necessary condition is satisfied. In general, $D_{bw} \subseteq D_{bv}$ and $D_{bv} - D_{bw}$ may be called a "grey zone", which is a result of an additional uncertainty caused by the fact that $W(b) \neq V(b)$. The condition $V(b)$ may be determined as a negation of a sufficient condition that the system is not GAS, i.e. such a property $V_{neg}(b)$ that

$$V_{neg}(b) \rightarrow \text{there exists } c_n \text{ satisfying (10.69)}$$

$$\text{such that (10.68) is not GAS.} \tag{10.72}$$

To estimate the probability P_s according to (10.70), it is necessary to determine the conditions $W(b)$ and $V(b)$. The sufficient conditions for the uncertain system under consideration may have forms presented in the previous section, based on a general form (10.3). It is not possible to determine an analogous general necessary condition $V(b)$ or a sufficient condition of non-stability $V_{neg}(b)$. Particular forms of necessary conditions are proposed here by the following theorems.

Theorem 10.12. If the system (10.68), (10.68) is GAS then

$$\bigvee_{c \in \Delta_c} \bigvee_{x \in X} \lambda_{\min}[A^{\mathrm{T}}(c,b,x)A(c,b,x)] < 1 \tag{10.73}$$

where $\lambda_{\min}$ is the minimum eigenvalue of the matrix $A^{\mathrm{T}}A$.

Proof. It is easy to show that if (10.73) is not satisfied, i.e. if

$$\bigwedge_{c \in \Delta_c} \bigwedge_{x \in X} [\lambda_{\min}[A^{\mathrm{T}}(c,b,x)A(c,b,x)] \ge 1] \tag{10.74}$$

then the system (10.68), (10.69) is not GAS. It follows from (10.74) that $A^{\mathrm{T}}A - I$ is a non-negative definite matrix. Then

$$\bigwedge_{c \in \Delta_c} \bigwedge_{x \in X} [x_n{}^{\mathrm{T}}A^{\mathrm{T}}(c_n,b,x_n)A(c_n,b,x_n)x_n \ge 0]$$

and $||x_{n+1}||_2 \geq ||x_n||_2$ for every n. Consequently, for every c_n satisfying (10.69) the system (10.68) is not GAS and, according to Definition 10.2 and the statement (10.72), the system (10.68), (10.69) is not GAS, $V_{neg}(b)$ is defined by (10.74) and $V(b) = \neg V_{neg}(b)$ is defined by (10.73). □

Theorem 10.13. Assume that the entries of the matrix $A(c,b,x)$ are non-negative for every $c \in \Delta_c$ and every $x \in X$. If the system (10.68), (10.69) is GAS then

$$\bigvee_{c \in \Delta_c} \bigvee_{x \in X} [\min_j \sum_{i=1}^{k} a_{ij}(c,b,x) < 1] \tag{10.75}$$

and

$$\bigvee_{c \in \Delta_c} \bigvee_{x \in X} [\min_i \sum_{j=1}^{k} a_{ij}(c,b,x) < 1]. \tag{10.76}$$

Proof. It is easy to show that if (10.75) or (10.76) is not satisfied, i.e. if

$$\bigwedge_{c \in \Delta_c} \bigwedge_{x \in X} [\min_j \sum_{i=1}^{k} a_{ij}(c,b,x) \geq 1] \tag{10.77}$$

or

$$\bigwedge_{c \in \Delta_c} \bigwedge_{x \in X} [\min_i \sum_{j=1}^{k} a_{ij}(c,b,x) \geq 1] \tag{10.78}$$

then the system (10.68), (10.69) is not GAS. Assume that $x_0 > \overline{0}$ (components of x_0 and consequently of x_n for $n > 0$ are positive). Then it follows from (10.77) that

$$\sum_{i=1}^{k} x_{n+1}^{(i)} = \sum_{j=1}^{k} [\sum a_{ij}(c_n,b,x_n)] x_n^{(j)} \geq \sum_{j=1}^{k} x_n^{(j)}, \tag{10.79}$$

i.e. $||x_{n+1}||_\infty \geq ||x_n||_\infty$ for every n. Consequently, for every sequence c_n satisfying (10.69) and for $x_0 > \overline{0}$, x_n does not converge to $\overline{0}$. This means, according to Definitions 10.1, 10.2 and the statement (10.72), that (10.75) is true, i.e. $V_{neg}(b)$ is defined by (10.77) and $V(b) = \neg V_{neg}(b)$ is defined by (10.75). The proof for (10.76) is analogous. □

Theorem 10.14. Consider the linear, time-varying system

$$x_{n+1} = A(c_n, b) x_n. \tag{10.80}$$

If the system (10.80), (10.69) is GAS then

$$\bigwedge_{c \in \Delta_c} [\max_i | \lambda_i[A(c,b)] | < 1] \tag{10.81}$$

where $\lambda_i(A)$ are the eigenvalues of the matrix A $(i = 1, 2, \ldots, k)$.
Proof. It is enough to note that if there exists $c \in \Delta_c$ such that

$$\max_i |\lambda_i[A(c,b)]| \geq 1 \tag{10.82}$$

and $c = \hat{c}$ satisfies (10.82) then for $c_n = \text{const.} = \hat{c}$ the known necessary and sufficient stability condition for the linear time-invariant system is not satisfied, i.e. the system (10.80) is not GAS. Thus, $V(b)$ is defined by (10.72). □

Remark 10.9. The determination of $D_{bv} = \{b \in B\colon\ V(b)\}$ directly from the necessary conditions $V(b)$ defined by (10.73), (10.75) and (10.76) is rather inconvenient. It is easier to determine

$$D_{b,neg} = \{\, b \in B\colon V_{neg}(b)\,\}$$

using (10.74), (10.77) or (10.78) and $D_{bv} = B - D_{b,neg}$. □

Remark 10.10. The estimation of P_s using (10.70) is better when the sufficient conditions $W(b)$ and/or $V_{neg}(b)$ are less conservative. If $V_{neg}(b)$ is less conservative then the necessary condition $V(b)$ is stronger, the set D_{bv} is smaller and the "grey zone" is smaller. It is easy to see that in general the condition (10.81) is stronger than the condition (10.73), (10.75) or (10.76) for the linear system. It follows from the fact that the condition (10.75), (10.77) or (10.78) assures non-stability *for every sequence* c_n satisfying (10.69), and according to the statement (10.72) it is sufficient that *there exists* c_n satisfying (10.69) such that the system is not GAS. Besides, the condition (10.75), (10.77) or (10.78) assures the inequality $\| x_{n+1} \| \geq \| x_n \|$ for every n, which is not necessary for non-stability. □

Let us consider one of the typical cases of uncertain systems (10.68), (10.69), when

$$\Delta_c = \{c \in C\colon \bigwedge_{x \in X} [\underline{A}(b) \leq A(c,b,x) \leq \overline{A}(b)]\}, \tag{10.83}$$

$\underline{A}(b)$ and $\overline{A}(b)$ are given matrices and the inequality in (10.83) denotes the inequalities for the entries:

$$\underline{a}_{ij}(b) \leq a_{ij}(c,b,x) \leq \overline{a}_{ij}(b). \tag{10.84}$$

The definition (10.83) of the set Δ_c means that if c_n satisfies (10.69) then for every $n \geq 0$

$$\underline{A}(b) \leq A(c_n,b,x_n) \leq \overline{A}(b).$$

If we introduce the notation

$$A(b) = \frac{1}{2}[\underline{A}(b) + \overline{A}(b)],\ A(c,b,x) = A(b) + \overline{A}(c,b,x)$$

then the inequality in (10.83) may be replaced by

$$\overline{A}^{+}(c,b,x) \le A_M(b) \tag{10.85}$$

where $\overline{A}^{+}$ is the matrix obtained by replacing the entries of $\overline{A}$ by their absolute values and $A_M(b) = \overline{A}(b) - A(b)$. Then the inequality (10.85) corresponds to the form (10.17) with $A_M(b)$ in place of A_M. Consequently, we can use the sufficient conditions (10.24) and (10.33):

$$\alpha(b) + \| M^{-1}(b)]^{+} A_M(b) M^{+}(b) \| < 1$$

where $\alpha(b) = \max | \lambda_i[A(b)] |$, and

$$\| A^{+}(b) + A_M(b) \| < 1$$

which for $A(b) \ge 0$ (all entries of $A(b)$ are non-negative) is reduced to

$$\| \overline{A}(b) \| < 1 . \tag{10.86}$$

Let us now turn to the necessary conditions for our case.

Corollary 6. Assume that $\underline{A}(b) \ge 0$. If the system (10.68), (10.83) is GAS then

$$\bigvee_{j} [\sum_{i=1}^{k} \underline{a}_{ij}(b) < 1] \tag{10.87}$$

and

$$\bigvee_{i} [\sum_{j=1}^{k} \underline{a}_{ij}(b) < 1] . \tag{10.88}$$

Proof. It follows from Theorem 10.13 that if (10.77) or (10.78) is satisfied then the system (10.68), (10.69) is not GAS. From (10.83) it follows that the condition

$$\bigwedge_{j} [\sum_{i=1}^{k} \underline{a}_{ij}(b) \ge 1] \tag{10.89}$$

implies (10.77) and the condition

$$\bigwedge_{i} [\sum_{j=1}^{k} \underline{a}_{ij}(b) \ge 1] \tag{10.90}$$

implies (10.78). Consequently, (10.89) or (10.90) is the sufficient condition that the system (10.68), (10.83) is not GAS. Thus inequalities (10.87) and (10.89) are necessary conditions of GAS. □

By choosing different sufficient and necessary conditions we may obtain the different estimations of the probability P_s. For example, if we choose the condition (10.86) with the norm $\|\cdot\|_\infty$ (see (10.8)) and the condition (10.87), i.e. the negation of (10.89), then

$$\int_{D_{bw}} f_b(b)db \le P_s \le \int_{D_{bv}} f_b(b)db$$

where

$$D_{bw} = \{b \in B: \bigwedge_j [\sum_{i=1}^{k} \overline{a}_{ij}(b) < 1]\}, \qquad (10.91)$$

$$D_{bv} = B - D_{b,neg}, \quad D_{b,neg} = \{b \in B: \bigwedge_j [\sum_{i=1}^{k} \underline{a}_{ij}(b) \ge 1]\}. \qquad (10.92)$$

Example 10.3.
Consider an uncertain system (10.68) where $k = 2$ and

$$A(c_n, b, x_n) = \begin{bmatrix} a_{11}(c_n, x_n) + b & a_{12}(c_n, x_n) \\ a_{21}(c_n, x_n) & a_{22}(c_n, x_n) + b \end{bmatrix}$$

with the uncertainty (10.84), i.e. non-linearities and the sequence c_n are such that

$$\bigwedge_{c \in \Delta_c} \bigwedge_{x \in X} [\underline{a}_{ij} \le a_{ij}(c, x) \le \overline{a}_{ij}]. \qquad (10.93)$$

Assume that $b \ge 0$ and $\underline{a}_{ij} \ge 0$. Applying the condition (10.86) with the norm $\|\cdot\|_\infty$ in (10.8) yields

$$\overline{a}_{11} + b + \overline{a}_{21} < 1, \qquad \overline{a}_{12} + \overline{a}_{22} + b < 1$$

and D_{bw} in (10.91) is defined by

$$b < 1 - \max(\overline{a}_{11} + \overline{a}_{21}, \overline{a}_{12} + \overline{a}_{22}). \qquad (10.94)$$

Applying the condition (10.89) yields

$$\underline{a}_{11} + b + \underline{a}_{21} \ge 1, \qquad \underline{a}_{12} + \underline{a}_{22} + b \ge 1.$$

Then $D_{b,neg}$ in (10.92) is determined by

$$b \ge 1 - \min(\underline{a}_{11} + \underline{a}_{21}, \underline{a}_{12} + \underline{a}_{22})$$

and the necessary condition (10.87) defining the set $D_{bv} = B - D_{b,neg}$ is as follows:

$$b < 1 - \min(\underline{a}_{11} + \underline{a}_{21}, \underline{a}_{12} + \underline{a}_{22}). \tag{10.95}$$

For the given probability density $f_b(b)$ we can determine

$$P_w = \int_0^{b_w} f_b(b)db\,, \quad P_v = \int_0^{b_v} f_b(b)db$$

where

$$b_w = 1 - \max(\overline{a}_{11} + \overline{a}_{21}, \overline{a}_{12} + \overline{a}_{22}),$$

$$b_v = 1 - \min(\underline{a}_{11} + \underline{a}_{21}, \underline{a}_{12} + \underline{a}_{22}).$$

For numerical data $a_{11} \in [0.3\,, 0.4]$, $a_{21} \in [0.2, 0.3]$, $a_{12} \in [0.3\,, 0.5]$, $a_{22} \in [0.2\,, 0.3]$ we obtain $b_w = 1 - \max(0.7, 0.8) = 0.2$, $b_v = 1 - \min(0.5, 0.5) = 0.5$.

Assume that b has exponential probability density $f(b) = \lambda e^{-\lambda b}$. Then the estimation of the probability P_s is the following:

$$1 - e^{-\lambda b_w} \le P_s \le 1 - e^{-\lambda b_v}.$$

Substituting $b_w = 0.2$ and $b_v = 0.5$ we obtain

$$0.55 \le P_s \le 0.86 \text{ for } \lambda = 4,$$

$$0.80 \le P_s \le 0.98 \text{ for } \lambda = 8.$$

Assume now that a_{ij} do not depend on x and apply the necessary condition (10.81). In our case

$$\lambda_{1,2} = \frac{1}{2}[-(a_{11} + a_{22} + b) \pm \sqrt{(a_{11} - a_{22})^2 + 4a_{12}a_{21}}\,]$$

and under the assumption $b \ge 0$, $\underline{a}_{ij} \ge 0$, the condition (10.81) is as follows:

$$b < 1 - \frac{1}{2}(a_{11} + a_{22} + \sqrt{(a_{11} - a_{22})^2 + 4a_{12}a_{21}}) \tag{10.96}$$

for every $a_{11} \in [\underline{a}_{11}, \overline{a}_{11}]$, $a_{12} \in [\underline{a}_{12}, \overline{a}_{12}]$, $a_{21} \in [\underline{a}_{21}, \overline{a}_{21}]$, $a_{22} \in [\underline{a}_{22}, \overline{a}_{22}]$. Then

$$b_v = 1 - \frac{1}{2}\max(a_{11} + a_{22} + \sqrt{(a_{11} - a_{22})^2 + 4\overline{a}_{12}\overline{a}_{21}})$$

where max denotes the maximum value for the pairs $(\underline{a}_{11}, \underline{a}_{22})$, $(\underline{a}_{11}, \overline{a}_{22})$, $(\overline{a}_{11}, \underline{a}_{22})$, $(\overline{a}_{11}, \overline{a}_{22})$.

For the numerical data we obtain $b_v = 0.26$ and

$$P_v = 1 - e^{-\lambda \cdot 0.26}.$$

Using the condition (10.81) we obtain the better condition $V(b)$ and consequently, better estimation of P_s :

$$0.55 \le P_s \le 0.64 \quad \text{for } \lambda = 4,$$

$$0.80 \le P_s \le 0.87 \quad \text{for } \lambda = 8.$$

It is worth noting that the conditions obtained for b and the estimations of P_s may be applied to different forms of $a_{ij}(c, x)$ in (10.93). In the linear time-varying system $a_{ij} = c_{ij,n}$, $c^{\mathrm{T}} = [c_{11}, c_{12}, c_{21}, c_{22}]$ and

$$\Delta_c = \{c \in C : \underline{a}_{ij} \le c_{ij} \le \overline{a}_{ij}\}.$$

In the non-linear time-invariant system

$$x_{n+1}^{(1)} = f_1^{(1)}(x_n^{(1)}) + b + f_2^{(1)}(x_n^{(2)}),$$

$$x_{n+1}^{(2)} = f_1^{(2)}(x_n^{(1)}) + f_2^{(2)}(x_n^{(2)}) + b,$$

i.e.

$$a_{ij}(x^{(j)}) = \frac{f_j^{(i)}(x^{(j)})}{x^{(j)}}$$

and the uncertainties concerning the non-linearities are as follows:

$$\bigwedge_{-\infty < x^{(j)} < \infty} [\underline{a}_{ij} \le \frac{f_j^{(i)}(x^{(j)})}{x^{(j)}} \le \overline{a}_{ij}].$$

In the non-linear time-varying system with $f_j^{(i)}(c_{ij,n}, x_n^{(j)})$ the non-linearities and the sequences $c_{ij,n}$ are such that

$$\bigwedge_{n \ge 0} \; \bigwedge_{-\infty < x^{(j)} < \infty} [\underline{a}_{ij} \le \frac{f_j^{(i)}(c_{ij,n}, x^{(j)})}{x^{(j)}} \le \overline{a}_{ij}].$$

For example

$$f_1^{(1)}(x^{(1)}, c_{11}) = \begin{cases} c_{11}[1 - \exp(-2x^{(1)})] & \text{for } x^{(1)} \ge 0 \\ c_{11}[\exp(2x^{(1)}) - 1] & \text{for } x^{(1)} < 0, \end{cases}$$

$$\bigwedge_{n \geq 0} \quad 0 \leq c_{11,n} \leq \frac{1}{2}\bar{a}_{11}.$$

(see the corresponding cases in Example 10.1). □

10.6 An Approach Based on Uncertain Variables

The considerations for the description based on uncertain variables are analogous to those presented in the previous section [52, 56]. Assume that the parameters c_n and b in Equation (10.68) are unknown and the uncertainties concerning c_n and b are formulated as follows:

1.
$$\bigwedge_{n \geq 0} (c_n \in \Delta_c) \tag{10.97}$$

where Δ_c is a given set in C.

2. b is a value of an uncertain variable $\bar{b}$ described by the certainty distribution $h_b(b)$ given by an expert.

Denote by v_s the certainty index that the uncertain system (10.68), (10.97) is GAS. The problem considered here consists in the determination of an estimation of v_s. Using the sets D_{bw} and D_{bv} introduced in the previous section, one obtains

$$v_w \leq v_s \leq v_g$$

where

$$v_w = \max_{b \in D_{bw}} h_b(b), \qquad v_g = \max_{b \in D_{bv}} h_b(b),$$

v_w is the certainty index that the sufficient condition is satisfied and v_g is the certainty index that the necessary condition is satisfied. Precisely speaking, they are the certainty indexes that the respective conditions are satisfied for approximate value of b, i.e. are "approximately satisfied". Choosing different sufficient and necessary conditions presented in the previous sections, we may obtain different estimations of v_s. For example, if we choose the condition (10.86) with the norm $\|\cdot\|_\infty$ (see (10.8)) and the condition (10.87), i.e. the negation of (10.89), then

$$\max_{b \in D_{bw}} h_b(b) \le v_s \le \max_{b \in D_{bv}} h_b(b),$$

where

$$D_{bw} = \{b \in B : \bigwedge_j [\sum_{i=1}^{k} \overline{a}_{ij}(b) < 1]\},$$

$$D_{bg} = B - D_{b,neg}, \qquad D_{b,neg} = \{b \in B : \bigwedge_j [\sum_{i=1}^{k} \underline{a}_{ij}(b) \ge 1]\}.$$

Example 10.4.
Consider an uncertain system the same as in Example 10.3. Assume that b is a value of an uncertain variable $\overline{b}$. For the given certainty distribution $h_b(b)$ we can determine

$$v_w = \max_{0 \le b \le b_w} h_b(b), \quad v_g = \max_{0 \le b \le b_g} h_b(b) \tag{10.98}$$

where

$$b_w = 1 - \max(\overline{a}_{11} + \overline{a}_{21}, \overline{a}_{12} + \overline{a}_{22}),$$

$$b_g = 1 - \min(\underline{a}_{11} + \underline{a}_{21}, \underline{a}_{12} + \underline{a}_{22}).$$

Assume that $h_b(b)$ has triangular form presented in Fig. 10.1. The results obtained from (10.98) for the different cases are as follows:

$$v_w = \begin{cases} 1 & \text{for} \quad b_w \ge d \\ (b_w - d)\gamma^{-1} + 1 & \text{for} \quad d - \gamma \le b_w \le d \\ 0 & \text{otherwise}. \end{cases}$$

$$v_g = \begin{cases} 1 & \text{for} \quad b_g \ge d \\ (b_g - d)\gamma^{-1} + 1 & \text{for} \quad d - \gamma \le b_g \le d \\ 0 & \text{otherwise}. \end{cases}$$

For example, if $d - \gamma \le b_w \le d$ and $b_g \le d$ then the certainty index v_s that the system is GAS satisfies the inequality

$$(b_w - d)\gamma^{-1} + 1 \le v_s \le (b_g - d)\gamma^{-1} + 1. \qquad \square$$

Example 10.5.
To show the role of the shape of $h_b(b)$ let us assume that in Example 10.4 $h_b(b)$ has a parabolic form (Fig. 10.2):

$$h_b(b) = \begin{cases} -\gamma^2 (b-d)^2 + 1 & \text{for } d-\gamma \le b \le d+\gamma \\ 0 & \text{otherwise} \end{cases},$$

$0 < \gamma < d$.

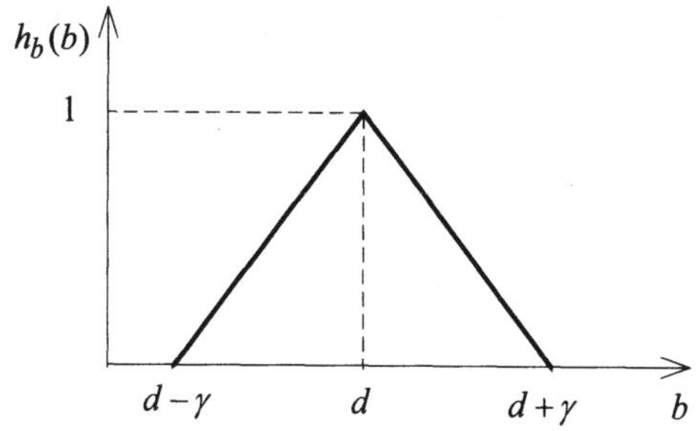

Figure 10.1. Certainty distribution in the example

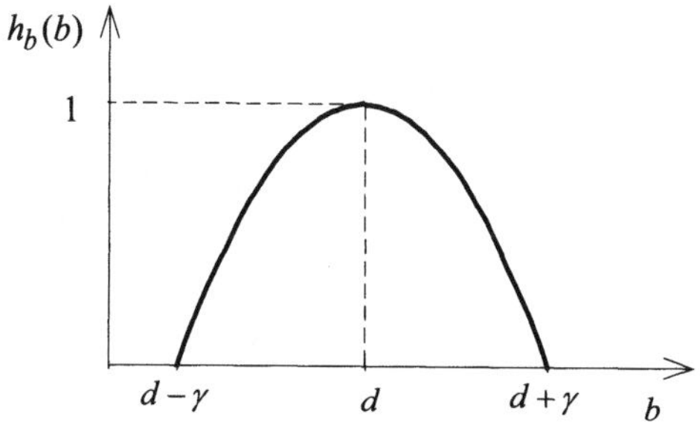

Figure 10.2. Certainty distribution in Example 10.5

The results obtained from (10.98) for the different cases are as follows:

1. For $b_w \ge d$ $\quad v_w = v_g = 1$.

2. For $d - \gamma \le b_w \le d$

$$v_w = -\gamma^{-2}(b_w - d)^2 + 1,$$

$v_g = 1$ for $b_g \ge d$ and $v_g = -\gamma^{-2}(b_g - d)^2 + 1$

for $b_g \le d$.

3. For $b_w \le d-\gamma \quad v_w = 0$,

$$v_g = \begin{cases} 1 & \text{for} \quad b_g \ge d \\ -\gamma^{-2}(b_g - d)^2 + 1 & \text{for} \quad d-\gamma \le b_g \le d \\ 0 & \text{for} \quad b_g \le d-\gamma . \end{cases}$$

For example, if $d-\gamma \le b_w \le d$ and $b_g \le d$ then the certainty index v_s that the system is globally asymptotically stable satisfies the inequality

$$-\gamma^{-2}(b_w - d)^2 + 1 \le v_s \le -\gamma^{-2}(b_g - d)^2 + 1 .$$

For numerical data $\overline{a}_{11} = 0.2$, $\underline{a}_{11} = 0.1$, $\overline{a}_{21} = 0.3$, $\underline{a}_{21} = 0.2$, $\overline{a}_{12} = 0.2$, $\underline{a}_{12} = 0.1$, $\overline{a}_{22} = 0.2$, $\underline{a}_{22} = 0.1$, $d = 0.9$, $\gamma^{-2} = 3$, we obtain $v_w = 0.52$, $v_g = 0.97$ and $0.52 \le v_s \le 0.97$. □

10.7 Stabilization

Consider a system

$$x_{n+1} = A(c_n, x_n, e) \tag{10.99}$$

where $e \in E$ is a vector of parameters which may be chosen by a designer. Now the condition (10.9) has the form

$$\bigwedge_{n \ge 0} \bigwedge_{x \in X} A(c_n, x, e) \in \mathcal{A}$$

and the set $\mathcal{A}_u(e)$ depends on e, i.e. for the fixed e

$$\bigwedge_{n \ge 0} \bigwedge_{x \in X} A(c_n, x, e) \in \mathcal{A}_u(e) .$$

The parameter e should be chosen from the set

$$D_e = \{e \in E : \mathcal{A}_u(e) \subseteq \mathcal{A}\} \tag{10.100}$$

and $D_e \ne \varnothing$ (empty set) is the sufficient *condition of stabilizability*. The set $\mathcal{A}$ may also depend on e when in (10.10) we use $\| P^{-1}\hat{A}P \|$ instead of $\| \hat{A} \|$ and P depends on e. In the formulation with D_c (10.12) and Δ_c (10.14), the set $D_c(e)$ depends on e, i.e. for the fixed e

$$D_c(e) = \{c \in C : \bigwedge_{x \in X} A(c,x,e) \in \mathcal{A}\}$$

and e should be chosen from the set

$$D_e = \{e \in E : \Delta_c \subseteq D_c(e)\}. \tag{10.101}$$

Consider now a closed-loop system (Fig. 10.3) described by

$$x_{n+1} = f_x(c_{xn}, x_n, u_n), \qquad y_n = \eta_x(c_{yn}, x_n) \tag{10.102}$$

$$s_{n+1} = f_s(c_{sn}, s_n, y_n, e_s), \quad u_n = \eta_s(c_{un}, s_n, e_u) \tag{10.103}$$

where in Equations (10.102) x_n, u_n and y_n are the state, input and output vectors, respectively, and in Equations (10.103) s_n, y_n and u_n are the state, input and output vectors, respectively. By substituting u_n from (10.103) into (10.102) and y_n from (10.102) into (10.103), the description (10.102), (10.103) may be reduced to

$$z_{n+1} = f(c_n, z_n, e) \tag{10.104}$$

where $z_n^{\mathrm{T}} = [x_n^{\mathrm{T}} \; s_n^{\mathrm{T}}]$, $c_n^{\mathrm{T}} = [c_{xn}^{\mathrm{T}} \; c_{yn}^{\mathrm{T}} \; c_{sn}^{\mathrm{T}} \; c_{un}^{\mathrm{T}}]$, $e^{\mathrm{T}} = [e_s^{\mathrm{T}} \; e_u^{\mathrm{T}}]$. Under the assumptions for the system (10.104) the same as for the system (10.1) we can consider

$$z_{n+1} = A(c_n, z_n, e) z_n \tag{10.105}$$

and apply to the system (10.105) all former statements and conditions for the system (10.99) with z_n in place of x_n. Let us assume that Equations (10.102) have the form

$$x_{n+1} = E(c_{x1n}, x_n)x_n + F(c_{x2n}, u_n)u_n, \quad y_n = G(c_{yn}, x_n)x_n \tag{10.106}$$

and apply the linear feedback (10.103):

$$s_{n+1} = H(e_{s1})s_n + K(e_{s2})y_n, \qquad u_n = L(e_u)s_n \tag{10.107}$$

where E, F, G, H, K and L are matrices of appropriate dimensions. Then $c^{\mathrm{T}} = [c_{x1}^{\mathrm{T}} \; c_{x2}^{\mathrm{T}} \; c_y^{\mathrm{T}}]$, $e^{\mathrm{T}} = [e_{s1}^{\mathrm{T}} \; e_{s2}^{\mathrm{T}} \; e_u^{\mathrm{T}}]$ and

$$A(c_n, z_n, e) = \begin{bmatrix} E(c_{x1n}, x_n) & F[c_{x2n}, L(e_u)s_n]L(e_u) \\ \\ H(e_{s1}) & K(e_{s2})G(c_{yn}, x_n) \end{bmatrix}. \tag{10.108}$$

The uncertainties may be formulated separately for E, F and G by $\mathcal{E}_u$, $\mathcal{F}_u$ and $\mathcal{G}_u$, respectively, and for c_{x1}, c_{x2} and c_y by Δ_{cx1}, Δ_{cx2} and Δ_{cx2}, respectively. Then

$$\mathcal{A}_u = \{A:\ E \in \mathcal{E}_u \wedge F \in \mathcal{F}_u \wedge G \in \mathcal{G}_u\},$$

$$\Delta_c = \Delta_{cx1} \times \Delta_{cx2} \times \Delta_{cy}\,.$$

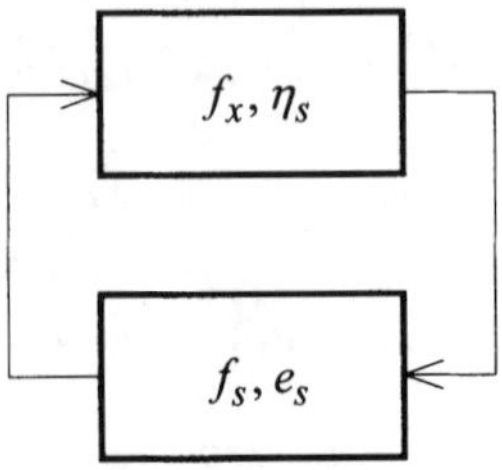

Figure 10.3. Closed-loop system under consideration

Consider a system

$$x_{n+1} = A(c_n, b, x_n, e)x_n \tag{10.109}$$

where $b \in B$ is a vector of unknown parameters. Now the conditions $W(b, e)$ and $V(b, e)$ considered in Sects 10.5 and 10.6 depend on e and, consequently, the probabilities $P_w(e)$, $P_v(e)$ in (10.7) and the certainty indexes v_v, v_g introduced in Sect. 10.6 depend on e. In the deterministic case (i.e. for the known value of b) the stabilization problem consists in the determination of such a set D_e that for every $e \in D_e$ the condition $W(b, e)$ is satisfied. To satisfy the sufficient condition

$$\bigwedge_{c \in \Delta_c} \bigwedge_{x \in X} [\| A(c,b,x,e) \| < 1]$$

(see (10.3)), the parameter e should be chosen from the set

$$D_e(b) = \{e \in E : W(b,e)\} = \{e \in E : \bigwedge_{c \in D_c} \bigwedge_{x \in X} \| A(c,b,x,e) \| < 1\}.$$

In the probabilistic case, a designer may have an influence on the values of P_w and P_v by choosing the value e. The problem of choosing e may be formulated in the following ways:

1. Choose e maximizing $P_w(e)$.
2. Choose e maximizing $P_w(e)$, subject to the constraint $P_v(e) \le \alpha$ where $0 < \alpha < 1$ is given.
3. Choose e to maximizing $P_w(e)$, subject to the constraint $P_v(e) - P_w(e) \le \beta$ where $0 < \beta < 1$ is given.

In the first case we try to maximize the probability that the condition assuring the stability is satisfied. To evaluate the grey zone one should determine $P_v(e^*)$ where

$e^* = \arg\max P_w(e)$. In the second and third case the grey zone is included into the optimization problem in two different ways.

The stabilization consisting in maximizing $P_w(e)$ may be performed by applying a feedback. To illustrate the idea, let us consider a simple case described by

$$x_{n+1} = A(c_n, b, x_n)x_n + du_n, \quad u_n = e^{\mathrm{T}} x_n \tag{10.110}$$

where $u_n \in R^1$ and $d \in R^k$ is a vector of the known parameters. From (10.110)

$$x_{n+1} = \hat{A}(c_n, b, x_n, e)x_n \tag{10.111}$$

where

$$\hat{A}(c_n, b, x_n, e) = A(c_n, b, x_n) + de^{\mathrm{T}},$$

and we may apply to the matrix $\hat{A}$ the same statements and approaches as to the matrix A in (10.109).

The similar problems of choosing the parameter e may be formulated in the case of uncertain variables, i.e. for $v_w(e)$ and $v_g(e)$ instead of $P_w(e)$ and $P_v(e)$.

Example 10.6.

Consider an uncertain system (10.110) where $k = 2$, the matrix $A(c_n, b, x_n)$ is the same as in Example 10.3, $d^{\mathrm{T}} = [1, 0]$, $e^{\mathrm{T}} = [\bar{e}, -\bar{e}]$. The matrix $\hat{A}$ in (10.111) is then as follows:

$$\hat{A}(c_n, b, x_n, \bar{e}) = \begin{bmatrix} a_{11}(c_n, x_n) + b + \bar{e} & a_{12}(c_n, x_n) - \bar{e} \\ a_{21}(c_n, x_n) & a_{22}(c_n, x_n) + b \end{bmatrix}.$$

Assume that $0 \le \bar{e} \le \underline{a}_{12}$. Applying to the matrix $\hat{A}$ the same conditions as (10.94) and (10.95) for the matrix A yields

$$b_w(\bar{e}) = 1 - \max(\bar{a}_{11} + \bar{a}_{21} + e,\ \bar{a}_{12} + \bar{a}_{22} - e), \tag{10.112}$$

$$b_v(\bar{e}) = 1 - \min(\underline{a}_{11} + \underline{a}_{21} + e,\ \underline{a}_{12} + \underline{a}_{22} - e).$$

It is easy to see that $\arg\max P_w(e) = \arg\max b_w(e)$. Using (10.112), one obtains

$$\bar{e}^* = \begin{cases} \hat{e} & \text{if} \quad \bar{a}_{12} + \bar{a}_{22} \ge \bar{a}_{11} + \bar{a}_{21} \ \text{ and } \ \hat{e} \le \underline{a}_{12} \\ \bar{a}_{12} + \bar{a}_{22} + \underline{a}_{12} & \text{if} \quad \bar{a}_{12} + \bar{a}_{22} \ge \bar{a}_{11} + \bar{a}_{21} \ \text{ and } \ \hat{e} > \underline{a}_{12} \\ 0 & \text{if} \quad \bar{a}_{12} + \bar{a}_{22} < \bar{a}_{11} + \bar{a}_{21} \end{cases}$$

where

$$\bar{e}^* = \arg\max[b_w(\bar{e}) : 0 \le \bar{e} \le \underline{a}_{12}]$$

and $\hat{e}$ is the solution of the equation $\overline{a}_{11}+\overline{a}_{21}+e= \overline{a}_{12}+\overline{a}_{22}-e$, i.e.

$$\hat{e}=\frac{1}{2}(\overline{a}_{12}+\overline{a}_{22}-\overline{a}_{11}-\overline{a}_{21}).$$

For the numerical data $\overline{e}^*=\hat{e}=0.05$, $b_w(\overline{e}^*)=0.25$, $b_v(\overline{e}^*)=0.55$. For the exponential probability density the estimation of P_s is as follows:

$$0.63 \le P_s \le 0.89 \quad \text{for } \lambda=4,$$

$$0.86 \le P_s < 1 \quad \text{for } \lambda=8.$$

The probability P_w is now greater than in Example 10.3, i.e. for $\overline{e}=0$ where $P_w=0.55$ for $\lambda=4$ and $P_w=0.80$ for $\lambda=8$. Applying for $a_{ij}(c)$ (i.e. for a linear time-varying system) the necessary condition (10.96) with $a_{11}+\overline{e}$, and $a_{12}-\overline{e}$, one can obtain $b_v(\overline{e}^*)\approx 0.2$, i.e. $P_w \approx P_v$ (the grey zone is approximately equal to 0), $P_s \approx 0.63$ for $\lambda=4$ and $P_s \approx 0.86$ for $\lambda=8$.

11 Learning Systems

This chapter is concerned with plants described by a knowledge representation in the form of relations with unknown parameters. The learning process consists here in *step by step* knowledge validation and updating [20, 28, 32, 34, 37, 51, 54]. At each step one should prove if the current observation "belongs" to the knowledge representation determined before this step (*knowledge validation*) and if not – one should modify the current estimation of the parameters in the knowledge representation (*knowledge updating*). The results of the successive estimation of the unknown parameters are used in the current determination of the decisions in a learning decision making system. This approach may be considered as an extension of the known idea of adaptation via identification for the plants described by traditional mathematical models (see e.g. [14]). We shall consider two versions of learning systems. In the first version the knowledge validation and updating is concerned with the knowledge of the plant (i.e. the relation R describing the plant), and in the second version – with the knowledge of the decision making (i.e. the set of decisions D_u). In both versions the learning algorithms based on the knowledge validation and updating will be presented.

11.1 Learning System Based on Knowledge of the Plant

Consider a static plant described by a relation

$$R(u, y; c) \subset U \times Y \tag{11.1}$$

where $c \in C$ is a vector parameter (a vector of parameters). As was said in Chapter 2, the relation R may be described by a set of equalities and/or inequalities concerning the components of u and y, e.g.

$$u^{\mathrm{T}} u + y^2 \le c^2 \tag{11.2}$$

where u is a column vector, c and y are one-dimensional variables, or

$$a^{\mathrm{T}} u \le y \le b^{\mathrm{T}} u$$

where a, b are vectors of parameters, i.e. $c = (a, b)$. As a solution of the decision problem for the given $D_y \subset Y$ (see Sect. 2.3) we obtain a set of decisions

$$D_u(c) = \{u \in U: \ D_y(u;c) \subseteq D_y\} \tag{11.3}$$

where

$$D_y(u;c) = \{y \in Y : (u, y) \in R(u, y; c)\}$$

is a set of possible outputs for the given u, and $D_u(c)$ is the largest set such that the implication $u \in D_u \rightarrow y \in D_y$ is satisfied. For example, if in the case (11.2) we have the requirement $y^2 \le \alpha^2$ then the solution $D_u(c)$ is determined by the inequality

$$c^2 - \alpha^2 \le u^{\mathrm{T}} u \le c^2 .$$

For the further considerations we assume that $R(u,y;c)$ for every $c \in C$ is a continuous and closed domain in $U \times Y$. Assume now that the parameter c in the relation R has the value $c = \overline{c}$ and $\overline{c}$ is unknown.

11.1.1 Knowledge Validation and Updating

Let us assume that the sequence of observations

$$(u_1, y_1),\ (u_2, y_2),\ \ldots,\ (u_n, y_n),\ \ \bigwedge_{i \in \overline{1,n}} [(u_i, y_i) \in R(u, y; \overline{c})]$$

is available and may be used for the estimation of $\overline{c}$. For the value u_i at the input, the corresponding value y_i is "generated" by the plant. The greater the variation of (u_i, y_i) inside R, the better the estimation that may be obtained. Let us introduce the set

$$D_c(n) = \{c \in C : \bigwedge_i [(u_i, y_i) \in R(u, y; c)]\} . \tag{11.4}$$

It is easy to note that $D_c(n)$ is a closed set in C. The boundary $\Delta_c(n)$ of the set $D_c(n)$ may be proposed as the estimation of $\overline{c}$. In the example (11.2) the set $D_c(n) = [c_{\min,n}, \infty)$ and $\Delta_c(n) = \{c_{\min,n}\}$ (a singleton) where

$$c_{\min,n}^2 = \max_i (u_i^{\mathrm{T}} u_i + y_i^2) . \tag{11.5}$$

Assume that the points (u_i, y_i) occur randomly from $R(u, y; \overline{c})$ with probability density $f(u, y)$, i.e. that (u_i, y_i) are the values of random variables $(\widetilde{u}, \widetilde{y})$ with probability density $f(u, y)$.

Theorem 11.1. If $f(u,y)>0$ for every $(u,y)\in R(u,y;\overline{c})$ and for every $c\neq\overline{c}$ $R(u,y;c)\neq R(u,y;\overline{c})$ then $\Delta_c(n)$ converges to $\{\overline{c}\}$ with probability 1.

Proof: From (11.4)

$$D_c(n+1)=\{c\in C:\bigwedge_{i\in\overline{1,n}}[(u_i,y_i)\in R(u,y;c)]\wedge(u_{n+1},y_{n+1})\in R(u,y;c)\}.$$

Then $D_c(n+1)\subseteq D_c(n)$, which means that $D_c(n)$ is a convergent sequence of sets. We shall show that $D_c=\overline{D}_c$ with probability 1, where

$$D_c=\lim_{n\to\infty}D_c(n)=\{c\in C:\bigwedge_{i\in\overline{1,\infty}}[(u_i,y_i)\in R(u,y;c)]\},\qquad(11.6)$$

$$\overline{D}_c=\{c\in C:R(u,y;\overline{c})\subseteq R(u,y;c)\}.\qquad(11.7)$$

Assume that $D_c\neq\overline{D}_c$, i.e. there exists $\hat{c}\in D_c$ such that $R(u,y;\overline{c})\not\subseteq R(u,y;\hat{c})$. There exists then the subset of $R(u,y;\overline{c})$

$$R(u,y;\overline{c})-R(u,y,\hat{c})\stackrel{\Delta}{=}D_R\qquad(11.8)$$

such that for every $i\in\overline{1,\infty}$ $(u_i,y_i)\notin D_R$. The probability of this property is the following:

$$\lim_{n\to\infty}p^n\stackrel{\Delta}{=}P_\infty$$

where

$$p=P[(\widetilde{u},\widetilde{y})\in U\times Y-D_R]=\int_{U\times Y-D_R}f(u,y)dudy.$$

From the assumption about $f(u,y)$ it follows that $p<1$ and $P_\infty=0$. Then $D_c=\overline{D}_c$ with probability 1. From (11.6)

$$\lim_{n\to\infty}\Delta_c(n)\stackrel{\Delta}{=}\Delta_c$$

where Δ_c is the boundary of D_c. Using the assumption about R it is easy to note from (11.7) that $\overline{\Delta}_c=\{\overline{c}\}$ where $\overline{\Delta}_c$ is the boundary of $\overline{D}_c$. Then with probability 1

$$\lim_{n\to\infty}\Delta_c(n)=\overline{\Delta}_c=\{\overline{c}\}.\qquad\square$$

The determination of $\Delta_c(n)$ may be presented in the form of the following recursive algorithm:

1. **Knowledge validation.** Prove if

$$\bigwedge_{c \in D_c(n-1)} [(u_n, y_n) \in R(u, y; c)]. \tag{11.9}$$

If yes then $D_c(n) = D_c(n-1)$ and $\Delta_c(n) = \Delta_c(n-1)$. If not then one should determine the new $D_c(n)$ and $\Delta_c(n)$, i.e. update the knowledge.

2. **Knowledge updating.**

$$D_c(n) = \{c \in D_c(n-1): \quad (u_n, y_n) \in R(u, y; c)\} \tag{11.10}$$

and $\Delta_c(n)$ is the boundary of $D_c(n)$. For $n = 1$

$$D_c(1) = \{c \in C: \quad (u_1, y_1) \in R(u, y; c)\}.$$

The successive estimations may be used in current updating of the solution of the decision problem in the open-loop learning system, in which the set $D_u(c_n)$ is determined by putting c_n in (11.3), where c_n is chosen randomly from $\Delta_c(n)$. For the random choice of c_n a generator of random numbers is required.

11.1.2 Learning Algorithm for Decision Making in a Closed-loop System

The successive estimations of $\bar{c}$ may be performed in a closed-loop learning system where u_i is the sequence of the decisions. For the successive decision u_n and its result y_n, knowledge validation and updating should be performed by using the algorithm presented in the first part of this section. The next decision u_{n+1} is based on the updated knowledge and is chosen randomly from $D_u(c_n)$. Finally, the decision making algorithm in the closed-loop learning system is the following:

1. Put u_n at the input of the plant and measure y_n.
2. Prove the condition (11.9), determine $D_c(n)$ and $\Delta_c(n)$. If (11.9) is not satisfied, then knowledge updating according to (11.10) is necessary.
3. Choose randomly c_n from $\Delta_c(n)$.
4. Determine $D_u(c_n)$ according to (11.3) with $c = c_n$.
5. Choose randomly u_{n+1} from $D_u(c_n)$.

For $n = 1$ one should choose randomly u_1 from U and determine $D_c(1)$. If for all $n < p$ the value u_n is such that y_n does not exist (i.e. u_n does not belong to the projection of $R(u, y; \bar{c})$ on U), then the estimation starts from $n = p$. If $D_u(c_n)$ is an empty set (i.e. for $c = c_n$ the solution of the decision problem does not exist) then u_{n+1} should be chosen randomly from U. The block scheme of the learning system is presented in Fig. 11.1. For the random choice of c_n and u_n the generators G_1 and G_2 are required. The probability distributions should be determined currently for $\Delta_c(n)$ and $D_u(c_n)$.

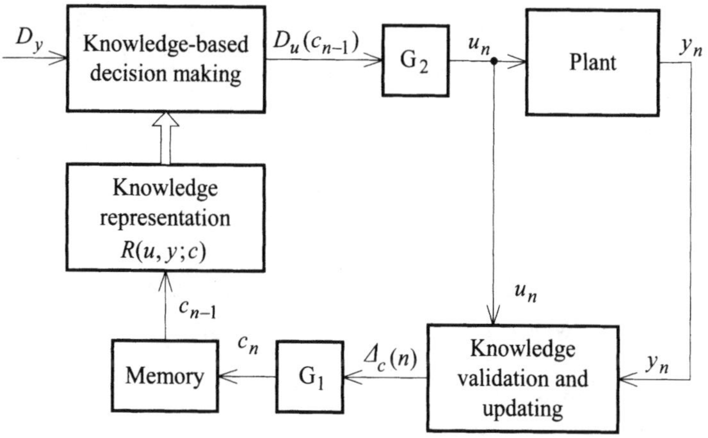

Figure 11.1. Learning system based on the knowledge of the plant

Assume that the points c_n are chosen randomly from $\Delta_c(n)$ with probability density $f_{cn}(c)$, the points u_n are chosen randomly from $D_u(c_{n-1})$ with probability density $f_u(u \mid c_{n-1})$ and the points y_n "are generated" randomly by the plant with probability density $f_y(y \mid u_n; \overline{c})$ from the set $D_y(u;c) = \{y \in Y : (u,y) \in R(u,y;c)\}$ where $u = u_n$ and $c = \overline{c}$. It means that (c_i, u_{i+1}, y_{i+1}) are the values of random variables $(\widetilde{c}_i, \widetilde{u}_{i+1}, \widetilde{y}_{i+1})$ with probability density $f_{ci}(c_i) \cdot f_u(u_{i+1} \mid c_i) \cdot f_y(y_{i+1} \mid u_{i+1}; \overline{c})$.

11.2 Learning System Based on Knowledge of Decisions

In this version the validation and updating directly concerns $D_u(c)$, i.e. the knowledge of the decision making. When the parameter $\overline{c}$ is unknown then for the fixed value u it is not known if u is a correct decision, i.e. if $u \in D_u(\overline{c})$ and consequently $y \in D_y$. Our problem may be considered as a classification problem with two classes. The point u should be classified to class $j = 1$ if $u \in D_u(\overline{c})$ and to class $j = 2$ if $u \notin D_u(\overline{c})$. Assume that we can use the learning sequence

$$(u_1, j_1), (u_2, j_2), \ldots, (u_n, j_n) \stackrel{\Delta}{=} S_n$$

where $j_i \in \{1, 2\}$ are the results of the correct classification given by an external trainer or obtained by testing the property $y_i \in D_y$ at the output of the plant. Let us assume for the further considerations that $D_u(c)$ is a continuous and closed domain in U, and consider the approaches analogous to those presented in the previous section.

11.2.1 Knowledge Validation and Updating

Let us denote by $\overline{u}_i$ the subsequence for which $j_i = 1$, i.e. $\overline{u}_i \in D_u(\overline{c})$ and by $\hat{u}_i$ the subsequence for which $j_i = 2$, and introduce the following sets in C:

$$\overline{D}_c(n) = \{c \in C : \overline{u}_i \in D_u(c) \text{ for every } \overline{u}_i \text{ in } S_n\}, \tag{11.11}$$

$$\hat{D}_c(n) = \{c \in C : \hat{u}_i \in U - D_u(c) \text{ for every } \hat{u}_i \text{ in } S_n\}. \tag{11.12}$$

It is easy to see that $\overline{D}_c$ and $\hat{D}_c$ are closed sets in C. The set

$$\overline{D}_c(n) \cap \hat{D}_c(n) \stackrel{\Delta}{=} \overline{\Delta}_c(n)$$

may be proposed as the estimation of $\overline{c}$. For example, if $D_u(\overline{c})$ is described by the inequality $u^{\mathrm{T}} u \le \overline{c}^2$ then

$$\overline{D}_c(n) = [c_{\min,n}, \infty), \qquad \hat{D}_c(n) = [0, c_{\max,n}), \qquad \overline{\Delta}_c(n) = [c_{\min,n}, c_{\max,n})$$

where

$$c_{\min,n}^2 = \max_i \overline{u}_i^{\mathrm{T}} \overline{u}_i, \qquad c_{\max,n}^2 = \min_i \hat{u}_i^{\mathrm{T}} \hat{u}_i.$$

Assume that the points u_i are chosen randomly from U with probability density $f(u)$.

Theorem 11.2. If $f(u) > 0$ for every $u \in U$ and $D_u(c) \ne D_u(\overline{c})$ for every $c \ne \overline{c}$ then $\overline{\Delta}_c(n)$ converges to $\{\overline{c}\}$ with probability 1 (w.p.1).

Proof: In the same way as for Theorem 11.1 we can prove that

$$\lim_{n\to\infty} \overline{D}_c(n) = \overline{D}_c, \qquad \lim_{n\to\infty} \hat{D}_c(n) = \hat{D}_c \tag{11.13}$$

w.p.1 where

$$\overline{D}_c = \{c \in C : D_u(\overline{c}) \subseteq D_u(c)\}, \quad \hat{D}_c = \{c \in C : D_u(c) \subseteq D_u(\overline{c})\}. \quad (11.14)$$

From (11.13) one can derive that $\overline{\Delta}_c(n)$ converges to $\overline{D}_c \cap \hat{D}_c \triangleq \overline{\Delta}_c$ (the boundary of $\overline{D}_c$) w.p.1. Using the assumption about D_u it is easy to note that $\overline{\Delta}_c = \{\overline{c}\}$. □

The determination of $\overline{\Delta}_c(n)$ may be presented in the form of the following recursive algorithm:

If $j_n = 1$ ($u_n = \overline{u}_n$).

1. **Knowledge validation** for $\overline{u}_n$. Prove if

$$\bigwedge_{c \in \overline{D}_c(n-1)} [u_n \in D_u(c)].$$

If yes then $\overline{D}_c(n) = \overline{D}_c(n-1)$. If not then one should determine the new $\overline{D}_c(n)$, i.e. update the knowledge.

2. **Knowledge updating** for $\overline{u}_n$

$$\overline{D}_c(n) = \{c \in \overline{D}_c(n-1) : u_n \in D_u(c)\}.$$

Put $\hat{D}_c(n) = \hat{D}_c(n-1)$.

If $j_n = 2$ ($u_n = \hat{u}_n$).

3. **Knowledge validation** for $\hat{u}_n$. Prove if

$$\bigwedge_{c \in \hat{D}_c(n-1)} [u_n \in U - D_u(c)].$$

If yes then $\hat{D}_c(n) = \hat{D}_c(n-1)$. If not then one should determine the new $\hat{D}_c(n)$, i.e. update the knowledge.

4. **Knowledge updating** for $\hat{u}_n$

$$\hat{D}_c(n) = \{c \in \hat{D}_c(n-1) : u_n \in U - D_u(c)\}.$$

Put $\overline{D}_c(n) = \overline{D}_c(n-1)$ and $\overline{\Delta}_c(n) = \overline{D}_c(n) \cap \hat{D}_c(n)$.

For $n = 1$, if $u_1 = \overline{u}_1$ determine

$$\overline{D}_c(1) = \{c \in C : u_1 \in D_u(c)\},$$

if $u_1 = \hat{u}_1$ determine

$$\hat{D}_c(1) = \{c \in C : u_1 \in U - D_u(c)\}.$$

If for all $i \le p$ $u_i = \overline{u}_i$ (or $u_i = \hat{u}_i$), put $\hat{D}_c(p) = C$ (or $\overline{D}_c(p) = C$).

11.2.2 Learning Algorithm for Decision Making in a Closed-Loop System

The successive estimation of $\bar{c}$ may be performed in a closed-loop learning system where u_i is the sequence of the decisions. The decision making algorithm is as follows:

1. Put u_n at the input of the plant and measure y_n.
2. Test the property $y_n \in D_y$, i.e. determine j_n.
3. Determine $\bar{\Delta}_c(n)$ using the estimation algorithm with knowledge validation and updating.
4. Choose randomly c_n from $\bar{\Delta}_c(n)$, put c_n into $R(u,y;c)$ and determine $D_u(c)$, or put c_n directly into $D_u(c)$ if the set $D_u(c)$ may be determined from R in an analytical form.
5. Choose randomly u_{n+1} from $D_u(c_n)$.

At the beginning of the learning process u_i should be chosen randomly from U. The block scheme of the learning system in the case when c_n is put directly into $D_u(c)$ is presented in Fig. 11.2, and in the case when $D_u(c_n)$ is determined from $R(u,y;c_n)$ is presented in Fig. 11.3. The blocks G_1 and G_2 are the generators of random variables for the random choosing of c_n from $\bar{\Delta}_c(n)$ and u_{n+1} from $D_u(c_n)$, respectively.

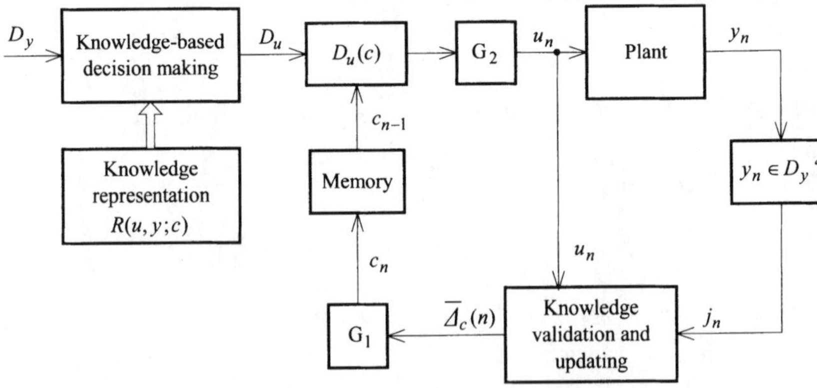

Figure 11.2. Learning system in the first version

Assume that the points c_n are chosen randomly from $\bar{\Delta}_c(n)$ with probability density $f_{cn}(c)$ and the points u_n are chosen randomly from $D_u(c_{n-1})$ with probability density $f_u(u \mid c_{n-1})$, i.e. (c_i, u_{i+1}) are the values of random variables

$(\widetilde{c}_i, \widetilde{u}_{i+1})$.

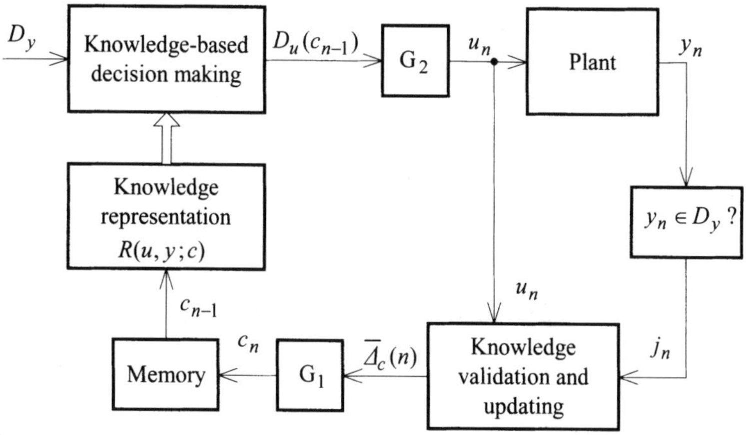

Figure 11.3. Learning system in the second version

Theorem 11.3. If

$$\bigwedge_{i} \bigwedge_{c \in \overline{\Delta}_c(i)} f_{ci}(c) > 0, \quad \bigwedge_{c \in C} \bigwedge_{u \in D_u(c)} f_u(u \mid c) > 0 \tag{11.15}$$

and for every

$$c \neq \overline{c}, \quad D_u(c) \neq D_u(\overline{c}) \tag{11.16}$$

then $\overline{\Delta}_c(n)$ converges to $\{\overline{c}\}$ w.p.1.

Proof: From (11.11) it is easy to note that $\overline{D}_c(n+1) \subseteq \overline{D}_c(n)$, which means that $\overline{D}_c(n)$ is a convergent sequence of sets. We shall show that $D_c = \overline{D}_c$ w.p.1. where

$$D_c = \lim_{n \to \infty} \overline{D}_c(n)$$

and $\overline{D}_c$ is defined in (11.14). Assume that $D_c \neq \overline{D}_c$, i.e. there exists $\hat{c} \in D_c$ such that $D_u(\overline{c}) \not\subseteq D_u(\hat{c})$. There exists then the subset of $D_u(\overline{c})$

$$D_u(\overline{c}) - D_u(\hat{c}) \triangleq D_R$$

such that $\overline{u}_i \notin D_R$ for every $\overline{u}_i$ in S_∞. The probability of this property is the following:

$$\lim_{n \to \infty} \prod_{i=1}^{n} p_i \triangleq P_\infty$$

where

$$p_i = P(\widetilde{u}_i \in U - D_R) = \int\limits_{U-D_R} f_{ui}(u)du\,,$$

$$f_{ui}(u) = \int\limits_{\overline{\Delta}_c(i)} f_u(u\,|\,c) f_{ci}(c)\,dc\,. \qquad (11.17)$$

Since $\overline{c} \in \overline{\Delta}_c(i)$ for every i then from (11.15) and (11.17) it follows that $f_{ui}(u) > 0$ for every $u \in D_u(\overline{c})$ and consequently $f_{ui}(u) > 0$ for every $u \in D_R$. Thus, $p_i < 1$ for every i and $P_\infty = 0$. Then $D_c = \overline{D}_c$ w.p.1. In the same way it may be proved that

$$\lim_{n\to\infty} \hat{D}_c(n) = \hat{D}_c \text{ w.p.1}$$

where $\hat{D}_c$ is defined in (11.14). Consequently, $\overline{\Delta}_c(n)$ converges w.p.1 to $\overline{D}_c \cap \hat{D}_c = \overline{\Delta}_c$ (the boundary of $\overline{D}_c$). Using (11.16) it is easy to note that $\overline{\Delta}_c = \{\overline{c}\}$. □

Remark 11.1. Let us note that the decisions in a closed-loop learning system may be based on j_n given by an external trainer, i.e. $j_n = 1$ if $u_n \in D_u(\overline{c})$ and $j_n = 2$ if $u_n \notin D_u(\overline{c})$, or may be obtained by testing the property $y_n \in D_y$, illustrated in Figs. 11.2 and 13.3. In this case,

if $y_n \notin D_y$ then $j_n = 2$ and $u_n \notin D_u(\overline{c})$,

if $y_n \in D_y$ then $j_n = 1$ and $u_n \in \widetilde{D}_u(\overline{c})$

where

$$\widetilde{D}_u(\overline{c}) = \{u \in U : D_y(\overline{c}) \cap D_y \neq \varnothing\}$$

and

$$D_y(\overline{c}) = \{y \in Y : (u, y) \in R(u, y; \overline{c})\}\,.$$

Consequently, in (11.11) and in the first part of the recursive algorithm presented in Sect. 11.2.1 for $\overline{u}_n$, one should use $\widetilde{D}_u(c)$ instead of $D_u(c)$.

It is worth noting that Theorem 11.3 concerns the case with an external trainer. □

Example 11.1.

Consider the single-output plant described by the inequality

$$0 \le y \le \frac{1}{c} u^{\mathrm{T}} P u$$

where P is a positive definite matrix. For the requirement $y \le \overline{y}$ we obtain

$$D_u(\bar{c}) = \{u \in U: u^{\mathrm{T}} P u \le \bar{c}\, \bar{y}\}. \tag{11.18}$$

According to (11.11) and (11.12)

$$\overline{D}_c(n) = [c_{\min,n}, \infty), \qquad \hat{D}_c(n) = [0, c_{\max,n}), \qquad \overline{\Delta}_c(n) = [c_{\min,n}, c_{\max,n})$$

where

$$c_{\min,n} = \bar{y}^{-1} \cdot \max_i \bar{u}_i^{\mathrm{T}} P \bar{u}_i, \qquad c_{\max,n} = \bar{y}^{-1} \cdot \min_i \hat{u}_i^{\mathrm{T}} P \hat{u}_i.$$

The decision making algorithm in the closed-loop learning system is the following:
1. Put u_n at the input, measure y_n and determine j_n.
2. For $j_n = 1$ ($u_n = \bar{u}_n$), prove if

$$\bar{y}^{-1} u_n^{\mathrm{T}} P u_n \le c_{\min,n-1}.$$

If yes then $c_{\min,n} = c_{\min,n-1}$. If not, determine new $c_{\min,n}$

$$c_{\min,n} = \bar{y}^{-1} u_n^{\mathrm{T}} P u_n.$$

Put $c_{\max,n} = c_{\max,n-1}$.
3. For $j_n = 2$ ($u_n = \hat{u}_n$), prove if

$$\bar{y}^{-1} u_n^{\mathrm{T}} P u_n \ge c_{\max,n-1}.$$

If yes then $c_{\max,n} = c_{\max,n-1}$. If not, determine new $c_{\max,n}$

$$c_{\max,n} = \bar{y}^{-1} u_n^{\mathrm{T}} P u_n.$$

Put $c_{\min,n} = c_{\min,n-1}$, $\overline{\Delta}_c(n) = [c_{\min,n}, c_{\max,n})$.
4. Choose randomly c_n from $\overline{\Delta}_c(n)$ and put $\bar{c} = c_{n-1}$ in (11.18).
5. Choose randomly u_n from $D_u(c_n)$. □

The example may be easily extended for the case when $D_u(c)$ is a domain closed by a hypersurface $F(u) = c$ for one-dimensional c and a given function F. The simulations showed the significant influence of the shape of $D_u(c)$ and the probability distributions $f_c(c)$, $f_u(u \mid c)$, on the convergence of the learning process and the quality of the decisions.

11.3 Learning Algorithms for a Class of Dynamical Systems [36]

The considerations for a dynamical plant described by the relations (8.2) are analogous to those for the static plant and are based on the solution of the decision problem described in Sect. 8.1. Assume that in relations $R_{\rm I}$ and $R_{\rm II}$ there are unknown vector parameters b and c, respectively, i.e. $b=\overline{b}$, $c=\overline{c}$, $\overline{b}$ and $\overline{c}$ are unknown. Consequently, $D_{s,n+1}(c)$ in (8.5) depends on c and $D_{un}(b,c)$ depends on b and c. Let us note that our knowledge representation (i.e. the form of $R_{\rm I}$ and $R_{\rm II}$) does not depend on n. To denote in a different way the variables u_n, s_n, y_n and their values in the sequence of observations, let us introduce the notation

$$R_{\rm I}(u,s,\overline{s};b)\,, \qquad R_{\rm II}(s,y;c)$$

where u, s, y denote the current input, state and output, respectively, and $\overline{s}$ denotes the next state. We assume further that $R_{\rm I}(u,s,\overline{s};b)$ and $R_{\rm II}(s,y;c)$ are continuous and closed domains in $U\times S\times S$ and $S\times Y$, respectively.

11.3.1 Knowledge Validation and Updating

Using the sequences of observations

$$u_i,\ i=0,1,...,n-1\,;\quad s_i,\ i=0,1,...,n\,;\quad y_i,\ i=1,2,...,n \tag{11.19}$$

we can propose an estimation of the unknown values $\overline{b}$ and $\overline{c}$. For the separate estimation of $\overline{b}$ and $\overline{c}$ it is convenient to form two separate sequences from the sequences (11.19)

$$(u_0,s_0,s_1),(u_1,s_1,s_2),...,(u_{n-1},s_{n-1},s_n),\quad \bigwedge_{i\in\overline{1,n}}[(u_{i-1},s_{i-1},s_i)\in R_{\rm I}(u,s,\overline{s};b)]\,, \tag{11.20}$$

$$(s_1,y_1),(s_2,y_2),...,(s_n,y_n),\quad \bigwedge_{i\in\overline{1,n}}[(s_i,y_i)\in R_{\rm II}(s,y;c)]\,. \tag{11.21}$$

In the second part of this section the current *step by step* estimation of $\overline{b}$ and $\overline{c}$ will be described. In each step, one should prove if the current observation "belongs" to the knowledge representation determined before this step (*knowledge validation*) and if not – one should modify the current estimation of the parameters in the knowledge representation (*knowledge updating*). Let us introduce the set

$$D_b(n)=\{b\in B:\bigwedge_{i\in\overline{1,n}}[(u_{i-1},s_{i-1},s_i)\in R_{\rm I}(u,s,\overline{s};b)]\}\,. \tag{11.22}$$

It is easy to see that $D_b(n)$ is a closed set in B. The boundary $\Delta_b(n)$ of the set $D_b(n)$ is proposed here as the estimation of $\overline{b}$. In the same way we can define the estimation of $\overline{c}$ using the sequence (11.21) and introducing the set

$$D_c(n) = \{c \in C : \bigwedge_{i \in \overline{1,n}} [(s_i, y_i) \in R_{II}(s, y; c)]\} . \tag{11.23}$$

The boundary $\Delta_c(n)$ of the set $D_c(n)$ may be proposed as the estimation of $\bar{c}$. For example, for a plant with k state variables, one output and R_{II} of the form

$$(s^{(1)})^2 + (s^{(2)})^2 + ... + (s^{(k)})^2 + y^2 \le c^2$$

where $s = [s^{(1)}, ..., s^{(k)}]^T$ we have $D_c(n) = [c_{min}, \infty)$ and $\Delta_c(n) = \{c_{min}\}$ where

$$c_{min}^2 = \max_{i \in \overline{1,n}} (s_i^T s_i + y_i^2) .$$

It is easy to see that the definitions of the estimations are the same as for the static plant described in the previous sections. Instead of (u, y) for the static plant we now have $(u, s, \bar{s})$ in the first part and (s, y) in the second part of the knowledge representation. Assume that the points (u_i, s_i, y_{i+1}) occur randomly with probability density $f(u_i, s_i, s_{i+1}, y_{i+1})$, i.e. that (u_i, s_i, y_{i+1}) is the sequence of values of random variables $(\tilde{u}_i, \tilde{s}_i, \tilde{y}_{i+1})$, $i = 0, 1, ..., n$ (discrete stochastic process) and there exists the probability density $f(u_i, s_i, s_{i+1}, y_{i+1})$. Denote the marginal probability densities by $f_I(u_i, s_i, s_{i+1})$ and $f_{II}(s_{i+1}, y_{i+1})$. The following theorem concerning the convergence of $\Delta_b(n)$ and $\Delta_c(n)$ may be proved:

Theorem 11.4. Assume that:
(i) $f_I(u_i, s_i, s_{i+1}) > 0$ for every $(u_i, s_i, s_{i+1}) \in R_I(u, s, \bar{s}; b)$,
(ii) $f_{II}(s_i, y_i) > 0$ for every $(s_i, y_i) \in R_{II}(s, y; c)$,
(iii) for every $b \ne \bar{b}$ $R_I(u, s, \bar{s}; b) \ne R_I(u, s, \bar{s}; \bar{b})$, and for every $c \ne \bar{c}$ $R_{II}(s, y; c) \ne R_{II}(s, y; \bar{c})$.

Then $\Delta_b(n)$ converges to $\{\bar{b}\}$ with probability 1 and $\Delta_c(n)$ converges to $\{\bar{c}\}$ with probability 1.

The proof is analogous to that of Theorem 11.1 Sect. 11.1. □

The idea of the determination of $\Delta_b(n)$ and $\Delta_c(n)$ may be presented in the form of a *recursive algorithm*. The algorithm for $\Delta_b(n)$ is the following:

Knowledge validation
For $n > 1$ one should prove if

$$\bigwedge_{b \in D_b(n-1)} [(u_{n-1}, s_{n-1}, s_n) \in R_I(u, s, \bar{s}; b)] . \tag{11.24}$$

If yes then $D_b(n) = D_b(n-1)$ and $\Delta_b(n) = \Delta_b(n-1)$. If not then one should determine the new $D_b(n)$ and $\Delta_b(n)$, i.e. update the knowledge:

Knowledge updating

$$D_b(n) = \{b \in D_b(n-1) : (u_{n-1}, s_{n-1}, s_n) \in R_{\text{I}}(u, s, \overline{s}; b)\} \tag{11.25}$$

and $\Delta_b(n)$ is the boundary of $D_b(n)$.

For $n = 1$

$$D_b(1) = \{b \in B : (u_0, s_0, s_1) \in R_{\text{I}}(u, s, \overline{s}; b)\} . \tag{11.26}$$

The algorithm for $\Delta_c(n)$ has an analogous form:

Knowledge validation

For $n > 1$ one should prove if

$$\bigwedge_{c \in D_c(n-1)} [(s_n, y_n) \in R_{\text{II}}(s, y; c)] . \tag{11.27}$$

If yes then $D_c(n) = D_c(n-1)$ and $\Delta_c(n) = \Delta_c(n-1)$. If not then one should determine the new $D_c(n)$ and $\Delta_c(n)$:

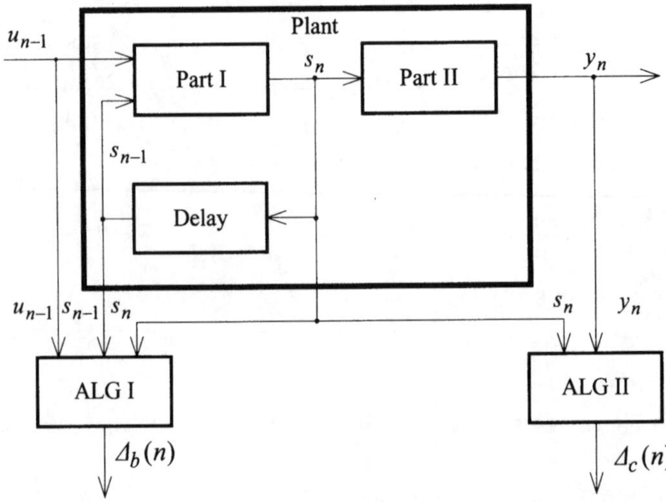

Figure 11.4. Illustration of knowledge validation and updating

Knowledge updating

$$D_c(n) = \{c \in D_c(n-1) : (s_n, y_n) \in R_{\text{II}}(s, y; c)\} \tag{11.28}$$

and $\Delta_c(n)$ is the boundary of $D_c(n)$.

For $n = 1$

$$D_c(1) = \{c \in C : (s_1, y_1) \in R_{\text{II}}(s, y; c)\}\,. \tag{11.29}$$

The algorithms for $\Delta_b(n)$ and $\Delta_c(n)$ (ALG I and ALG II, respectively) may be executed concurrently, using the current data of observations (u_i, s_i, y_i) (Fig. 11.4).

11.3.2 Learning Control System

The idea of the closed-loop learning system presented here consists in the following: (1) For the successive decision u_{n-1} and the results of observations s_{n-1}, s_n, y_n, knowledge validation and updating should be performed. (2) Then the next decision u_n is based on the updated knowledge, i.e. is chosen randomly from $D_{un}(b_n, c_n)$ with a probability density $f_{un}(u)$ where b_n and c_n are the values of b and c chosen randomly from $\Delta_b(n)$ and $\Delta_c(n)$ with probability densities $f_{bn}(b)$ and $f_{cn}(c)$, respectively. The forms of the probability distributions are fixed but their parameters should be changed according to current parameters of D_{un}, $\Delta_b(n)$ and $\Delta_c(n)$. If $D_{un}(b_n, c_n) = \varnothing$ (the controllability condition is not satisfied for $b = b_n$ and $c = c_n$) then u_n is chosen randomly from U without the restriction to $D_{un}(b_n, c_n)$. Finally, the *control algorithm* with knowledge validation and updating in the closed-loop learning system is the following:

1. Put u_{n-1} at the input of the plant.
2. Measure s_n and y_n, put s_n into the memory.
3. Take s_{n-1} from the memory and prove the condition (11.24) (knowledge validation for part I).
4. Determine $D_b(n)$ and $\Delta_b(n)$. If (11.24) is not satisfied, then knowledge updating according to (11.25) is necessary.
5. Prove the condition (11.27) (knowledge validation for part II).
6. Determine $D_c(n)$ and $\Delta_c(n)$. If (11.27) is not satisfied, then knowledge updating according to (11.28) is necessary.
7. Choose randomly b_n from $\Delta_b(n)$ and c_n from $\Delta_c(n)$.
8. Determine $D_{s,n+1}(c_n)$ according to (8.5), with $c = c_n$.
9. Determine $D_{un}(b_n, c_n)$ according to (8.6), with $b = b_n$ and $c = c_n$.
10. Choose randomly u_n from $D_{un}(b_n, c_n)$.
11. If $D_{un}(b_n, c_n) = \varnothing$, choose randomly u_n from U.

For $n = 1$, choose randomly u_0 from U and determine $D_b(1)$ and $D_c(1)$ in (11.26), (11.29), respectively. A block scheme of the learning control system is presented in Fig. 11.5. For the random choice of b_n, c_n and u_n the generators G_1, G_2 and G of the random numbers are required. Their probability distributions are determined currently for $\Delta_b(n)$, $\Delta_c(n)$ and $D_{un}(b_n, c_n)$.

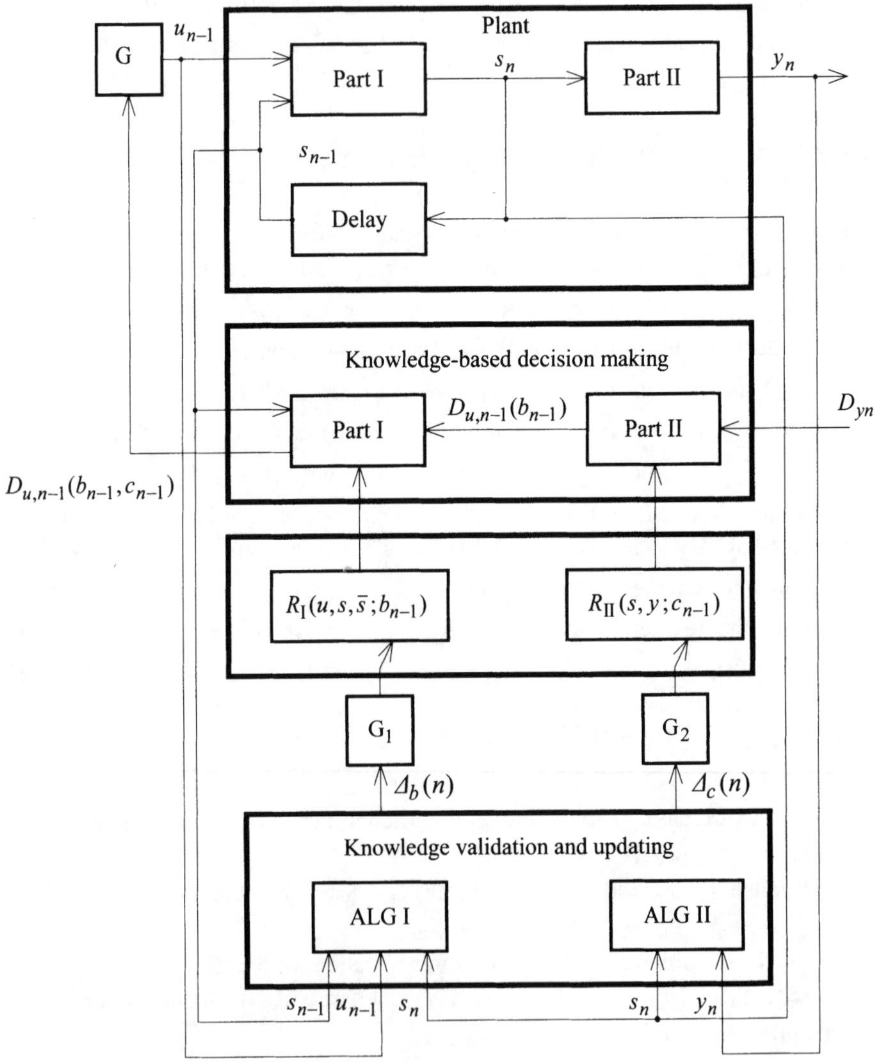

Figure 11.5. Learning control system with dynamical plant

11.3.3 Example

Consider a very simple example with the first-order single-input and single-output plant described by the following relations R_{I} and R_{II} :

$$\alpha s_n + b^{(1)} u_n \le s_{n+1} \le \alpha s_n + b^{(2)} u_n , \quad b^{(2)} > b^{(1)} > 0 , \quad b = (b^{(1)}, b^{(2)}) ,$$

$$c^{(1)} s_n \le y_n \le c^{(2)} s_n , \quad c^{(2)} > c^{(1)} > 0 , \quad c = (c^{(1)}, c^{(2)}) .$$

Let the requirement $D_{yn} = D_y = [y_{\min}, y_{\max}]$, $y_{\min}, y_{\max} > 0$. It is easy to note that

$$D_s(c) = [\frac{y_{\min}}{c^{(1)}}, \frac{y_{\max}}{c^{(2)}}] ,$$

$$D_{u,n-1}(b,c;s_{n-1}) = [\frac{y_{\min}}{b^{(1)}c^{(1)}} - \frac{\alpha s_{n-1}}{b^{(1)}}, \frac{y_{\max}}{b^{(2)}c^{(2)}} - \frac{\alpha s_{n-1}}{b^{(2)}}] \qquad (11.30)$$

and the solution exists if

$$\frac{y_{\min}}{b^{(1)}c^{(1)}} - \frac{\alpha s_{n-1}}{b^{(1)}} \le \frac{y_{\max}}{b^{(2)}c^{(2)}} - \frac{\alpha s_{n-1}}{b^{(2)}} .$$

In our case the sets $D_b(n)$ in (11.22) and $D_c(n)$ in (11.23) are determined by

$$b_{n1}^{(1)} \le b^{(1)} \le b_{n2}^{(1)} , \qquad b_{n1}^{(2)} \le b^{(2)} \le b_{n2}^{(2)} ,$$

and

$$c_{n1}^{(1)} \le c^{(1)} \le c_{n2}^{(1)} , \qquad c_{n1}^{(2)} \le c^{(2)} \le c_{n2}^{(2)} ,$$

respectively, where

$$b_{n1}^{(1)} = \max_i \, (\frac{s_i - \alpha s_{i-1}}{u_{i-1}} : u_{i-1} < 0) , \qquad b_{n2}^{(1)} = \min_i \, (\frac{s_i - \alpha s_{i-1}}{u_{i-1}} : u_{i-1} > 0) ,$$

$$b_{n1}^{(2)} = \max_i \, (\frac{s_i - \alpha s_{i-1}}{u_{i-1}} : u_{i-1} > 0) , \qquad b_{n2}^{(2)} = \min_i \, (\frac{s_i - \alpha s_{i-1}}{u_{i-1}} : u_{i-1} < 0) ,$$

$$c_{n1}^{(1)} = \max_i \, (\frac{y_i}{s_i} : s_i < 0) , \qquad c_{n2}^{(1)} = \min_i \, (\frac{y_i}{s_i} : s_i > 0) ,$$

$$c_{n1}^{(2)} = \max_i \, (\frac{y_i}{s_i} : s_i > 0) , \qquad c_{n2}^{(2)} = \min_i \, (\frac{y_i}{s_i} : s_i < 0) .$$

Consequently, we obtain

$$\Delta_b(n) = \{b_{n1}, b_{n2}, b_{n3}, b_{n4}\} , \qquad \Delta_c(n) = \{c_{n1}, c_{n2}, c_{n3}, c_{n4}\}$$

where

$$b_{n1}=(b_{n1}^{(1)},b_{n1}^{(2)}),\quad b_{n2}=(b_{n2}^{(1)},b_{n1}^{(2)}),\quad b_{n3}=(b_{n1}^{(1)},b_{n2}^{(2)}),\quad b_{n4}=(b_{n2}^{(1)},b_{n2}^{(2)}),$$

$$c_{n1}=(c_{n1}^{(1)},c_{n1}^{(2)}),\quad c_{n2}=(c_{n2}^{(1)},c_{n1}^{(2)}),\quad c_{n3}=(c_{n1}^{(1)},c_{n2}^{(2)}),\quad c_{n4}=(c_{n2}^{(1)},c_{n2}^{(2)}).$$

In our example u_n is chosen randomly according to the rectangular probability density

$$f_{un}(u)=\begin{cases}\dfrac{1}{\gamma-\beta} & \text{for } \beta\le u\le\gamma,\\ 0 & \text{otherwise.}\end{cases} \tag{11.31}$$

As the set U we assume $[-\bar{\gamma},\bar{\gamma}]$ with $\bar{\gamma}>0$ sufficiently large. The control algorithm in the learning system is the following.

1. Put u_{n-1} at the input of the plant.
2. Measure s_n and y_n, put s_n into the memory.
3. Knowledge validation and updating for part I:

Introduce the notation

$$A_1=s_n-\alpha s_{n-1},\quad A_2=b_{n-1,1}^{(1)}u_{n-1},\quad A_3=b_{n-1,2}^{(1)}u_{n-1},$$

$$A_4=b_{n-1,1}^{(2)}u_{n-1},\quad A_5=b_{n-1,2}^{(2)}u_{n-1},\quad A_6=\frac{s_n-\alpha s_{n-1}}{u_{n-1}}.$$

Then the knowledge validation and updating may be described by IF-THEN rules as follows:

If $u_{n-1}>0$ then

$$(A_3\le A_1)\wedge(A_1\le A_4)\Rightarrow b_{n1}^{(1)}=b_{n-1,1}^{(1)},\ b_{n2}^{(1)}=b_{n-1,2}^{(1)},\ b_{n1}^{(2)}=b_{n-1,1}^{(2)},\ b_{n2}^{(2)}=b_{n-1,2}^{(2)},$$

$$(A_3> A_1)\wedge(A_1\le A_4)\Rightarrow b_{n1}^{(1)}=b_{n-1,1}^{(1)},\ b_{n2}^{(1)}=A_6,\ b_{n1}^{(2)}=b_{n-1,1}^{(2)},\ b_{n2}^{(2)}=b_{n-1,2}^{(2)},$$

$$(A_3\le A_1)\wedge(A_1> A_4)\Rightarrow b_{n1}^{(1)}=b_{n-1,1}^{(1)},\ b_{n2}^{(1)}=b_{n-1,2}^{(1)},\ b_{n1}^{(2)}=A_6,\ b_{n2}^{(2)}=b_{n-1,2}^{(2)}.$$

If $u_{n-1}<0$ then

$$(A_2\le A_1)\wedge(A_1\le A_5)\Rightarrow b_{n1}^{(1)}=b_{n-1,1}^{(1)},\ b_{n2}^{(1)}=b_{n-1,2}^{(1)},\ b_{n1}^{(2)}=b_{n-1,1}^{(2)},\ b_{n2}^{(2)}=b_{n-1,2}^{(2)},$$

$$(A_2> A_1)\wedge(A_1\le A_5)\Rightarrow b_{n1}^{(1)}=A_6,\ b_{n2}^{(1)}=b_{n-1,2}^{(1)},\ b_{n1}^{(2)}=b_{n-1,1}^{(2)},\ b_{n2}^{(2)}=b_{n-1,2}^{(2)},$$

$$(A_2\le A_1)\wedge(A_1> A_5)\Rightarrow b_{n1}^{(1)}=b_{n-1,1}^{(1)},\ b_{n2}^{(1)}=b_{n-1,2}^{(1)},\ b_{n1}^{(2)}=b_{n-1,1}^{(2)},\ b_{n2}^{(2)}=A_6.$$

4. Knowledge validation and updating for part II:
Introduce the notation

$$B_1 = y_n, \quad B_2 = c_{n-1,1}^{(1)} s_n, \quad B_3 = c_{n-1,2}^{(1)} s_n,$$

$$B_4 = c_{n-1,1}^{(2)} s_n, \quad B_5 = c_{n-1,2}^{(2)} s_n, \quad B_6 = \frac{y_n}{s_n}.$$

Then the knowledge validation and updating may be described by IF-THEN rules as follows:
If $s_n > 0$ then

$$(B_3 \le B_1) \wedge (B_1 \le B_4) \Rightarrow c_{n1}^{(1)} = c_{n-1,1}^{(1)}, \; c_{n2}^{(1)} = c_{n-1,2}^{(1)}, \; c_{n1}^{(2)} = c_{n-1,1}^{(2)}, \; c_{n2}^{(2)} = c_{n-1,2}^{(2)},$$

$$(B_3 > B_1) \wedge (B_1 \le B_4) \Rightarrow c_{n1}^{(1)} = c_{n-1,1}^{(1)}, \; c_{n2}^{(1)} = B_6, \; c_{n1}^{(2)} = c_{n-1,1}^{(2)}, \; c_{n2}^{(2)} = c_{n-1,2}^{(2)},$$

$$(B_3 \le B_1) \wedge (B_1 > B_4) \Rightarrow c_{n1}^{(1)} = c_{n-1,1}^{(1)}, \; c_{n2}^{(1)} = c_{n-1,2}^{(1)}, \; c_{n1}^{(2)} = B_6, \; c_{n2}^{(2)} = c_{n-1,2}^{(2)}.$$

If $s_n < 0$ then

$$(B_2 \le B_1) \wedge (B_1 \le B_5) \Rightarrow c_{n1}^{(1)} = c_{n-1,1}^{(1)}, \; c_{n2}^{(1)} = c_{n-1,2}^{(1)}, \; c_{n1}^{(2)} = c_{n-1,1}^{(2)}, \; c_{n2}^{(2)} = c_{n-1,2}^{(2)},$$

$$(B_2 > B_1) \wedge (B_1 \le B_5) \Rightarrow c_{n1}^{(1)} = B_6, \; c_{n2}^{(1)} = c_{n-1,2}^{(1)}, \; c_{n1}^{(2)} = c_{n-1,1}^{(2)}, \; c_{n2}^{(2)} = c_{n-1,2}^{(2)},$$

$$(B_2 \le B_1) \wedge (B_1 > B_5) \Rightarrow c_{n1}^{(1)} = c_{n-1,1}^{(1)}, \; c_{n2}^{(1)} = c_{n-1,2}^{(1)}, \; c_{n1}^{(2)} = c_{n-1,1}^{(2)}, \; c_{n2}^{(2)} = B_6.$$

5. Choose randomly b_n from $\Delta_b(n)$ and c_n from $\Delta_c(n)$.
6. Determine the set D_{un} according to (11.30), i.e.

$$u_{\min,n} = \frac{y_{\min}}{b_n^{(1)} c_n^{(1)}} - \frac{a s_{n-1}}{b_n^{(1)}}, \qquad u_{\max,n} = \frac{y_{\max}}{b_n^{(2)} c_n^{(2)}} - \frac{a s_{n-1}}{b_n^{(2)}}.$$

7. If $u_{\min,n} < u_{\max,n}$ – choose randomly u_n according to (11.31) with $\beta = u_{\min,n}$, $\gamma = u_{\max,n}$.
8. If $u_{\min,n} > u_{\max,n}$ – choose randomly u_n according to (11.31) with $\beta = -\bar{\gamma}, \gamma = \bar{\gamma}$ where $\bar{\gamma} >> \max\,(|\, x_{\min,n} \,|, |\, x_{\max,n} \,|\,)$.

The example shows that the learning algorithms may be rather complicated, even for simple plants.

11.4 Learning Algorithms for a Class of Knowledge-based Assembly Systems

As an example of the application of the learning idea to dynamical plants, let us take into consideration the knowledge-based assembly process described in Sect. 8.5, i.e. a specific dynamical plant in which the states s_n and the decisions u_n belong to finite sets [48].

11.4.1 Knowledge Validation and Updating

Let us consider the knowledge representation with time-varying unknown parameters in the second part, i.e. in the relation R_{II} (see Sect. 8.5.1). The relation $R_{\mathrm{II}}(j_n, y_n; c_n)$, and consequently the sets $D_{y,n}(j_n; c_n)$ in (8.34) and the set $D_{j,n}(c_n)$ in (8.35) rewritten with n in place of n+1, depend on the vector parameter $c_n \in C_n$. Assume that the parameter c_n has the value $c_n = \bar{c}_n$ and $\bar{c}_n$ is unknown. We shall present the algorithm of a *step by step* estimation of the unknown value $\bar{c}_n$ based on the results of observations. In each step one should prove if the current observation "belongs" to the knowledge representation determined before this step (*knowledge validation*) and if not – one should modify the current estimation of the parameters in the knowledge representation (*knowledge updating*). The successive estimations will be used in the determination of the decision concerning the assembly operations, based on the current knowledge in the learning system. According to the general approach, the validation and updating may concern the knowledge in the form of $D_{y,n}(j_n; c_n)$ or directly the knowledge in the form of $D_{j,n}(c_n)$. The second version is used here, taking into account the specific form of the knowledge representation for the assembly process. Let us assume that the assembly process consisting of N operations is repeated in successive cycles and denote by i_{pn}, j_{pn}, y_{pn} the variables in the p-th cycle. One cycle corresponds to one step of the estimation process. When the parameter $\bar{c}_n$ is unknown then for the fixed value i_{n-1} it is not known if i_{n-1} is a correct decision, i.e. if $j_n \in D_{j,n}(c_n)$ and consequently $y_n \in D_{y,n}$. Our problem may be considered as a classification problem with two classes. The index j_n should be classified to class $k=1$ if $j_n \in D_{j,n}(c_n)$ or to class $k=2$ if $j_n \notin D_{j,n}(c_n)$. Assume that we can use the *learning sequence*

$$(j_{1n}, k_{1n}), (j_{2n}, k_{2n}), \ldots, (j_{vn}, k_{vn}) \triangleq S_{vn}$$

where $k_{pn} \in \{1, 2\}$ are the results of the correct classification given by an external trainer or obtained by testing the property $y_{pn} \in D_{y,n}$ at the output of the assembly plant. Let us denote the index j_{pn} by $\bar{j}_{pn}$ if $k_{pn} = 1$ and by $\hat{j}_{pn}$ if $k_{pn} = 2$, and introduce the following sets in C_n:

$$\overline{D}_{c,n}(p) = \{c_n \in C_n : j_{pn} \in D_{j,n}(c_n) \text{ for every } j_{pn} = \bar{j}_{pn} \text{ in } S_{vn}\}, \quad (11.32)$$

$$\hat{D}_{c,n}(p) = \{c_n \in C_n : j_{pn} \notin D_{j,n}(c_n) \text{ for every } j_{pn} = \hat{j}_{pn} \text{ in } S_{vn}\}. \quad (11.33)$$

The set

$$\overline{D}_{c,n}(p) \cap \hat{D}_{c,n}(p) \stackrel{\Delta}{=} \overline{\Delta}_{c,n}(p) \quad (11.34)$$

is proposed as the estimation of $\bar{c}_n$ after p assembly cycles. The value c_n may be chosen randomly from $\overline{\Delta}_{c,n}(p)$ and put into $D_{j,n}(c_n)$. The determination of $\overline{\Delta}_{c,n}(p)$ may be presented in the form of the following recursive algorithm:

If $k_{pn} = 1$:

prove whether

$$\bigwedge_{c_n \in \overline{D}_{c,n}(p-1)} [j_{pn} \in D_{j,n}(c_n)]$$

(*knowledge validation*).

If yes then $\overline{D}_{c,n}(p) = \overline{D}_{c,n}(p-1)$. If not – determine new $\overline{D}_{c,n}(p)$ (*knowledge updating*)

$$\overline{D}_{c,n}(p) = \{c_n \in \overline{D}_{c,n}(p-1) : \ j_{pn} \in D_{j,n}(c_n)\}.$$

Put

$$\hat{D}_{c,n}(p) = \hat{D}_{c,n}(p-1).$$

If $k_{pn} = 2$:

prove whether

$$\bigwedge_{c_n \in \hat{D}_{c,n}(p-1)} [j_{pn} \notin D_{j,n}(c_n)]$$

(*knowledge validation*).

If yes then $\hat{D}_{c,n}(p) = \hat{D}_{c,n}(p-1)$. If not – determine new $\hat{D}_{c,n}(p)$ (*knowledge updating*)

$$\hat{D}_{c,n}(p) = \{c_n \in \hat{D}_{c,n}(p-1): \ j_{pn} \notin D_{j,n}(c_n)\}.$$

Put

$$\overline{D}_{c,n}(p) = \overline{D}_{c,n}(p-1)$$

and

$$\overline{\Delta}_{c,n}(p) = \overline{D}_{c,n}(p) \cap \hat{D}_{c,n}(p).$$

Let us note that for the determination of $\overline{\Delta}_{c,n}(p)$ it is necessary to observe the states j_{pn}.

Example 11.2.
Let the sets $D_{y,n}(j_n)$ be described by the inequalities

$$c_n a(j_n) \le y_n^{\mathrm{T}} y_n \le 2c_n a(j_n), \quad c_n > 0, \quad a(j_n) > 0, \tag{11.35}$$

given by an expert. It is then known that the value $y_n^{\mathrm{T}} y_n$ (where y_n is a vector of the features that characterizes the assembly plant) satisfies the inequality (11.35) and the bounds are the coefficients a depending on the state, i.e. for the different assembly operations i_n and consequently – the different states j_n we have the different bounds for the value $y_n^{\mathrm{T}} y_n$ which denotes a quality index. The requirement concerning the quality (i.e. the set $D_{y,n}$) is the following:

$$\beta \le y_n^{\mathrm{T}} y_n \le \alpha, \quad \beta > 0, \quad \alpha \ge 2\beta.$$

Then, according to (8.35)

$$D_{j,n}(c_n) = \{ j_n \in M_n : \beta \le c_n a(j_n) \le \frac{\alpha}{2} \}.$$

Using (11.32) and (11.33), we obtain

$$\overline{D}_{c,n}(p) = \{c_n : \frac{\beta}{a(j_{pn})} \le c_n \le \frac{\alpha}{2a(j_{pn})} \quad \text{for every} \quad j_{pn} = \bar{j}_{pn} \quad \text{in} \quad S_{vn}\},$$

$$\hat{D}_{c,n}(p) = \{c_n : \ c_n < \frac{\beta}{a(j_{pn})} \quad \text{or} \quad c_n > \frac{\alpha}{2a(j_{pn})} \quad \text{for every} \ j_{pn} = \hat{j}_{pn} \quad \text{in} \quad S_{vn}\}.$$

Hence,

$$\overline{D}_{c,n}(p) = [\,\bar{c}_{\min,p}, \bar{c}_{\max,p}\,] \tag{11.36}$$

where

$$\bar{c}_{\min,p} = \max_{p} \frac{\beta}{a(j_{pn})},$$

$$\bar{c}_{\max,p} = \min_{p} \frac{\alpha}{2a(j_{pn})}$$

for all p such that $j_{pn} = \bar{j}_{pn}$, and

$$\hat{D}_{c,n}(p) = (0, c_{\max,p}) \cup (c_{\min,p}, \infty) \tag{11.37}$$

i.e.

$$0 < c_n < c_{\max,p} \quad \text{or} \quad c_n > c_{\min,p}$$

where

$$c_{\max,p} = \min_{p} \frac{\beta}{a(j_{pn})}, \qquad c_{\min,p} = \max_{p} \frac{\alpha}{2a(j_{pn})}$$

for all p such that $j_{pn} = \hat{j}_{pn}$. Finally c_n should be chosen randomly from the set

$$\bar{\Delta}_{c,n}(p) = \bar{D}_{c,n}(p) \cap \hat{D}_{c,n}(p)$$

where $\bar{D}_c$ and $\hat{D}_c$ are given by (11.36) and (11.37).

11.4.2 Learning Algorithm for Decision Making in a Closed-loop System

The estimations of $\bar{c}_n$ may be performed for the successive cycles p and used for the proper choosing of the assembly operations i_{pn} in a closed-loop learning system. Before the cycle (p+1) the values c_{pn} (n = 1, 2, ... , N) should be chosen randomly from the sets $\bar{\Delta}_{c,n}(p)$ obtained after the cycle p and put into the sets $D_{j,n}(c_n)$ in place of c_n. The decisions i_n in the cycle (p+1) should be chosen randomly from the sets $D_{i,n}(c_n)$ determined according to (8.36) for $c_n = c_{pn}$. The decision making (control) algorithm in the learning system is then the following:

1. Execute the assembly operations i_{pn} (n = 0, 1, ... , N – 1) and measure y_{pn} (n = 1, 2, ... , N).
2. Determine $\bar{\Delta}_{c,n}(p)$ (n = 1, 2, ... , N) using the procedure presented in Sect. 11.4.1.

3. Choose randomly c_{pn} from $\overline{\Delta}_{c,n}(p)$ and put c_{pn} into $D_{j,n}(c_n)$.

4. Determine the sets $D_{i,n}(c_{pn}) \stackrel{\Delta}{=} D_{i,p+1,n}$ $(n = 0, 1, \dots, N-1)$ using $D_{j,n}(c_{pn})$ and the procedure presented in Sect. 8.5.1.
5. Choose randomly the assembly operations $i_{p+1,n}$ $(n = 0, 1, \dots, N-1)$ from the sets $D_{i,p+1,n}$.

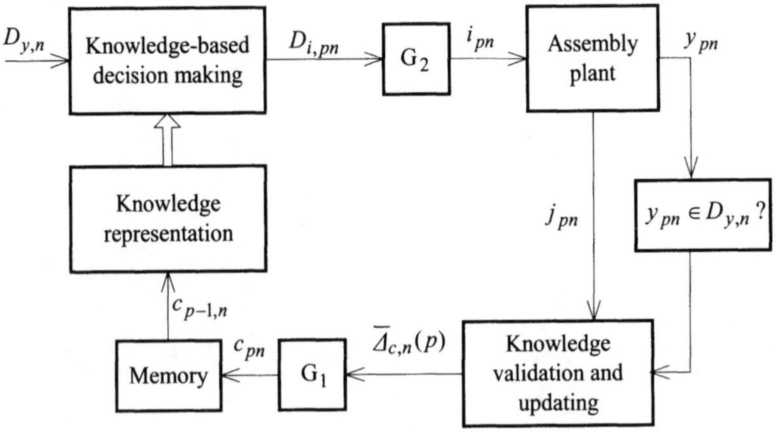

Figure 11.6. Learning control system

The sequence $i_{p+1,0}, i_{p+1,1}, \dots, i_{p+1,N-1}$ forms the assembly plan for the cycle $(p+1)$. The operations i_{pn} may be chosen randomly from the set $D_{i,pn}$ with the same probability for each operation equal to $[d(p,n)]^{-1}$ where $d(p,n)$ denotes the number of operations in the set $D_{i,pn}$. Assume that the points c_{pn} are chosen randomly from $\overline{\Delta}_{c,n}(p)$ with probability density $f_{pn}(c)$. It may be shown that if $f_{pn}(c) > 0$ for every $c \in \overline{\Delta}_{c,n}(p)$ then $\overline{\Delta}_{c,n}(p)$ converges with probability 1 to $\{\overline{c}_n\}$, i.e. to the value of the unknown parameter, and the sequence $D_{i,pn}$ converges to the set $D_{i,n}$ obtained for $c_n = \overline{c}_n$. The proof is based on the theorem concerning the convergence of the learning process, presented in Sect. 11.2.2. The block scheme of the closed-loop control system with learning is presented in Fig.11.6 where i_{pn}, y_{pn}, c_{pn} etc. denote the respective sequences of variables for different n in the cycle p, e.g. i_{pn} denotes the sequence of operations $i_{p0}, \dots, i_{p,N-1}$ and c_{pn} denotes the sequence $c_{p1}, \dots, c_{pN}$. Two generators of random variables are used in the system: G_1 for the random choosing of c_{pn} from $\overline{\Delta}_{c,n}(p)$ and G_2 for the random choosing of i_{pn} from $D_{i,pn}$.

12 Complex Problems and Systems

In the previous chapters we have considered a single decision plant with one unified form of the uncertainty based on random, uncertain or fuzzy variables. Complex problems arise when there are different forms of uncertainty in the knowledge representation of a single plant or when there are different levels of uncertainty concerning a single plant. If the relational knowledge representation presented in Chapter 2 is considered as a basic level, then the description of unknown parameters in the relations, using random or uncertain variables, forms the second (the upper) level of uncertainty or the second-order uncertainty, as was explained in Sect. 3.6 for the relational and random levels. The first part of this chapter concerns this kind of complex problems.

The second part of the chapter deals with selected cases of complex uncertain systems, i.e. the complex decision plants consisting of single plants considered in the previous chapters.

12.1 Decision Problems for Plants with Uncertain and Random Parameters

The purpose of this section is to present the application of uncertain variables to decision making for decision plants containing two kinds of unknown parameters in the relational knowledge representation: *uncertain parameters* described by certainty distributions and *random parameters* described by probability distributions [58]. The considerations are limited to static (memoryless) plants and open-loop decision (control) systems, but the basic idea may be extended to more complicated cases.

Let us recall the decision problem for the plant described by the relation $R(u,y;x) \subset U \times Y$ where $x \in X$ is an unknown vector parameter which is assumed to be a value of an uncertain variable $\bar{x}$ described by $h(x)$ given by an expert (Sect. 5.5, version II). The set of all possible outputs is as follows:

$$D_y(u;x) = \{y \in Y : (u,y) \in R(u,y;x)\}. \tag{12.1}$$

For the requirement $y \in D_y$ we can formulate and solve the following decision problem: given R, $h(x)$ and D_y, find u^* maximizing the certainty index

$$v(u) = v[D_y(u;\bar{x}) \,\tilde{\subseteq}\, D_y] = v[u \,\tilde{\in}\, D_u(\bar{x})] \tag{12.2}$$

where

$$D_u(x) = \{u \in U : D_y(u;x) \subseteq D_y\} . \tag{12.3}$$

The optimal decision u^* maximizes the certainty index that a set of possible outputs approximately belongs to D_y. It is easy to note that

$$v(u) = v[\bar{x} \,\tilde{\in}\, D_x(u)] = \max_{x \in D_x(u)} h(x) \tag{12.4}$$

where $D_x(u) = \{x \in X : D_y(u;x) \subseteq D_y\} = \{x \in X : u \in D_u(x)\}$.

Denote by x^* the value maximizing $h(x)$. Then the set of the optimal decisions $D_u = \{u \in U : x^* \in D_x(u)\}$ and $v(u^*) = 1$. If $\bar{x}$ is considered as a *C*-uncertain variable then, according to (4.59), one should determine u_c^* maximizing

$$v_c(u) = \frac{1}{2}\{v[\bar{x} \,\tilde{\in}\, D_x(u)] + 1 - v[\bar{x} \,\tilde{\in}\, X - D_x]\} . \tag{12.5}$$

Consider now a plant described by $R(u, y; z, x, w) \subset U \times Y$ where $z \in Z$ is a vector of external disturbances which may be measured, $x \in X$ is a value of an uncertain variable $\bar{x}$ described by the certainty distribution $h(x)$, and $w \in W$ is a value of a random variable $\tilde{w}$ described by a probability density $f(w)$. In general w is a vector and W is a vector space. Now the sets in (12.1) and (12.3) depend on z and w:

$$D_y(u; z, x, w) = \{y \in Y : (u, y) \in R(u, y; z, x, w)\} , \tag{12.6}$$

$$D_u(z, x, w) = \{u \in U : D_y(u; z, x, w) \subseteq D_y\} . \tag{12.7}$$

Then, v and v_c in (12.2), (12.4) and (12.5) depend on z and w:

$$v(u; z, w) = v[D_y(u; z, \bar{x}, w) \,\tilde{\subseteq}\, D_y] = v[u \,\tilde{\in}\, D_u(z, \bar{x}, w)] ,$$

$$v(u; z, w) = \max_{x \in D_x(u; z, w)} h(x)$$

where

$$D_x(u; z, w) = \{x \in X : D_y(u; z, x, w) \subseteq D_y\} = \{x \in X : u \in D_u(z, x, w)\} ,$$

and

$$v_c(u;z,w) = \frac{1}{2}\{ \max_{x \in D_x(u;z,w)} h(x) + 1 - \max_{x \in X - D_x(u;z,w)} h(x)\}. \qquad (12.8)$$

Consequently, the optimal decisions u^* and u_c^* depend on z and w:

$$u^*(z,w) = \arg\max_{u \in U} v(u;z,w), \qquad (12.9)$$

$$u_c^*(z,w) = \arg\max_{u \in U} v_c(u;z,w).$$

Now two versions of the decision problem may be proposed:

Decision problem – version I: Given $R, h(x), f(w)$ and D_y, find u_1 or u_{c1} maximizing the expected value of $v(u;z,w)$ or $v_c(u;z,w)$, respectively.
Then

$$u_1 = \arg\max_{u \in U} \mathrm{E}_w[v(u;z,\tilde{w})] \stackrel{\Delta}{=} \Psi_1(z)$$

where

$$\mathrm{E}_w[v(u;z,\tilde{w})] = \int_W v(u;z,w) f(w) dw \stackrel{\Delta}{=} G(u,z). \qquad (12.10)$$

The decision $u_{c1} = \Psi_{c1}(z)$ is determined in an analogous form, with $v_c(u;z,w)$ in place of $v(u;z,w)$.

Decision problem – version II: Given $R, h(x), f(w)$ and D_y, find u_2 or u_{c2} as the expected value of $u^*(z,w)$ or $u_c^*(z,w)$, respectively.
Then

$$u_2 = \mathrm{E}_w[u^*(z,\tilde{w})] = \int_W u^*(z,w) f(w) dw \stackrel{\Delta}{=} \Psi_2(z)$$

$$u_{c2} = \mathrm{E}_w[u_c^*(z,\tilde{w})] = \int_W u_c^*(z,w) f(w) dw \stackrel{\Delta}{=} \Psi_{c2}(z).$$

The results in versions I and II may be different, i.e. in general $u_1 \neq u_2$ and $u_{c1} \neq u_{c2}$ (see Example 12.1). They have different practical interpretations. In version I u_1 (or u_{c1}) is a decision maximizing the mean value of the certainty index that the requirement $y \in D_y$ is approximately satisfied, where the mean certainty index is understood in a probabilistic sense as an expected value. In version II as an optimal decision u_2 (or u_{c2}) we accept the mean value of the decision maximizing the certainty index that the requirement is approximately satisfied for the given w. We may say that $\mathrm{KD} = < u^*(z,w) >$ is a *knowledge of decision making* (or a *random decision algorithm* in an open-loop system), obtained from the *knowledge of the plant*

$$\mathrm{KP} = < R(u,y;z,x,w),\ h(x) >$$

given by an expert. In versions I and II different forms of determinization consisting in replacing an uncertain description by a corresponding deterministic model have been applied. In version I the function $v(u;z,w)$ is replaced by $G(u,z)$ and in version II the random decision algorithm $u^*(z,w)$ is replaced by the deterministic algorithm $\Psi_2(z)$. Two versions of the knowledge-based decision making in open-loop systems are illustrated in Figs. 12.1 and 12.2.

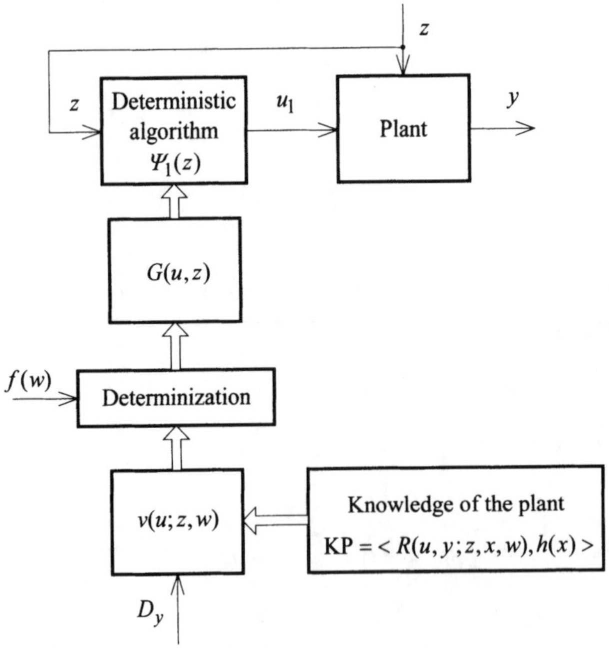

Figure 12.1. Open-loop knowledge-based system – version I

Example 12.1.
To illustrate the presented approach let us consider a simple example of a plant without disturbances, described by the inequality

$$xu \le y \le \frac{xu}{w}$$

where u, y, x, w are one-dimensional positive variables and $w < 1$. For

$D_y = [y_1, y_2]$, according to (12.7), the set $D_u(x,w)$ is determined by the inequality

$$\frac{y_1}{x} \le u \le \frac{wy_2}{x}$$

and the set $D_x(u;w)$ is determined by the inequality

$$\frac{y_1}{u} \le x \le \frac{wy_2}{u}.$$

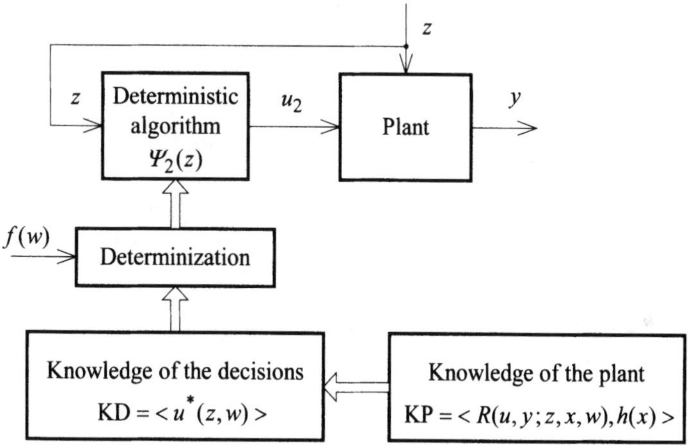

Figure. 12.2. Open-loop knowledge-based system – version II

Assume that x is a value of an uncertain variable $\bar{x}$ with triangular certainty distribution: $h = 2x$ for $0 \le x \le \frac{1}{2}$, $h = -2x + 2$ for $\frac{1}{2} \le x \le 1$, $h = 0$ otherwise. Using (12.8) it is easy to obtain the following results for the given w (see Example 5.6):

$$v_c(u;w) = \begin{cases} \frac{y_2 w}{u} & \text{for} \quad u \ge y_1 + wy_2 \\ 1 - \frac{y_1}{u} & \text{for} \quad y_1 \le u \le y_1 + wy_2 \\ 0 & \text{for} \quad u \le y_1 \end{cases},$$

$$u_c^*(w) = y_1 + wy_2. \tag{12.11}$$

Now assume that w is a value of a random variable with rectangular probability density: $f(w) = \beta^{-1}$ for $0 \le w \le \beta$ and $f(w) = 0$ otherwise, $\beta < 1$. Then, according to (12.10) with $v_c(u;w)$ instead of $v(u;w)$, we may obtain the following formula for $G(u) = \mathrm{E}_w[v_c(u;\widetilde{w})]$:

$$G(u) = \begin{cases} 0 & \text{for} \quad u \le y_1 \\ \dfrac{1}{2\beta}\dfrac{y_2}{u}(\dfrac{u-y_1}{y_2})^2 + \dfrac{1}{\beta}(1-\dfrac{y_1}{u})(\beta - \dfrac{u-y_1}{y_2}) & \text{for} \quad y_1 \le u \le \beta y_2 + y_1 \\ \dfrac{\beta\, y_2}{2u} & \text{for} \quad u \ge \beta y_2 + y_1 \, . \end{cases}$$

To obtain u_{c1} the function $G(u)$ should be maximized with respect to u. In version II, using (12.11), we have $u_{c2} = y_1 + \mathrm{E}_w(\widetilde{w})y_2 = y_1 + \frac{\beta}{2}y_2$. For the numerical data the functions $G(u)$ are presented in Fig. 12.3 and the results are as follows:

Data	Results
$\beta = 0.5,\ y_1 = 1,\ y_2 = 4$	$u_{c1} = 3,\ G(u_{c1}) = 0.33,\ u_{c2} = 2$
$\beta = 0.5,\ y_1 = 2,\ y_2 = 12$	$u_{c1} = 8,\ G(u_{c1}) = 0.38,\ u_{c2} = 5$
$\beta = 0.8,\ y_1 = 2,\ y_2 = 12$	$u_{c1} = 7,\ G(u_{c1}) = 0.47,\ u_{c2} = 6.8$

□

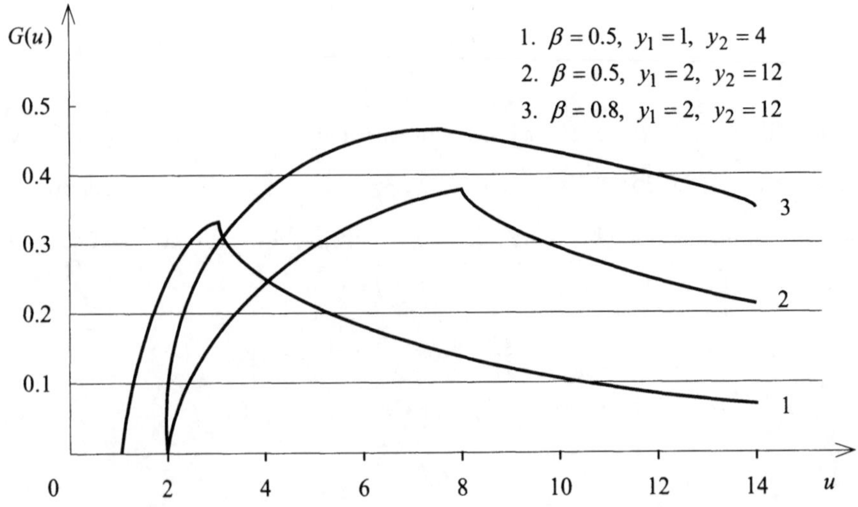

Figure 12.3. Relationship between the expected certainty index and u

12.2 Other Formulations. Three-level Uncertainty

It is possible to invert the order of the considerations concerning the uncertain and random parameters and to introduce another approach to the decision problem, in which the certainty index that $D_y(u;z,w,\bar{x}) \tilde{\subseteq} D_y$ will be replaced by the probability

$$P[D_y(u;z,x,\tilde{w}) \subseteq D_y] = \int_{D_w(u;z,x)} f(w)dw \stackrel{\Delta}{=} p(u;z,x) \qquad (12.12)$$

where

$$D_w(u;z,x) = \{w \in W : D_y(u;z,x,w) \subseteq D_y\} = \{w \in W : u \in D_u(z,x,w)\},$$

$D_y(u;z,x,w)$ and $D_u(z,x,w)$ are determined by (12.6) and (12.7), respectively. The probability (12.12) has been introduced in the probabilistic (random) form of a decision problem, described in Sect. 3.5 (see (3.33)).

Decision problem – version I: Given R, $h(x)$, $f(w)$ and D_y, find u_1 maximizing the mean value of $p(u;z,x)$.
Then

$$u_1 = \arg\max_{u \in U} \mathrm{M}_x[p(u;z,\bar{x})] \stackrel{\Delta}{=} \Psi_1(z) \qquad (12.13)$$

where

$$\mathrm{M}_x[p(u;z,\bar{x})] = \int_X p(u;z,x)h(x)dx \cdot [\int_X h(x)dx]^{-1} \stackrel{\Delta}{=} G(u,z).$$

Decision problem – version II: Given R, $h(x)$, $f(w)$ and D_y, find u_2 as a mean value of $u^*(z,x)$ where $u^*(z,x)$ is the decision maximizing the probability (12.12), i.e.

$$u^*(z,x) = \arg\max_{u \in U} p(u;z,x).$$

Then

$$u_2 = \mathrm{M}_x[u^*(z,\bar{x})] = \int_X u^*(z,x)h(x)dx \cdot [\int_X h(x)dx]^{-1}. \qquad (12.14)$$

The considerations for C-uncertain variables have an analogous form. The results in versions I and II have different interpretations. In version I u_1 is a decision maximizing the mean value of the probability that for the given z the set of possible inputs belongs to D_y. In version II as an optimal decision u_2 we accept the mean value of the decision maximizing the probability (12.11). The practical meaning of these interpretations is much less clear than for the approach in Sect. 12.1.

Up until now we have considered the decision problem for a plant with two unknown parameters: uncertain and random, i.e. with a mixed form of the uncertainty, based on uncertain and random variables. It may be interesting and useful to take into account two other versions with uncertain and random variables:

a) The plant is described by a relation $R(u,y;z,x)$ with the uncertain parameter described by $h(x;w)$ where w is a value of a random variable $\widetilde{w}$ described by $f(w)$.

b) The plant is described by a relation $R(u,y;z,w)$ with the random parameter described by $f(w;x)$ where x is a value of an uncertain variable described by $h(x)$.

In both cases we can speak about the second-order uncertainty or about two levels of uncertainty concerning the unknown parameters. In fact, we have three levels of uncertainty concerning the plant: the relational level described by R, and two additional levels.

In case a):

1. Uncertain (lower) level described by $h(x;w)$.
2. Random (upper) level described by $f(w)$.

In case b):

1. Random (lower) level described by $f(w;x)$.
2. Uncertain (upper) level described by $h(x)$.

The formulations and solutions of the decision problems are similar to those presented for $R(u,y;z,x,w)$.

In case a) one should determine $v(u;z,w)$:

$$v(u;z,w) = \max_{x \in D_x(u;z)} h(x;w)$$

where

$$D_x(u;z) = \{x \in X : D_y(u;z,x) \subseteq D_y\},$$

$$D_y(u;z,x) = \{y \in Y : (u,y) \in R(u,y;z,x)\}.$$

Then two versions presented in Sect. 12.1 may be applied to determine the decisions u_1 and u_2.

In case b) one should determine $p(u;z,x)$:

$$p(u;z,x) = \int_{D_w(u;z)} f(w;x)dw$$

where

$$D_w(u;z) = \{w \in W : D_y(u;z,w) \subseteq D_y\},$$

$$D_y(u;z,w) = \{y \in Y : (u,y) \in R(u,y;z,w)\}.$$

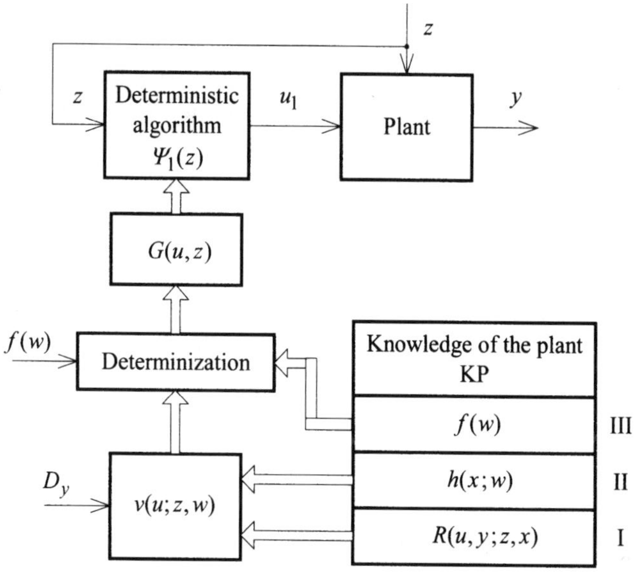

Figure 12.4. Decision system with three-level uncertainty in case a); I – relational level, II – uncertain level, III – random level

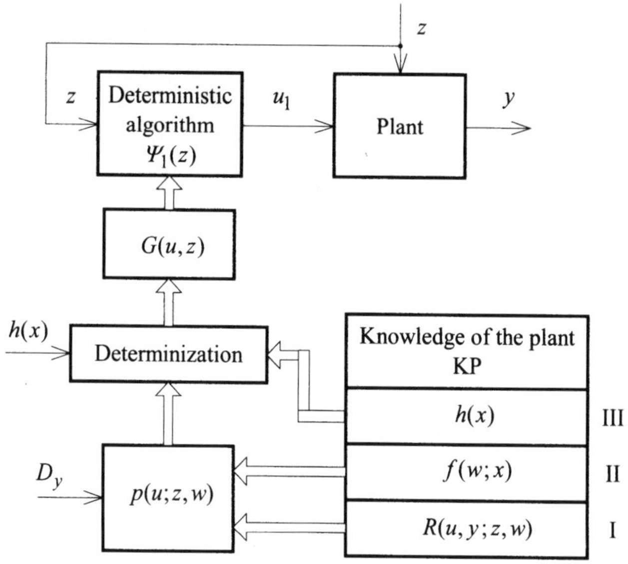

Figure 12.5. Decision system with three-level uncertainty in case b); I – relational level, II – random level, III – uncertain level

Then two versions presented in the first part of this section may be applied to the determination of u_1 and u_2. The similar approaches may be used in the case of C-uncertain variables. The knowledge-based decision systems with three levels of uncertainty in cases a) and b) are illustrated in Figs. 12.4 and 12.5, respectively. The decisions are determined according to version I, i.e. the deterministic algorithm is obtained via the determinization of $v(u;z,w)$ in case a) or the determinization of $p(u;z,x)$ in case b), and then – the maximization of the function $G(u,z)$ obtained as a result of the determinization. The systems presented in Figs. 12.4 and 12.5 correspond to the system presented in Fig. 12.1. The figures in version II, corresponding to Fig. 12.2, have analogous forms.

12.3 Complex Systems with Distributed Knowledge

12.3.1 Complex Relational System

As an example of a complex system (a complex plant) let us consider the two-level system presented in Fig. 12.6, described by a relational knowledge representation with uncertain parameters where $u_i \in U_i$, $x_i \in X_i$, $y \in Y$ [32, 40, 50, 51, 54]. For example, it may be a production system containing k parallel operations (production units) in which x_i is a vector of variables characterizing the product (e.g. the amounts of some components), u_i is a vector of variables characterizing the raw material which are accepted as the control variables. The block P may denote an additional production unit or an evaluation of a vector y of global variables characterizing the system as a whole.

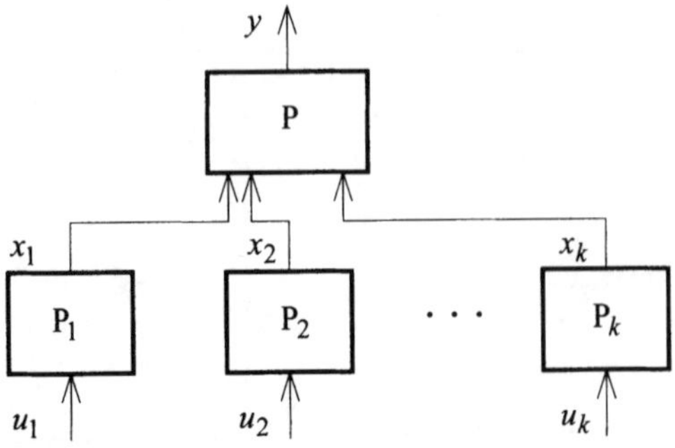

Figure 12.6. Material flow in the manufacturing process

The lower level subsystems are described by the relations

$$R_j(u_j, x_j; c_j) \subseteq U_j \times X_j, \quad j \in \overline{1,k} \tag{12.15}$$

where $c_j \in C_j$ is a parameter (real number vector), and the upper level subsystem is described by a relation

$$R(x_1, \ldots, x_k, y; b) \subseteq X_1 \times \ldots \times X_k \times Y \tag{12.16}$$

where $b \in B$ is a parameter (real number vector).

The set of relations forms the knowledge representation of our system (*distributed relational knowledge representation)*. Each relation may be presented as a set of inequalities and/or equalities concerning the components of the respective vectors.

We can formulate the following decision problem, adequately for the form of the knowledge representation: for the given knowledge representation and the set $D_y \subset Y$ find the largest set $D_u \subset U$ such that the implication $\overline{u} \in D_u \Rightarrow y \in D_y$ is satisfied. In this notation $\overline{u} = (u_1, \ldots, u_k)$ is the input of the system as a whole, $U = U_1 \times U_2 \times \ldots \times U_k$ and $y \in D_y$ is a user's requirement. The decision problem for the system as a whole may be decomposed according to the decomposition of the knowledge representation in the distributed system.

1. The decision problem for the upper level: given D_y, find the sets $D_{xj} \subset X_j$ such that the implication

$$(x_1 \in D_{x1}) \wedge \ldots \wedge (x_k \in D_{xk}) \Rightarrow y \in D_y$$

is satisfied.

2. The decision problem for the lower level: given D_{xj}, find the largest set D_{uj} such that the implication $u_j \in D_{uj} \Rightarrow x_j \in D_{xj}$ is satisfied. In general, we do not have a unique solution of the upper level decision problem, and the solutions for the system as a whole and via the decomposition are not equivalent. The method of problem-solving is the same as for a simple one-unit plant, described in Chapter 2. Now it is important to note that the results depend on the parameters in the knowledge representation, i.e. we obtain $D_{xj}(b)$ as the results for the upper level and $D_{uj}(c_j, b)$ as the results for the lower level $(j = 1, \ldots, k)$. The sets D_{uj} depend on b because the sets $D_{xj}(b)$, i.e. the data for the lower level, depend on b. We may say that the relations (12.15), (12.16) describe the knowledge concerning the system and the sets D_{uj} form the knowledge concerning the decision making.

As a generalization of the system presented in Fig. 12.6 let us consider a complex manufacturing process containing k parallel cascades of operations. The material flow in the process is presented in Fig. 12.7. The product of the operation $\mathrm{P}_{m,s}$ is used as a raw material in the operation $\mathrm{P}_{m,s+1}$, $y^{(m,s)} \in Y_{m,s}$ is a vector of the variables characterizing the product of the operation $P_{m,s}$ (e.g. the amounts of some components) $x^{(m)} \in X_m$ is a vector of variables characterizing the raw material for the operation $P_{m,1}$ (which are assumed as the decision or control variables), and $z^{(m,s)} \in Z_{m,s}$ is a vector of disturbances which are measured. The system presented in Fig. 12.7 may be also considered as a distributed task system in which the operation $P_{m,s}$ denotes a separate task described by the data $y^{(m,s-1)}$, $z^{(m,s)}$ and the result $y^{(m,s)}$.

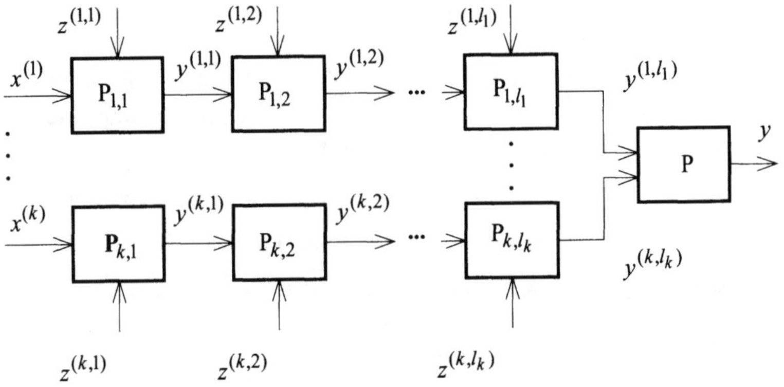

Figure 12.7. Material flow in the manufacturing process

The block P may denote an additional production unit or an evaluation of a vector y of global variables characterizing the system as a whole. If each block is described by a function (i.e. the relations (12.15) and (12.16) are reduced to functions) then the system as a whole is described by a function $y = F(x,z)$ where $x = (x^{(1)},...,x^{(k)})$ and z is a vector containing all variables in the vectors $z^{(m,s)}$ in all operations. Then the decision problem may by formulated as follows: given F, z and a desirable value y^*, find $x = \Phi(z)$ such that $y = y^*$. Assume that the system is described by a *relational knowledge representation*, which has a form of a set of relations:

$$\left.\begin{aligned} &R_{m,s}(y^{(m,s-1)}, y^{(m,s)}, z^{(m,s)}; c^{(m,s)}) \subset Y_{m,s-1} \times Y_{m,s} \times Z_{m,s} \\ &\qquad\qquad s = 1,2,...,l_m, \quad m = 1,2,...,k \\ &R_y(\bar{y}, y; c^{(k+1)}) \subset \bar{Y} \times Y \end{aligned}\right\} \tag{12.17}$$

where $y^{(m,0)} = x^{(m)}, c^{(m,s)}$ and $c^{(k+1)}$ are vectors of parameters, $y \in Y$,

$$\bar{y} = (y^{(1,l_1)}, y^{(2,l_2)}, ..., y^{(m,l_m)}) \in \bar{Y}.$$

Each relation may be presented as a set of inequalities and/or equalities concerning the components of the respective vectors. The relations (12.17) may be reduced to one relation $R(x,y,z;c) \in X \times Y \times Z$ where $x \in X$, $z \in Z$, and c is the vector of the parameters in all operations. Now the decision problem may be formulated as follows: given R, z and $D_y \subset Y$, find the largest set $D_x(z;c)$ such that the implication $x \in D_x \to y \in D_y$ is satisfied. In this formulation $y \in D_y$ denotes the user's requirement.

According to the notation in Chapter 2

$$D_x(z;c) = \{x \in X : D_y(x,z;c) \subseteq D_y\} \tag{12.18}$$

where

$$D_y(x,z;c) = \{y \in Y : (x,y,z) \in R(x,y,z;c)\}. \tag{12.19}$$

We assume for the further considerations that the solution exists, i.e. for every $z \in Z$ the set $D_x(z;c)$ is not empty.

We can consider a more complicated case when in the operation $\mathrm{P}_{m,s}$ an additional raw material is used and $x^{(m,s)}$ is a vector of variables characterizing this raw material (which are assumed as the control variables for this operation). Then the operation $\mathrm{P}_{m,s}$ is described by the relation

$$R_{m,s}\,(x^{(m,s)}, y^{(m,s-1)}, y^{(m,s)}, z^{(m,s)}; c^{(m,s)}).$$

If the variables $y^{(m,s)}$ may be measured then the decision problem may be formulated and solved for each operation separately, in the same way as for the system as a whole with the relation $R(x,y,z;c)$.

12.3.2 Complex System with Uncertain and Random Parameters

Let us come back to the system presented in Fig. 12.6 and assume that the system is described by a relational knowledge representation which has the form of a set of relations:

$$\left.\begin{array}{l} R_i(u_i, y_i, z_i; x_i) \subset U_i \times Y_i \times Z_i, \ i \in \overline{1,k} \\ R_y(\overline{y}, y; x_{k+1}) \subset \overline{Y} \times Y \end{array}\right\} \tag{12.20}$$

where $z_i \in Z_i$ are vectors of external disturbances, $x_i \in X_i$ ($i = 1, 2, ..., k+1$) are vectors of unknown parameters,

$$\overline{y} = (y_1, y_2, ..., y_k) \in \overline{Y}.$$

The unknown parameters x_i are assumed to be values of uncertain variables described by certainty distributions $h_{xi}(x_i)$ given by an expert. The relations (12.20) may be reduced to one relation

$$R(u,y,z;x) \in U \times Y \times Z$$

where

$$u = (u_1, u_2, ..., u_k) \subset U, \qquad z = (z_1, z_2, ..., z_k) \subset Z,$$

$$x = (x_1, x_2, ..., x_{k+1}) \subset X .$$

Now the decision problem may be formulated directly for the system as a whole, i.e. for the plant with input u, output y, disturbance z and uncertain vector parameter x. The formulation of the solution of the decision problem with the requirement $y \in D_y$ given by a user has been described in Sect. 5.5. If the uncertain variables $x_1, x_2, ..., x_{k+1}$ are independent then

$$h_x(x) = \min\{h_{x1}(x_1), h_{x2}(x_2), ..., h_{x,k+1}(x_{k+1})\} .$$

The direct solution of the decision problem for the system as a whole may be very complicated and it may be reasonable to apply a *decomposition*, i.e. to decompose our decision problem into separate subproblems for the block P and the blocks P_i .

1. The decision problem for the block P: For the given $R_y(\bar{y}, y; x_{k+1})$, $h_{x,k+1}(x_{k+1})$ and D_y find $\bar{y}$ maximizing the certainty index

$$v[D_y(\bar{y}; \bar{x}_{k+1}) \mathrel{\tilde{\subseteq}} D_y]$$

where

$$D_y(\bar{y}; x_{k+1}) = \{y \in Y : (\bar{y}, y) \in R_y(\bar{y}, y; x_{k+1})\} .$$

2. The decision problem for the blocks P_i $(i \in \overline{1,k})$: For the given $R_i(u_i, y_i, z_i; x_i)$, $h_{xi}(x_i)$ and D_{yi} find x_i^* maximizing the certainty index

$$v[D_{yi}(u_i, z_i; \bar{x}_i) \mathrel{\tilde{\subseteq}} D_{yi}]$$

where

$$D_{yi}(u_i, z_i; x_i) = \{y_i \in Y_i : (u_i, y_i, z_i) \in R_i(u_i, y_i, z_i; x_i)\} .$$

The decision problem for the block P_i with the given D_{yi} is then the same as the problem for the system as a whole with the given D_y . The sets D_{yi} are such that

$$D_{y1} \times D_{y2} \times ... \times D_{yk} \subseteq D_{\bar{y}} \qquad (12.21)$$

where $D_{\overline{y}}$ is the set of the solutions $\overline{y}$ of the decision problem for the block P.

In general, the results of the decomposition are not unique (the condition (12.21) may be satisfied by different sets D_{yi}) and differ from the results obtained from the direct approach for the system as a whole.

If there is no unknown parameter in the block P then the decomposition may have the following form:

1. The decision problem for the block P: For the given $R_y(\overline{y}, y)$ and D_y find the largest set $\hat{D}_{\overline{y}}$ such that the implication

$$\overline{y} \in \hat{D}_{\overline{y}} \rightarrow y \in D_y$$

is satisfied. This is a decision problem for the relational plant described in Sect. 2.2.

2. The decision problem for the block P_i is the same as in the previous formulation, with $\hat{D}_{\overline{y}}$ instead of $D_{\overline{y}}$.

A similar approach may be applied to the system with random parameters. The unknown parameters x_i in (12.20) are assumed to be values of random variables $\widetilde{x}_i$ described by probability densities $f_{xi}(x_i)$. The formulation of the solution of the decision problem for the system as a whole with the requirement $y \in D_y$ has been described in Sect. 3.5. In the case of the decomposition, the decision problem for the block P consists in maximizing the probability

$$P[D_y(\overline{y}; \widetilde{x}_{k+1}) \subseteq D_y],$$

and in the decision problem for the blocks P_i one should find the decisions x_i^* maximizing the probability

$$P[D_{yi}(u_i, z_i; \widetilde{x}_i) \subseteq D_{yi}].$$

Then the problems for the separate blocks are the same as the decision problem for the one block presented in Sect. 3.5.

12.4 Knowledge Validation and Updating

The learning processes described in Chapter 11 may be applied to the complex system under consideration.

12.4.1 Validation and Updating of the Knowledge Concerning the System

For the system presented in Fig. 12.6 assume that $b = \overline{b}$, $c_j = \overline{c}_j$ $(j = 1,\ldots,k)$, the values $\overline{b}$, $\overline{c}_j$ are unknown, but we have the sequence of observations

$$(u_{1i}, u_{2i}, \ldots, u_{ki}; x_{1i}, x_{2i}, \ldots, x_{ki}; y_i), \quad i = 1, 2, \ldots, n$$

which may be organized in the sets of data corresponding to the distribution of the knowledge representation:

For the upper level

$$(\overline{x}_1, y_1), (\overline{x}_2, y_2), \ldots, (\overline{x}_n, y_n), \tag{12.22}$$

$$\bigwedge_{i \in \overline{1,n}} (\overline{x}_i, y_i) \in R(\overline{x}, y; \overline{b})$$

where $\overline{x} = (x_1, \ldots, x_k)$.

For the lower level

$$(u_{j1}, x_{j1}), \ldots, (u_{jn}, x_{jn}), \tag{12.23}$$

$$\bigwedge_{i \in \overline{1,n}} (u_{ji}, x_{ji}) \in R(u_j, x_j; \overline{c}_j), \quad j = 1,2,\ldots,k .$$

Using the observations (12.22) one can estimate *step by step* the unknown parameter $\overline{b}$. In each step, one should prove if the current observation "belongs" to the knowledge representation determined before this step (*knowledge validation*) and if not, one should modify the current estimation of the parameters in the knowledge representation (*knowledge updating*). Let us introduce the set

$$D_b(n) = \{b \in B: \bigwedge_{i \in \overline{1,n}} (\overline{x}_i, y_i) \in R(\overline{x}, y; b)\} .$$

The boundary $\Delta_b(n)$ of the set $D_b(n)$ is proposed here as the estimation of $\overline{b}$. For example, if R is described by

$$\overline{x}^{\mathrm{T}} \overline{x} + y^2 \le \overline{b}^2$$

where $\overline{x}$ is a column vector, y and $\overline{b}$ are one-dimensional variables, then $D_b(n) = [b_{\min}, \infty)$ and $\Delta_b(n) = \{b_{\min}\}$ (a singleton) where

$$b_{\min}^2 = \max_i (\overline{x}_i^{\mathrm{T}} \overline{x}_i + y_i^2) .$$

The estimation of $\overline{c}_j$ using the observations (12.23) at the lower level has the analogous form with u_j, x_j, R_j, $\overline{c}_j$ in place of $\overline{x}$, y, R, $\overline{b}$, respectively. As the result we obtain the estimation $\Delta_{cj}(n)$ of $\overline{c}_j$, $j \in \overline{1,k}$. If the points $(\overline{u}_i, y_i)$ occur randomly then under some assumptions it may be proved that for $n \to \infty$ $\Delta_b(n)$ and $\Delta_{cj}(n)$ converge to $\{\overline{b}\}$ and $\{\overline{c}_j\}$, respectively, with probability one. It is an extension of Theorem 11.1 for one subsystem (or the centralized system).

The idea of the determination of $\Delta_b(n)$ may be presented in the form of the following recursive algorithm for $n > 1$.

Knowledge validation

Prove if

$$\bigwedge_{b \in D_b(n-1)} (\overline{x}_n, y_n) \in R(\overline{x}, y; b).$$

If yes then $D_b(n) = D_b(n-1)$ and $\Delta_b(n) = \Delta_b(n-1)$. If not then determine new $D_b(n)$ and $\Delta_b(n)$, i.e. update the knowledge.

Knowledge updating

$$D_b(n) = \{b \in D_b(n-1) : (\overline{x}_n, y_n) \in R(\overline{x}, y; b)\}$$

and $\Delta_b(n)$ is the boundary of $D_b(n)$. For $n = 1$

$$D_b(1) = \{b \in B : (\overline{x}_1, y_1) \in R(\overline{x}, y; b)\}.$$

The algorithms for the subsystems on the lower level have an analogous form. At the end of the validation and updating process (with the given "stop" condition) the values $\overline{b}$ and $\overline{c}_j$ may be chosen randomly from the sets $\Delta_b(n)$ and $\Delta_{cj}(n)$, respectively, and put into R, R_j or directly into the sets $D_{uj}(c_j, b)$, i.e. the sets forming the knowledge concerning the decision making.

12.4.2 Validation and Updating of the Knowledge Concerning the Decision Making

Let us assume that we have a learning sequence

$$(\overline{u}_1, j_1), (\overline{u}_2, j_2), \ldots, (\overline{u}_n, j_n)$$

where $\overline{u} = (u_1, \ldots, u_k)$, $j_i = 1$ if $y_i \in D_y$ (the requirement is satisfied), $j_i = 2$ if $y_i \notin D_y$. The values j may be given by a trainer or may be obtained when it is possible to measure y and verify the property $y \in D_y$. Let us denote by $\overline{u}_i^*$ the subsequence for which $j_i = 1$ and by $\hat{\bar{u}}_i$ the subsequence for which $j_i = 2$. It means that

$$\bar{u}_i^* \in D_u(\bar{e}), \quad \hat{\bar{u}}_i \notin D_u(\bar{e})$$

where

$$\bar{e} = (\bar{c}_1, ..., \bar{c}_k; \bar{b}) \in E, \qquad D_u(\bar{e}) = D_{u1} \times ... \times D_{uk}.$$

Let us introduce the following sets:

$$D_e^*(n) = \{e \in E: \text{ every } \bar{u}_i^* \in D_u(e)\}, \tag{12.24}$$

$$\hat{D}_e(n) = \{e \in E: \text{ every } \hat{\bar{u}}_i \in U - D_u(e)\}. \tag{12.25}$$

The set

$$D_e^*(n) \cap \hat{D}_e(n) \triangleq \bar{\Delta}_e(n) \tag{12.26}$$

is proposed here as the estimation of $\bar{e}$. Let us consider a simple example when $D_u(\bar{e})$ is a circle

$$u_1^2 + u_2^2 \le \bar{e}^{\,2}.$$

Then

$$D_e^*(n) = [e_{\min,n}, \infty), \qquad \hat{D}_e(n) = [0, e_{\max,n}),$$

$$\bar{\Delta}_e(n) = [e_{\min,n}, e_{\max,n})$$

where

$$e_{\min,n} = \max_i [(u_{1i}^*)^2 + (u_{2i}^*)^2],$$

$$e_{\max,n} = \min_i [(\hat{u}_{1i})^2 + (\hat{u}_{2i})^2].$$

If $\bar{u}$ is chosen randomly form U then under some assumptions it may be proved that $\bar{\Delta}_e(n)$ converges to $\{\bar{e}\}$ with probability 1. It is an extension of Theorem 11.2 for one subsystem. Consequently, we can use the following recursive algorithm for $n > 1$: Measure y_n. If $y_n \in D_y$

Knowledge validation for $\bar{u}_n = \bar{u}_n^*$

Prove if

$$\bigwedge_{e \in D_e^*(n-1)} \overline{u}_n \in D_u(e).$$

If yes then $D_e^*(n) = D_e^*(n-1)$. If not, determine new $D_e^*(n)$.

Knowledge updating for $\overline{u}_n = \overline{u}_n^*$

$$D_e^*(n) = \{e \in D_e^*(n-1) : \overline{u}_n \in D_u(e)\}.$$

Put $\hat{D}_e(n) = \hat{D}_e(n-1)$.

If $y_n \notin D_y$

Knowledge validation for $\overline{u}_n = \hat{\overline{u}}_n$

Prove if

$$\bigwedge_{e \in \hat{D}_e(n-1)} \overline{u}_n \in U - D_u(e).$$

If yes then $\hat{D}_e(n) = \hat{D}_e(n-1)$. If not, determine new $\hat{D}_e(n)$.

Knowledge updating for $\overline{u}_n = \hat{\overline{u}}_n$

$$\hat{D}_e(n) = \{e \in \hat{D}_e(n-1) : \overline{u}_n \in U - D_u(e)\}.$$

Put $D_e^*(n) = D_e^*(n-1)$.

Determine $\overline{\Delta}_e(n) = D_e^*(n) \cap \hat{D}_e(n)$.

We can also apply another version in which the estimation is performed independently in the particular subsystems on the lower level. For the j-th subsystem we can apply the procedure analogous to that for the system as a whole, with u_j, $D_{uj}(e_j)$, $D_{ej}^*(n)$ and $\hat{D}_{ej}(n)$ in place of $\overline{u}$, $D_u(e)$, $D_e^*(n)$ and $\hat{D}_e(n)$, respectively, where $e_j = (c_j, b)$. Now

$$u_{ji} = u_{ji}^*, \quad \text{if } y_i \in D_y, \quad \text{i.e.} \quad u_{ji}^* \in D_{uj}(\overline{e}_j),$$

$$u_{ji} = \hat{u}_{ji}, \quad \text{if } y_i \notin D_y, \quad \text{i.e.} \quad \hat{u}_{ji} \notin D_{uj}(\overline{e}_j).$$

Consequently, one obtains $\overline{\Delta}_{ej}(n)$ as the estimation of $\overline{e}_j$, and finally

$$\overline{\Delta}_e(n) = \overline{\Delta}_{e1}(n) \cap \ldots \cap \overline{\Delta}_{ek}(n).$$

If c_j and b are independent then $\overline{\Delta}_{ej}(n)$ may be reduced to the independent estimations $\overline{\Delta}_{cj}(n)$ and $\overline{\Delta}_b^{(j)}(n)$ of $\overline{c}_j$ and $\overline{b}$, respectively, where $\overline{\Delta}_b^{(j)}(n)$

denotes the estimation of $\bar{b}$ obtained from the j-th subsystem. Finally

$$\bar{\Delta}_b(n) = \bar{\Delta}_b^{(1)}(n) \cap \ldots \cap \bar{\Delta}_b^{(k)}(n)$$

where $\bar{\Delta}_b(n)$ denotes the final estimation of $\bar{b}$.

12.5 Learning System

The current validation and updating of the knowledge concerning the decision making may be used in the closed-loop learning system (Fig. 12.8) in which the successive estimations are used for the current determination of the decisions. The values of the parameters $e_n = (c_{1n}, c_{2n}, \ldots, c_{kn}; b_n)$ in the n-th step are chosen randomly from the set $\bar{\Delta}_e(n)$ and the decisions $\bar{u}_n = (u_{1n}, u_{2n}, \ldots, u_{kn})$ are chosen randomly from the set $D_u(e_n)$. Then, in the system under consideration two generators of random numbers are required: G_2 for the random choosing of e_n and G_1 for the random choosing of $\bar{u}_n$. The learning system may be considered as a closed-loop control system in which our distributed system is a control plant with inputs $u_1, \ldots, u_k$ and output y. Consequently, we have the following control algorithm (i.e. algorithm of the decision making with learning):

1. Choose randomly $\bar{u}_n$ from $D_u(e_{n-1})$.
2. Put $\bar{u}_n$ at the input of the plant, measure output and test the property $y_n \in D_y$.
3. Perform the knowledge validation and updating according to the procedure described in Sect. 12.4.2 and determine $\bar{\Delta}_e(n)$.
4. Choose randomly e_n from $\bar{\Delta}_e(n)$ and put into $D_u(e)$ in place of e.

The initial decision $\bar{u}_1$ is chosen randomly from U. The control algorithm with a trainer has the same form with $j_n = 1$, $j_n = 2$ instead of $y_n \in D_y$, $y_n \notin D_y$, respectively. In the second version described in Sect. 12.4.2 (with independent estimation in the particular subsystems on the lower level) the control algorithm has an analogous form.

In the closed-loop learning system one can consider another way of the estimation of the unknown parameters (the third version). It consists in the decomposition of the knowledge validation and updating into two levels, which corresponds to the decomposition of the decision making presented in Sect. 12.3.1. On the upper level,

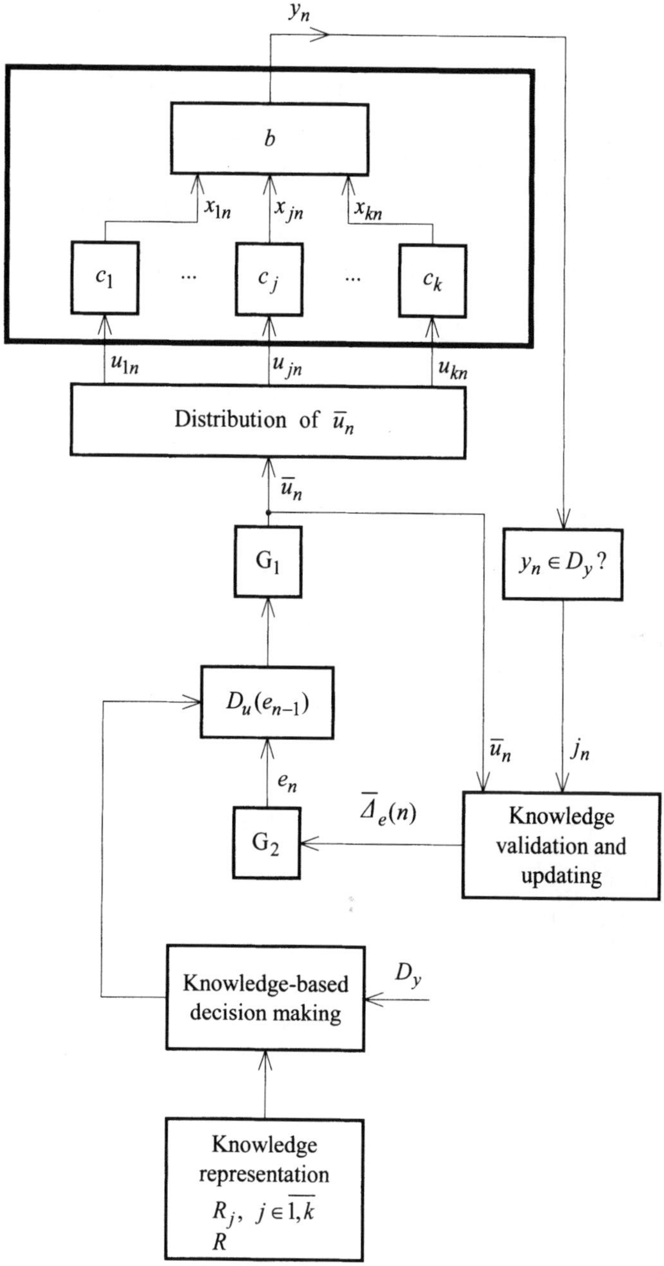

Figure 12.8. Closed-loop knowledge-based control system

for the estimation of $\overline{b}$ the measurements of x and testing the property $y \in D_y$ are used. The estimations of $\overline{c}_j$ on the lower level (independent for different j) are based on the measurements of u_j and testing the property $x_j \in D_{xj}(b)$. Hence, the estimation on the lower level may be performed after the estimation on the upper level, i.e. after the choosing of b_n which should be put into $D_{xj}(b)$ and $D_{uj}(c_j, b)$ in place of b. It is important to note that for the realization of the version under consideration the measurement of x is required.

Finally, the control algorithm in the closed-loop learning system has the following form:

1. Choose randomly u_{jn} from $D_{uj}(c_{j,n-1}, b_{n-1})$ $j = 1, 2, ..., k$.
2. Put u_{jn} at the input of the *j*-th subsystem.
3. Measure $x_{1n}, ..., x_{kn}$ and y_n.
4. Test the property $y_n \in D_y$.
5. If $y_n \in D_y$

Upper level knowledge validation for $x_n = x_n^*$

Prove if

$$\bigwedge_{b \in D_b^*(n-1)} x_{jn} \in D_{xj}(b), \quad j \in \overline{1,k}.$$

If yes then $D_b^*(n) = D_b^*(n-1)$. If not, determine new $D_b^*(n)$.

Upper level knowledge updating for $x_n = x_n^*$

$$D_b^*(n) = \{b \in D_b^*(n-1): \bigwedge_{j \in \overline{1,k}} x_{jn} \in D_{xj}(b)\}.$$

Put $\hat{D}_b(n) = \hat{D}_b(n-1)$.

6. If $y_n \notin D_y$

Upper level knowledge validation for $x_n = \hat{x}_n$

Prove if

$$\bigwedge_{b \in \hat{D}_b(n-1)} x_{jn} \in X - D_{xj}(b), \quad j \in \overline{1,k}.$$

If yes then $\hat{D}_b(n) = \hat{D}_b(n-1)$. If not, determine new $\hat{D}_b(n)$.

Upper level knowledge updating for $x_n = \hat{x}_n$

$$\hat{D}_b(n) = \{b \in \hat{D}_b(n-1): \bigwedge_{j \in \overline{1,k}} x_{jn} \in X_j - D_{xj}(b)\}.$$

Put $D_b^*(n) = D_b^*(n-1)$.

7. Choose randomly b_n from $\overline{\Delta}_b(n) = D_b^*(n) \cap \hat{D}_b(n)$ and put b_n into $D_{xj}(b)$ and $D_{uj}(c_j, b)$, $j = 1, ..., k$.

8. Test the property $x_{jn} \in D_{xj}(b_n)$.

9. If $x_{jn} \in D_{xj}(b_n)$

Lower level knowledge validation for $u_{jn} = u_{jn}^*$

Prove if

$$\bigwedge_{c_j \in D_{cj}^*(n-1)} u_{jn} \in D_{uj}(c_j, b_n).$$

If yes then $D_{cj}^*(n) = D_{cj}^*(n-1)$. If not, determine new $D_{cj}^*(n)$.

Lower level knowledge updating for $u_{jn} = u_{jn}^*$

$$D_{cj}^*(n) = \{c_j \in D_{cj}^*(n-1) : u_{jn} \in D_{uj}(c_j, b_n)\}.$$

Put $\hat{D}_{cj}(n) = \hat{D}_{cj}(n-1)$.

10. If $x_{jn} \notin D_{xj}(b_n)$

Lower level knowledge validation for $u_{jn} = \hat{u}_{jn}$

Prove if

$$\bigwedge_{c_j \in \hat{D}_{cj}(n-1)} u_{jn} \in U_j - D_{uj}(c_j, b_n).$$

If yes then $\hat{D}_{cj}(n) = \hat{D}_{cj}(n-1)$. If not, determine new $\hat{D}_{cj}(n)$.

Lower level knowledge updating for $u_{jn} = \hat{u}_{jn}$

$$\hat{D}_{cj}(n) = \{c_j \in \hat{D}_{cj}(n-1) : u_{jn} \in U_j - D_{uj}(c_j, b_n)\}.$$

Put $D_{cj}^*(n) = D_{cj}^*(n-1)$.

11. Choose randomly c_{jn} from $\overline{\Delta}_{cj}(n) = D_{cj}^*(n) \cap \hat{D}_{cj}(n)$ and put c_{jn} into $D_{uj}(c_j, b_n)$ in place of c_j.

Points 8, 9, 10, 11 should be performed for $j = 1, 2, ..., k$. The initial decisions u_{j1} are chosen randomly from U_j.

For the knowledge validation and updating on the upper and lower levels we used the same method as for the system as a whole, based on the formulas (12.24), (12.25) and (12.26).

In particular

$$x_i = x_i^* \quad \text{if} \quad y_i \in D_y ,$$

$$x_i = \hat{x}_i \quad \text{if} \quad y_i \notin D_y ,$$

$$u_{ji} = u_{ji}^* \quad \text{if} \quad x_{ji} \in D_{xj} ,$$

$$u_{ji} = \hat{u}_{ji} \quad \text{if} \quad x_{ji} \notin D_{xj} .$$

The block scheme of the learning system is presented in Fig. 12.9 where G_{j1}, G_{j2} and G_3 are the generators of random numbers, for random choosing of u_{jn}, c_{jn} and b_n, respectively. For the simplicity, only the determination of u_{jn} for the *j*-th subsystem is denoted in the figure and putting b_n into the block testing the property $x_{jn} \in D_{xj}$ is not denoted. Lower Level Knowledge and Upper Level Knowledge are denoted by LLK and ULK, respectively.

The decision making algorithm and the learning process have particular forms in a special case when the set D_u in the direct approach for the system as a whole or the sets D_{uj} and D_{xj} in the approach via the decomposition are domains closed by hypersurfaces (see the similar cases considered in Chapter 11). Let us explain it for the *j*-th subsystem and assume that $D_{uj}(c_j)$ is a domain closed by a hypersurface

$$G(u_j) \le c_j^2$$

where G is a function $G: U_j \to R^+$, e.g.

$$D_{uj}(c_j) = \{u_j \in U_j : u_j^{\mathrm{T}} u_j \le c_j^2\}$$

where u_j is the column vector. Let us assume that we have a learning sequence

$$(u_{j1}, j_1), \ldots, (u_{jn}, j_n)$$

and

$$u_{ji} = u_{ji}^* \quad \text{if} \quad j_i = 1, \quad \text{i.e.} \quad x_i \in D_{xj} ,$$

$$u_{ji} = \hat{u}_{ji} \quad \text{if} \quad j_i = 2, \quad \text{i.e.} \quad x_i \notin D_{xj} ,$$

$$D_{cj}^*(n) = \{c_j : \text{ every } u_{ji}^* \in D_{uj}(c_j)\} = [\underline{c}_{jn}, \infty) ,$$

$$\hat{D}_{cj}(n) = \{c_j : \text{ every } \hat{u}_{ji} \in U_j - D_{uj}(c_j)\} = [0, \overline{c}_{jn}) ,$$

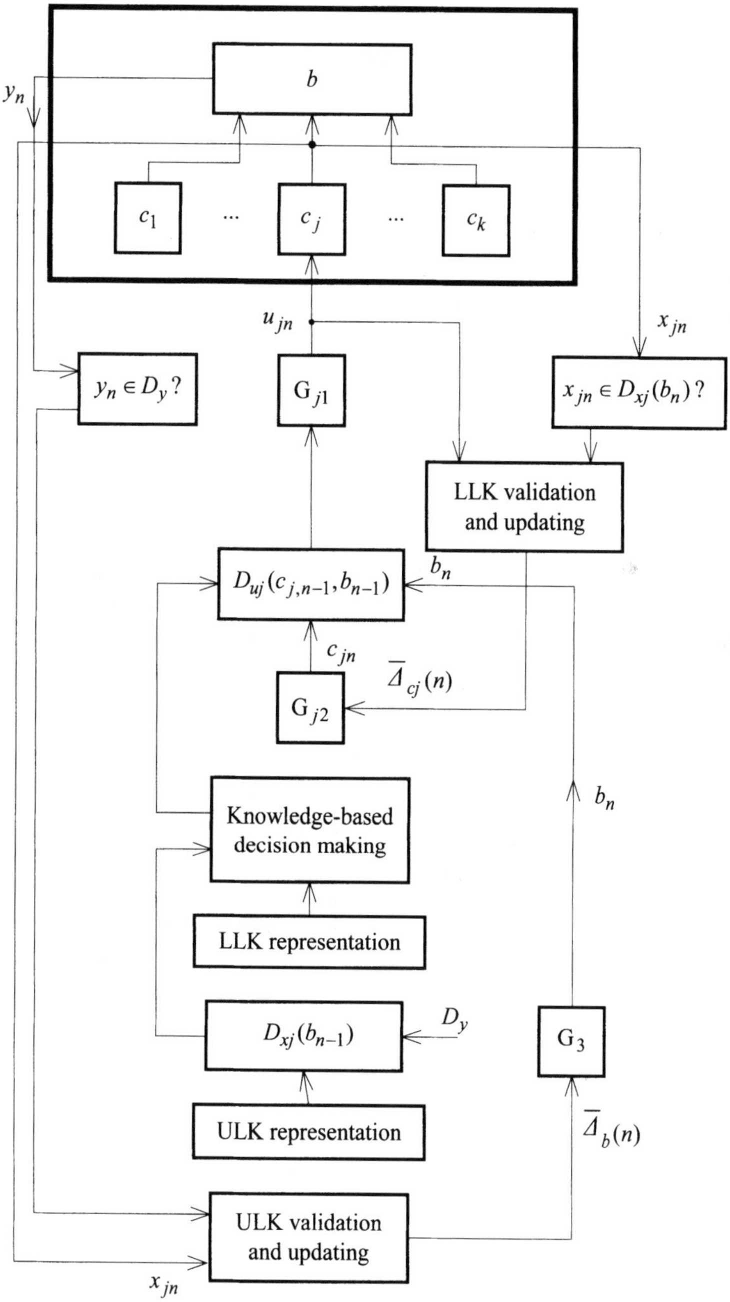

Figure 12.9. Another version of closed-loop learning system

$$\overline{\Delta}_{cj}(n) = D^{*}_{cj}(n) \cap \hat{D}_{cj}(n) = [\underline{c}_{jn}, \overline{c}_{jn})$$

where

$$\underline{c}^{2}_{jn} = \max_{i} \overset{*}{u}{}^{\mathrm{T}}_{ji} \overset{*}{u}_{ji}, \qquad \overline{c}^{2}_{jn} = \min_{i} \hat{u}^{\mathrm{T}}_{ji} \hat{u}_{ji}.$$

Another special case is concerned with the knowledge of the system (the decision making plant). In many practical applications the upper and lower level knowledge representations are described by linear inequalities. On the upper level the relation $R(x, y; b)$ is described by

$$x^{\mathrm{T}} b^{(1)} \le y \le x^{\mathrm{T}} b^{(2)}$$

where y is a one-dimensional variable and $b^{(1)}$, $b^{(2)}$ are the vectors of parameters, i.e. $b = (b^{(1)}, b^{(2)})$. On the lower level the relations $R_j(u_j, x_j; c_j)$ are described by

$$u_j^{\mathrm{T}} c_j^{(1)} \le x_j \le u_j^{\mathrm{T}} c_j^{(2)}$$

where x_j is a one-dimensional variable and $c_j = (c_j^{(1)}, c_j^{(2)})$. In this case the requirement $y \in D_y$ may have the form: $y \in [y_{\min}, y_{\max}]$ with the given numbers $y_{\min}$, $y_{\max}$. This is a typical description for a wide class of management problems, in particular for a project management or the management of a production process when the decisions u_j are concerned with task and/or resource allocations and the mathematical description of the plant is decomposed for particular subsystems which may be distributed in space.

For the special cases described above the computer programs (software systems) based on the presented methods of knowledge validation and updating have been written and used for simulations and for several practical applications e.g. [32, 51]. The main parts of the systems corresponding to the algorithms of the knowledge validation and learning presented above are the following:

1. The lower and upper level knowledge base (with two versions: relational and logical knowledge representation).
2. The procedure of the decision making in the form of D_{uj}.
3. The procedure of knowledge validation and updating.
4. The generators of random numbers.

For the decision making in the case of the logical knowledge representation a general purpose expert system CONTROL-LOG based on the logic-algebraic method [35] has been used.

Let us now consider two very simple examples to illustrate the general ideas, similar to the examples in Chapter 11.

Example 12.2.
At the upper level

$$b^{(1)} x_1 \le y \le b^{(2)} x_2,$$

at the lower level

$$x_1 \ge c_1 u_1, \quad x_2 \le c_2 u_2$$

where y, x_1, x_2, $b^{(1)}$, $b^{(2)}$, c_1, c_2 are one-dimensional positive variables.

For the requirement $y \in D_y = [y_{\min}, y_{\max}]$ it is easy to obtain the following results using the decomposition:

1. $D_{x1}(b^{(1)})$ and $D_{x2}(b^{(2)})$ are determined by

$$x_1 \ge \frac{y_{\min}}{b^{(1)}}, \quad x_2 \le \frac{y_{\max}}{b^{(2)}}.$$

2. $D_{u1}(c_1, b^{(1)})$ and $D_{u2}(c_2, b^{(2)})$ are determined by

$$u_1 \ge \frac{y_{\min}}{c_1 b^{(1)}}, \quad u_2 \le \frac{y_{\max}}{c_2 b^{(2)}}.$$

3. The estimations $\overline{\Delta}_{b1}(n)$ for $b^{(1)}$ and $\overline{\Delta}_{b2}(n)$ for $b^{(2)}$ are as follows:

$$\overline{\Delta}_{b1}(n) = [\frac{y_{\min}}{\min \overset{*}{x}_{1i}}, \frac{y_{\min}}{\max \hat{x}_{1i}}),$$

$$\overline{\Delta}_{b2}(n) = (\frac{y_{\max}}{\min \hat{x}_{2i}}, \frac{y_{\max}}{\max \overset{*}{x}_{2i}}].$$

4. The estimations $\overline{\Delta}_{c1}(n)$ for c_1 and $\overline{\Delta}_{c2}(n)$ for c_2 are the following:

$$\overline{\Delta}_{c1}(n) = [\frac{y_{\min}}{b_n^{(1)} \cdot \min \overset{*}{u}_{1i}}, \frac{y_{\min}}{b_n^{(1)} \cdot \max \hat{u}_{1i}}),$$

$$\overline{\Delta}_{c2}(n) = (\frac{y_{\max}}{b_n^{(2)} \cdot \min \hat{u}_{2i}}, \frac{y_{\max}}{b_n^{(2)} \cdot \max \overset{*}{u}_{2i}}]$$

where $b_n^{(1)}$ and $b_n^{(2)}$ are the values chosen randomly from the sets $\overline{\Delta}_{b1}(n)$ and $\overline{\Delta}_{b2}(n)$, respectively. □

Example 12.3.
Suppose that in the j-th subsystem u_j is a two-dimensional variable and for the

given value b_n D_{uj} is a circle

$$(u_j^{(1)})^2 + (u_j^{(2)})^2 \le c_j^2 . \tag{12.27}$$

Then

$$\overline{\Delta}_{cj}(n) = [\underline{c}_{jn}, \overline{c}_{jn})$$

where

$$\underline{c}_{jn}^2 = \max_i \, [\, (u_{ji}^{(1)*})^2 + (u_{ji}^{(2)*})^2 \,],$$

$$\overline{c}_{jn}^2 = \min_i [\, (\hat{u}_{ji}^{(1)})^2 + (\hat{u}_{ji}^{(2)})^2].$$

The knowledge updating and learning process has been simulated for the different numerical values c_j and different forms of probability densities f_u and f_c describing the random choosing of u_{jn} from $D_{uj}(c_{j,n-1})$ and c_{jn} from $\overline{\Delta}_{cj}(n)$, respectively. Fig. 12.10 presents the results for $c_j = 2$, the rectangular probability density f_c, and probability density f_u constant in the circle (12.27) and equal to zero outside this circle. The simulations for different cases showed the great influence of the shape of probability distributions on the convergence and quality of learning.

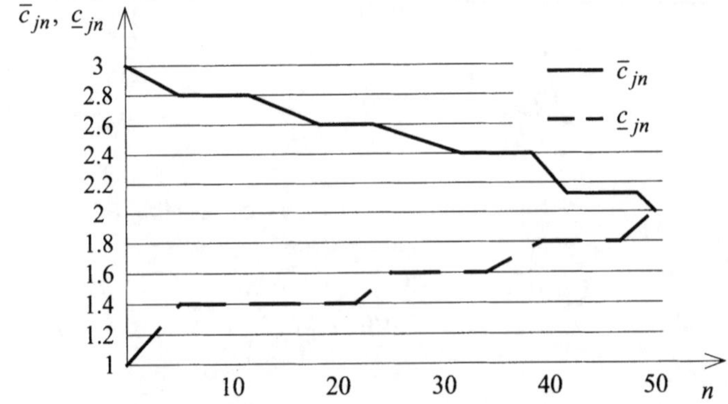

Figure 12.10. Results of simulations in Example 12.3

In the computer implementation of the learning algorithms it is possible to apply the parallel architecture with concurrent knowledge updating and decision making for the separate subsystems. The simulations showed the significant influence of the parameters in the presented algorithms on the convergence and quality of learning. It is worth noting that for the complicated forms of the knowledge representation the third version presented in Sect. 12.5 (with the decomposition of knowledge validation and updating into two levels) may be simpler and more convenient than the first and the second version.

13 Complex of Operations

There exists a great variety of decision problems concerning allocation and scheduling in a complex of operations (a complex operation system), with many applications to manufacturing and computer systems (see e.g. [8]). This chapter is concerned with the control of the complex of parallel operations containing unknown parameters in the relational knowledge representation. The complex of parallel operations is considered here as a specific uncertain decision plant. The control consists in a proper distribution of the given size of a task or the given amount of a resource, taking into account the execution time of the whole complex. It may mean the distribution of raw material in the case of a manufacturing process or a load distribution in a group of parallel computers. In the deterministic case where the operations are described by functions determining the relationship between the execution time and the size of the task or the amount of the resource, the optimization problem consisting in the determination of the distribution that minimizes the execution time of the complex may be formulated and solved (see e.g. [13]). In the case of uncertainty, various formulations of decision problems adequate for the different descriptions of the uncertainty may be considered [15, 16, 49, 51, 54].

13.1 Complex of Parallel Operations with Relational Knowledge Representation

In the deterministic case the complex of parallel operations is described by a set of functions

$$T_i = \varphi_i(u_i), \quad i = 1,2,...,k, \tag{13.1}$$

$$T = \max\{T_1, T_2, ..., T_k\} = \max_i \varphi_i(u_i) \triangleq \Phi(u) \tag{13.2}$$

where T_i is the execution time of the i-th operation, u_i for each i is the size of a task in the problem of tasks allocation or the amount of a resource in the problem of resources allocation, T is the execution time of the whole complex and

$$u = (u_1, u_2, ..., u_k) \in \overline{U}.$$

The set $\overline{U} \subset R^k$ is determined by the constraints

$$\bigwedge_{i \in \overline{1,k}} u_i \geq 0, \qquad \sum_{i=1}^{k} u_i = U \tag{13.3}$$

where U is the total size of the task or the total amount of the resource to be distributed among the operations. The function φ_i is an increasing function of u_i (and $\varphi_i(0) = 0$) in the case of tasks and a decreasing function of u_i in the case of resources. The complex of the parallel operations may be considered as a specific decision (control) plant (Fig. 13.1) with input vector u and a single output $y = T$, described by a function $\Phi(u)$ determined according to (13.2). If the functions φ_i are known, it is possible to formulate the following decision (control) problem, optimal from the execution time point of view: to determine the allocation (the distribution) u^* minimizing the execution time T, subject to constraints (13.3). It is easy to show that if all operations start at the same time then u^* satisfies the following equations:

$$T_1(u_1) = T_2(u_2) = \ldots = T_k(u_k). \tag{13.4}$$

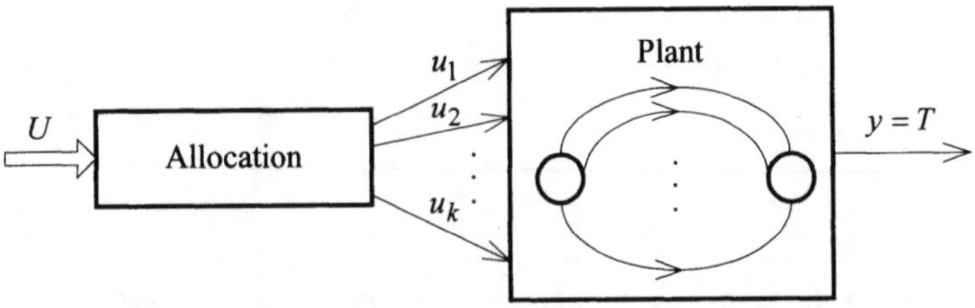

Figure 13.1. Complex of parallel operation as a decision plant

For example, if

$$T_i = c_i u_i, \quad c_i > 0, \quad i = 1, 2, \ldots, k, \tag{13.5}$$

then

$$u_i^* = \frac{U}{c_i} \left(\sum_{i=1}^{k} \frac{1}{c_i} \right)^{-1} \tag{13.6}$$

and

$$T^* = \arg\max_{u \in \overline{U}} T = U \left(\sum_{i=1}^{k} \frac{1}{c_i} \right)^{-1}. \tag{13.7}$$

If

$$T_i = \frac{c_i}{u_i}, \quad c_i > 0, \quad i = 1, 2, ..., k,$$

then

$$u_i^* = c_i U (\sum_{i=1}^{k} c_i)^{-1}, \qquad T^* = \frac{1}{U} \sum_{i=1}^{k} c_i .$$

Consider now an uncertain complex described by a set of inequalities

$$T_i \le \varphi_i(u_i) \tag{13.8}$$

where φ_i is a known function, increasing in the case of tasks and decreasing in the case of resources. The inequalities (13.10) together with the function $T = \max\{T_1, T_2, ..., T_k\}$ form a relational knowledge representation $R(u, y)$ in our case. For a user's requirement $T \le \alpha$ or $T \in [0, \alpha]$ one can formulate the decision problem in a way described in Sect. 2.2.

Decision problem: For the given φ_i $(i = 1, 2, ..., k)$ and α one should find the largest set $D_u \subset \overline{U}$ such that the implication

$$u \in D_u \to T \in [0, \alpha]$$

is satisfied.
Then

$$D_u = \{u \in \overline{U} : D_T(u) \subseteq [0, \alpha]\} \tag{13.9}$$

where $D_T(u)$ is the set of possible values T for the fixed u, i.e. $D_T(u)$ is determined by the inequality

$$T \le \max_i \varphi_i(u_i) .$$

Consequently,

$$D_u = \{u \in \overline{U} : \bigwedge_{i \in \overline{1,k}} [\varphi_i(u_i) < \alpha]\} ; \tag{13.10}$$

e.g. for (13.5)

$$D_u = \{u \in R^k : (\sum_{i=1}^{k} u_i = U) \wedge \bigwedge_{i \in \overline{1,k}} (u_i \ge 0) \wedge (c_i u_i \le \alpha)\} .$$

It is easy to note that the solution exists (i.e. D_u is not empty set) of $\alpha \ge T^*$ where T^* is the minimal execution time obtained for the optimal allocation u^*. In this case $u^* \in D_u$. For example, if $k = 2$ and

$$\alpha \geq \frac{c_1 c_2 U}{c_1 + c_2} = T^*$$

then D_u is determined by equality $u_2 = U - u_1$ and inequality $\beta \leq u_1 \leq \delta$ where

$$\beta = \frac{\alpha}{c_1}, \qquad \delta = U - \frac{\alpha}{c_2}$$

(see Fig. 13.2).

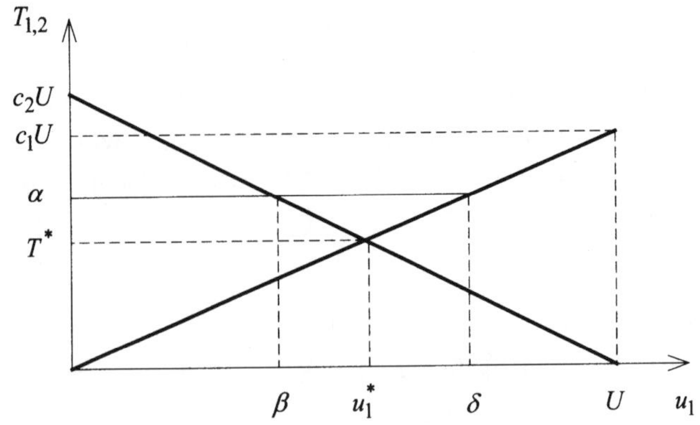

Figure 13.2. Illustration for the example

13.2 Application of Uncertain Variables [49, 54]

Let us consider a complex of parallel operations described by the inequalities

$$T_i \leq \varphi_i(u_i, x_i), \quad i = 1, 2, ..., k \tag{13.11}$$

where u_i is the size of a task assigned to the i-th operation, $x_i \in R^1$ is a parameter and φ_i is a known increasing function of u_i. The parameter x_i is unknown and is assumed to be a value of an uncertain variable $\bar{x}_i$ described by a certainty distribution $h_{xi}(x_i)$ given by an expert. Now the relational knowledge representation, consisting of (13.11) and the relationship $T = \max(T_1, T_2, ..., T_k)$, is completed by the functions $h_{xi}(x_i)$. We assume that $\bar{x}_1, \bar{x}_2, ..., \bar{x}_k$ are independent uncertain variables, i.e.

$$h_x(x) = \min_i h_{xi}(x_i)$$

where $x = (x_1, x_2, ..., x_k)$. The largest set of decisions $D_u(x)$ depends on x and is determined by (13.10) with $\varphi_i(u_i, x_i)$ in place of $\varphi_i(u_i)$. The description of the complex is analogous for the resource allocation problem. Then u_i is the amount of a resource assigned to the *i*-th operation, φ_i is a decreasing function of u_i and U denotes the total amount of the resource to be distributed.

According to the general formulation of the decision problem presented in Sect. 5.5 (version II), the allocation problem may be formulated as an optimization problem consisting in finding the optimal allocation u^* that maximizes the certainty index of the soft property: "u approximately belongs to $D_u(\overline{x})$" or "the set of possible values T approximately belongs to $[0,\alpha]$" (i.e. belongs to $[0,\alpha]$ for an approximate value of $\overline{x}$).

Optimal allocation (decision) problem: For the given φ_i , h_{xi} $(i \in \overline{1,k})$, U and α find

$$u^* = \arg\max_{u \in \overline{U}} v(u)$$

where

$$v(u) = v\{D_T(u;\overline{x}) \tilde{\subseteq} [0,\alpha]\} = v(T(u,\overline{x}) \tilde{\leq} \alpha) . \tag{13.12}$$

The soft property "$D_T(u;\overline{x}) \tilde{\subseteq} [0,\alpha]$" is denoted here by "$T(u,\overline{x}) \tilde{\leq} \alpha$", and $D_T(u;x)$ denotes the set of possible values T for the fixed u, determined by the inequality

$$T \leq \max_i \varphi_i(u_i, x_i) .$$

According to (13.12)

$$v(u) = v\{[T_1(u_1,\overline{x}_1) \tilde{\leq} \alpha] \wedge [T_2(u_2,\overline{x}_2) \tilde{\leq} \alpha] \wedge ... \wedge [T_k(u_k,\overline{x}_k) \tilde{\leq} \alpha]\} .$$

Then

$$u^* = \arg\max_{u \in \overline{U}} \min_i v_i(u_i) \tag{13.13}$$

where

$$v_i(u_i) = v[T_i(u_i,\overline{x}_i) \tilde{\leq} \alpha] = v[\varphi_i(u_i,\overline{x}_i) \tilde{\leq} \alpha] = v[\overline{x}_i \tilde{\in} D_{xi}(u_i)],$$

$$D_{xi}(u_i) = \{x_i \in R^1 : \varphi_i(u_i, x_i) \leq \alpha\} .$$

Finally

$$v_i(u_i) = \max_{x_i \in D_{xi}(u_i)} h_{xi}(x_i) \tag{13.14}$$

and

$$u^* = \arg\max_{u \in \overline{U}} \min_i \max_{x_i \in D_{xi}(u_i)} h_{xi}(x_i) \tag{13.15}$$

The value $v_i(u_i)$ denotes the certainty index that in the i-th operation an approximate value of the execution time is less than α. The procedure of finding the optimal allocation u^* is then the following:

1. To determine $v_i(u_i)$ using (13.14).
2. To determine u^* according to (13.13), subject to constraints (13.3).

Assume that $\varphi_i(u_i, x_i)$ is an increasing function of x_i. Then the set $D_{xi}(u_i)$ is determined by the inequality $x_i \le \hat{x}_i(u_i)$ where $\hat{x}_i(u)$ is the solution of the equation

$$\varphi_i(u_i, x_i) = \alpha \tag{13.16}$$

and

$$v_i(u_i) = \max_{x_i \le \hat{x}_i(u_i)} h_{xi}(x_i). \tag{13.17}$$

Let us denote by x_i^* the value maximizing $h_{xi}(x_i)$, i.e. $h_{xi}(x_i^*) = 1$. It is easy to see that in the determination of v_i according to (13.17) only the values of $h_{xi}(x_i)$ for $x_i \le x_i^*$ are taken into account.

We may use another version of the uncertain variables described in Sect. 4.3. Let us introduce an N-uncertain variable $\overline{x}$ for which the certainty index that $\overline{x} \,\tilde{\in}\, D_x$ is defined as follows:

$$v_n(\overline{x} \,\tilde{\in}\, D_x) = 1 - v(\overline{x} \,\tilde{\in}\, \overline{D}_x) = 1 - \max_{x \in \overline{D}_x} h_x(x)$$

where $\overline{D}_x$ is the complement of D_x, i.e. $\overline{D}_x = X - D_x$.

Using the description of the uncertain parameters x_i in the form of N-uncertain variables we obtain

$$v_n[T(u,\overline{x}) \,\tilde{\le}\, \alpha] = 1 - v[T(u,\overline{x}) \,\tilde{\ge}\, \alpha] \triangleq v_n(u)$$

and

$$v[T(u,\overline{x}) \,\tilde{\ge}\, \alpha] = v\{[T_1(u_1,\overline{x}_1) \,\tilde{\ge}\, \alpha] \vee \ldots \vee [T_k(u_k,\overline{x}_k) \,\tilde{\ge}\, \alpha]\} = \max_i \hat{v}_i(u_i)$$

where

$$\hat{v}_i(u_i) = v[T_i(u_i,\overline{x}_i) \,\tilde{\ge}\, \alpha] = \max_{x_i \in \overline{D}_{xi}(u_i)} h_{xi}(x_i).$$

Then the optimal decision u_N^* maximizing $v_n(u)$ is as follows:

$$u_N^* = \arg\max_{u \in \overline{U}} [1 - \max_i \hat{v}_i(u_i)] = \arg\min_{u \in \overline{U}} \max_i \hat{v}_i(u_i). \qquad (13.18)$$

In the case corresponding to (13.17)

$$\hat{v}_i(u_i) = \max_{x_i \geq \hat{x}_i(u_i)} h_{xi}(x_i). \qquad (13.19)$$

The value $\hat{v}_i(u_i)$ denotes the certainty index that in the i-th operation an approximate value of the execution time is greater than α.

Theorem 13.1. If $\varphi_i(u_i, x_i)$ is an increasing function of x_i for every $u_i \geq 0$ and $h_{xi}(x_i)$ is a decreasing function of x_i for $x_i \geq x_i^*$ then the optimal allocation u_N^* is determined by the equations

$$\left.\begin{aligned} \hat{v}_1(u_1) = \hat{v}_2(u_2) = \ldots = \hat{v}_k(u_k), \\ u_1 + u_2 + \ldots + u_k = U. \end{aligned}\right\} \qquad (13.20)$$

Proof: If $h_{xi}(x_i)$ is a decreasing function of x_i for $x_i \geq x_i^*$ then it follows from (13.19) that

$$\hat{v}_i(u_i) = \begin{cases} 1 & \text{for} \quad \hat{x}_i(u_i) \leq x_i^* \\ h_{xi}[\hat{x}_i(u_i)] & \text{for} \quad \hat{x}_i(u_i) \geq x_i^*. \end{cases}$$

The function $\hat{x}_i(u_i)$ as a solution of Equation (13.16) is a decreasing function of u_i where u_i is the size of a task or an increasing function of u_i where u_i is the amount of a resource. Consequently, $\hat{v}_i(u_i)$ is a non-decreasing (non-increasing) function of u_i and the result of minimization in (13.18) with the constraints (13.3) satisfies Equations (13.20). □

The determination of u_N^* from the formal point of view is analogous to the determination of the optimal allocation in the deterministic case, for the operations described by traditional functional models

$$T_i = \varphi_i(u_i, x_i), \quad i = 1, 2, \ldots, k,$$

i.e. the determination of the allocation u minimizing the execution time

$$T = \max_i [\varphi_1(u_1, x_1), \ldots, \varphi_k(u_k, x_k)]$$

with the constraints (13.3).

The procedure of finding the optimal allocation u_N^* based on (13.20) is then as follows:

1. To determine the inverse functions $u_i = g_i^{-1}(\hat{v}_i)$ where $\hat{v}_i(u_i) \stackrel{\Delta}{=} g_i(u_i)$.
2. To solve the equation

$$\sum_{i=1}^{k} g_i^{-1}(\hat{v}) = U \tag{13.21}$$

with respect to $\hat{v}$.
3. To determine

$$u_{Ni}^* = g_i^{-1}(\hat{v}^*)$$

where $\hat{v}^*$ is the solution of Equation (13.21).
If $\overline{x}$ is considered as a *C*-uncertain variable then the optimal decision u_c^* maximizing the certainty index

$$v_c[T(u,\overline{x}) \mathrel{\tilde{\le}} \alpha] = \frac{1}{2}[v(u) + v_n(u)] \stackrel{\Delta}{=} v_c(u)$$

is as follows:

$$u_c^* = \arg\max_{u \in \overline{U}} [\min_i v_i(u_i) + 1 - \max_i \hat{v}_i(u_i)] = \arg\max_{u \in \overline{U}} [\min_i v_i(u_i) - \max_i \hat{v}_i(u_i)]\,. \tag{13.22}$$

13.3 Special Cases and Examples

In many cases an expert gives the value x_i^* and the interval of the approximate values of $\overline{x}_i$: $x_i^* - d_i \le x_i \le x_i^* + d_i$. Then we assume that $h_{xi}(x_i)$ has a triangular form presented in Fig. 13.3 where $d_i \le x_i^*$. Let us consider the relation (13.11) in the form $T_i \le x_i u_i$ where $x_i > 0$ and u_i denotes the size of a task. In this case, using (13.17) and (13.19) it is easy to obtain the following formulas for the functions $v_i(u_i)$ and $\hat{v}_i(u_i)$:

$$v_i(u_i) = \begin{cases} 1 & \text{for} \quad u_i \le \dfrac{\alpha}{x_i^*} \\ \dfrac{1}{d_i}\left(\dfrac{\alpha}{u_i} - x_i^*\right) + 1 & \text{for} \quad \dfrac{\alpha}{x_i^*} \le u_i \le \dfrac{\alpha}{x_i^* - d_i} \\ 0 & \text{for} \quad u_i \ge \dfrac{\alpha}{x_i^* - d_i}\,, \end{cases} \tag{13.23}$$

$$\hat{v}_i(u_i) = \begin{cases} 0 & \text{for} \quad u_i \le \dfrac{\alpha}{x_i^* + d_i} \\ -\dfrac{1}{d_i}(\dfrac{\alpha}{u_i} - x_i^*) + 1 & \text{for} \quad \dfrac{\alpha}{x_i^* + d_i} \le u_i \le \dfrac{\alpha}{x_i^*} \\ 1 & \text{for} \quad u_i \ge \dfrac{\alpha}{x_i^*}. \end{cases} \quad (13.24)$$

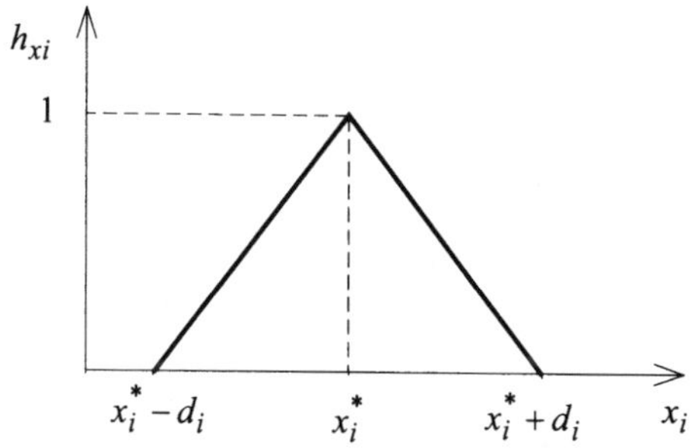

Figure 13.3. Example of the certainty distribution

For the relations $T_i \le x_i u_i^{-1}$ where u_i denotes the size of a resource, the functions $v_i(u_i)$ and $\hat{v}_i(u_i)$ have an analogous form with u_i^{-1} in place of u_i:

$$v_i(u_i) = \begin{cases} 0 & \text{for} \quad u_i \le \dfrac{x_i^* - d_i}{\alpha} \\ \dfrac{1}{d_i}(\alpha u_i - x_i^*) + 1 & \text{for} \quad \dfrac{x_i^* - d_i}{\alpha} \le u_i \le \dfrac{x_i^*}{\alpha} \\ 1 & \text{for} \quad u_i \ge \dfrac{x_i^*}{\alpha}, \end{cases} \quad (13.25)$$

$$\hat{v}_i(u_i) = \begin{cases} 1 & \text{for} \quad u_i \le \dfrac{x_i^*}{\alpha} \\ -\dfrac{1}{d_i}(\alpha u_i - x_i^*) + 1 & \text{for} \quad \dfrac{x_i^*}{\alpha} \le u_i \le \dfrac{x_i^* + d_i}{\alpha} \\ 0 & \text{for} \quad u_i \ge \dfrac{x_i^* + d_i}{\alpha}. \end{cases} \quad (13.26)$$

Example 13.1.
Let us consider the resource allocation for two operations ($k = 2$). Now in the maximization problem (13.13) the decision u_1^* may be found by solving the equation $v_1(u_1) = v_2(U - u_1)$ and $u_2^* = U - u_1^*$. Using (13.25), we obtain the

following result:
1. For

$$\alpha \le \frac{x_1^* - d_1 + x_2^* - d_2}{U} \tag{13.27}$$

$v(u) = 0$ for any u_1.
2. For

$$\frac{x_1^* - d_1 + x_2^* - d_2}{U} \le \alpha \le \frac{x_1^* + x_2^*}{U} \tag{13.28}$$

we obtain

$$u_1^* = \frac{\alpha d_1 U + x_1^* d_2 - x_2^* d_1}{\alpha(d_1 + d_2)}, \tag{13.29}$$

$$v(u^*) = \frac{1}{d_1}[\alpha u_1^* - x_1^*] + 1. \tag{13.30}$$

3. For

$$\alpha \ge \frac{x_1^* + x_2^*}{U} \tag{13.31}$$

we obtain $v(u^*) = 1$ for any u_1 satisfying the condition

$$\frac{x_1^*}{\alpha} \le u_1 \le U - \frac{x_2^*}{\alpha}.$$

In the case (13.27) α is too small (the requirement is too strong) and it is not possible to find the allocation for which $v(u)$ is greater than 0. In the case (13.28) we obtain one solution maximizing $v(u)$. For the numerical data $U = 9$, $\alpha = 0.5$, $x_1^* = 2$, $x_2^* = 3$, $d_1 = d_2 = 1$, using (13.29) and (13.30) we obtain $u_1^* = 3.5$, $u_2^* = 5.5$ and $v = 0.75$, which means that the requirement $T \le \alpha$ will be approximately satisfied with the certainty index 0.75. The solution of the optimization problem (13.18) based on (13.26) may be obtained in an analogous way:
1. For

$$\alpha \le \frac{x_1^* + x_2^*}{U} \tag{13.32}$$

$v_n(u) = 0$ for any u_1.
2. For

$$\frac{x_1^* + x_2^*}{U} \le \alpha \le \frac{x_1^* + d_1 + x_2^* + d_2}{U} \tag{13.33}$$

$u_{N1}^* = u_1^*$ in the formula (13.29) and

$$v_n(u^*) = \frac{1}{d_1}(\alpha u_{N1}^* - x_1^*). \tag{13.34}$$

3. For

$$\alpha \ge \frac{x_1^* + d_1 + x_2^* + d_2}{U}$$

we obtain $v_n(u^*) = 1$ for any u_1 satisfying the condition

$$\frac{x_1^* + d_1}{\alpha} \le u_1 \le U - \frac{x_2^* + d_2}{\alpha}.$$

For the numerical data we have the case (13.32) and $v_n(u) = 0$.

The optimization problem (13.22) for *C*-uncertain variables is much more complicated and should be considered in the different intervals of α introduced for v and v_n. For example, if

$$\frac{x_1^* + x_2^*}{U} \le \alpha \le \frac{x_1^* + d_1 + x_2^* + d_2}{U} \tag{13.35}$$

which means the combination of the cases (13.31) and (13.33), then $u_{c1}^* = u_{N1}^*$ and

$$v_c(u^*) = \frac{1}{2}[v(u^*) + 1 - v_n(u^*)].$$

Substituting $v(u^*) = 1$ and (13.34) yields

$$v_c(u^*) = 1 - \frac{1}{2d_1}(\alpha u_{c1}^* - x_1^*). \tag{13.36}$$

For the numerical data $U = 9$, $\alpha = 0.6$, $x_1^* = 2$, $x_2^* = 3$, $d_1 = d_2 = 1$ the inequality (13.35) is satisfied. Then, by using (13.29) and (13.36) we obtain $u_{c1}^* = 3.67$ and $v_c(u^*) = 0.9$. The results for these data in the case v and v_n are as follows: $u_{N1}^* = u_{c1}^* = 3.67$ and $v_n(u^*) = 0.2$; $v(u^*) = 1$ for any u_1 from the interval $[3.33, 4]$. □

Example 13.2.

Let us consider the task allocation for two operations. In the maximization problem (13.13) the decision u_1^* may be found by solving the equation $v_1(u_1) = v_2(U - u_1)$

and $u_2^* = U - u_1^*$. Using (13.23), we obtain the following result:

1. For

$$\alpha \le \frac{U(x_1^* - d_1)(x_2^* - d_2)}{x_1^* - d_1 + x_2^* - d_2} \tag{13.37}$$

$v(u) = 0$ for any u_1.

2. For

$$\frac{U(x_1^* - d_1)(x_2^* - d_2)}{x_1^* - d_1 + x_2^* - d_2} \le \alpha \le \frac{Ux_1^* x_2^*}{x_1^* + x_2^*} \tag{13.38}$$

u_1^* is a root of the equation

$$\frac{1}{d_1}(\frac{\alpha}{u_1} - x_1^*) = \frac{1}{d_2}(\frac{\alpha}{U - u_1} - x_2^*)$$

satisfying the condition

$$\frac{\alpha}{x_1^*} \le u_1^* \le \frac{\alpha}{x_1^* - d_1},$$

and $v(u^*) = v_1(u_1^*)$.

3. For

$$\alpha \ge \frac{Ux_1^* x_2^*}{x_1^* + x_2^*} \tag{3.39}$$

$v(u^*) = 1$ for any u_1 satisfying the condition

$$U - \frac{\alpha}{x_2^*} \le u_1 \le \frac{\alpha}{x_1^*}.$$

For example, if $U = 2$, $\alpha = 2$, $x_1^* = 2$, $x_2^* = 3$, $d_1 = d_2 = 1$ then using (13.38) yields $u_1^* = 1.25$, $u_2^* = 0.75$, $v(u^*) = 0.6$.

The result is simpler under the assumption

$$\frac{x_1^*}{d_1} = \frac{x_2^*}{d_2} \triangleq \gamma. \tag{13.40}$$

Then in the case (13.40)

$$u_1^* = \frac{Ux_2^*}{x_1^* + x_2^*}, \qquad u_2^* = \frac{Ux_1^*}{x_1^* + x_2^*},$$

$$v(u^*) = v_1(u_1^*) = \frac{1}{d_1}(\frac{\alpha}{u_1^*} - x_1^*) + 1 = \gamma[\frac{\alpha(x_1^* + x_2^*)}{U x_1^* x_2^*} - 1] + 1. \quad (13.41)$$

The formula (13.41) shows that $v(u^*)$ is a linear function of the parameter γ characterizing the expert's uncertainty. □

The result in point 3 of Example 13.2 may be easily generalized for k operations described by the inequalities $T_i \le x_i u_i$ and for any form of $h_{xi}(x_i)$. Let us denote by x_i^* the value maximizing $h_{xi}(x_i)$, i.e. $h_{xi}(x_i^*) = 1$.

Theorem 13.2. If

$$\alpha \ge \frac{U}{\sum_{i=1}^{k} (x_i^*)^{-1}} \quad (13.42)$$

then

$$D_u = \{u : (\bigwedge_{i \in \overline{1,k}} 0 \le u_i \le \frac{\alpha}{x_i^*}) \wedge \sum_{i=1}^{k} u_i = U\} \quad (13.43)$$

is the set of all allocations $u^* = (u_1^*, u_2^*, ..., u_k^*)$ such that $v(u^*) = 1$.

Proof: From (13.14) it follows that if

$$u_i \le (x_i^*)^{-1} \quad (13.44)$$

then $x_i^* \in D_{xi}(u_i)$ and consequently $v_i(u_i) = 1$. It is easy to see that under the assumption (13.42) there exists an allocation $u = (u_1, u_2, ..., u_k)$ such that (13.44) is satisfied for each $i \in \overline{1,k}$ and $u_1 + u_2 + ... + u_k = U$. All allocations satisfying these conditions form the set D_u described by (13.43), and if $u \in D_u$ then, according to (13.13), $v(u^*) = 1$. □

13.4 Decomposition and Two-level Control

The determination of the control decision u^* may be difficult for $k > 2$ because of the great computational difficulties. To decrease these difficulties we can apply the decomposition of the complex into two subcomplexes and consequently to obtain a two-level control system (Fig. 13.4). This approach is based on the idea of decomposition and two-level control presented for the deterministic case [13]. At the upper level the value U is divided into U_1 and U_2 assigned to the first and the second subcomplex, respectively, and at the lower level the allocation $u^{(1)}$, $u^{(2)}$

for the subcomplexes is determined. Let us introduce the following notation:
n, m – the number of operations in the first and the second complex, respectively, $n+m=k$, $T^{(1)}$, $T^{(2)}$ – the execution times in the subcomplexes, i.e.

$$T^{(1)} = \max(T_1, T_2, ..., T_n), \quad T^{(2)} = \max(T_{n+1}, T_{n+2}, ..., T_{n+m}),$$

$u^{(1)}$, $u^{(2)}$ – the allocations in the subcomplexes, i.e.

$$u^{(1)} = (u_1, ..., u_n), \quad u^{(2)} = (u_{n+1}, ..., u_{n+m}).$$

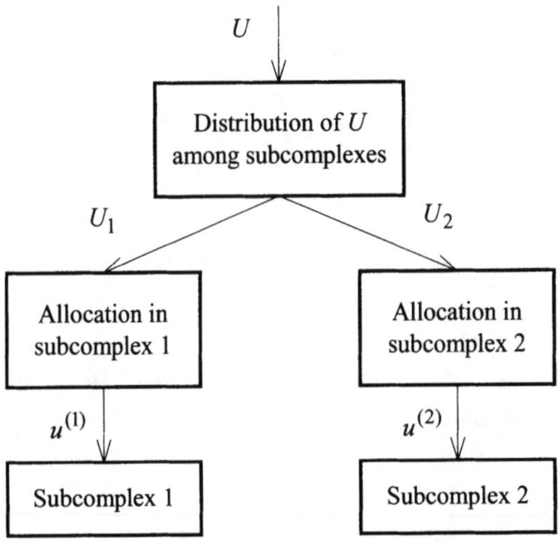

Figure 13.4. Two-level control system

The procedure of the determination of u^* is then the following:
1. To determine the allocation $u^{(1)*}(U_1)$, $u^{(2)*}(U_2)$ and the certainty indexes $v^{(1)*}(U_1)$, $v^{(2)*}(U_2)$ in the same way as u^*, v^* in Sect. 13.2, with U_1 and U_2 in place of U.
2. To determine U_1^*, U_2^* via the maximization of

$$v(T \,\tilde{\leq}\, \alpha) = v[(T^{(1)} \,\tilde{\leq}\, \alpha) \wedge (T^{(2)} \,\tilde{\leq}\, \alpha)] \stackrel{\Delta}{=} v(U_1, U_2).$$

Then

$$(U_1^*, U_2^*) = \arg \max_{U_1, U_2} \min\{v^{(1)*}(U_1), v^{(2)*}(U_2)\}$$

with the constraints: $U_{1,2} \geq 0$, $U_1 + U_2 = U$.

3. To find the values of $u^{(1)*}$, $u^{(2)*}$ and v^* putting U_1^* and U_2^* into the results $u^{(1)*}(U_1)$, $u^{(2)*}(U_2)$ obtained in point 1 and into $v(U_1,U_2)$ in point 2.

It may be shown that the result obtained via the decomposition is the same as the result of the direct approach presented in Sect. 13.2.

Example 13.3.

Let us consider the resource allocation problem the same as in Example 13.1 for $k=4$ and introduce the decomposition into two subcomplexes with $n=m=2$. Using the result obtained in Example 13.1 with $U^{(1)}$, $v^{(1)}$ in place of U, v, we have the following result for the first subcomplex:

1. For

$$U \le \frac{x_1^* - d_1 + x_2^* - d_2}{\alpha}$$

$v^{(1)*}(U_1) = 0$.

2. For

$$\frac{x_1^* - d_1 + x_2^* - d_2}{\alpha} \le U \le \frac{x_1^* + x_2^*}{\alpha}$$

we obtain

$$v^{(1)*}(U_1) = A_1 U_1 + B_1$$

where

$$A_1 = \frac{\alpha}{d_1 + d_2}, \quad B_1 = \frac{x_1^* d_2 - x_2^* d_1}{d_1(d_1 + d_2)} - \frac{x_1^*}{d_1} + 1\,.$$

3. For

$$U \ge \frac{x_1^* + x_2^*}{\alpha}$$

$v^{(1)*}(U_1) = 1$.

The relationship $v^{(2)*}(U_2)$ is the same with x_3, x_4, d_3, d_4, A_2, B_2 in place of x_1, x_2, d_1, d_2, A_1, B_1. The value U_1^* may be determined by solving the equation $v^{(1)*}(U_1) = v^{(2)*}(U - U_1)$ and $U_2^* = U - U_1^*$.

The result is as follows:

1. For

$$\alpha \le \frac{x_1^* - d_1 + x_2^* - d_2 + x_3^* - d_3 + x_4^* - d_4}{U}$$

$v(U_1, U_2) = 0$.

2. For

$$\frac{x_1^* - d_1 + x_2^* - d_2 + x_3^* - d_3 + x_4^* - d_4}{U} \le \alpha \le \frac{x_1^* + x_2^* + x_3^* + x_4^*}{U}$$

we obtain

$$U_1^* = \frac{A_2 U + B_2 - B_1}{A_1 + A_2}, \quad U_2^* = \frac{A_1 U + B_1 - B_2}{A_1 + A_2},$$

$$v(U_1^*, U_2^*) = \frac{A_1 A_2 U + A_1 B_2 + A_2 B_1}{A_1 + A_2}.$$

3. For

$$\alpha \ge \frac{x_1^* + x_2^* + x_3^* + x_4^*}{U}$$

we obtain $v(U_1^*, U_2^*) = 1$ for any U_1 satisfying the condition

$$\frac{x_1^* + x_2^*}{\alpha} \le U_1 \le U - \frac{x_3^* + x_4^*}{\alpha}.$$

For the numerical data $U = 20$, $\alpha = 0.5$, $x_1^* = 2$, $x_2^* = 3$, $x_3^* = 3$, $x_4^* = 4$, $d_1 = d_2 = 1$, $d_3 = d_4 = 2$ we obtain: $U_1^* = 8\frac{2}{3}$, $U_2^* = 11\frac{1}{3}$, $u_1^* = 3\frac{1}{3}$, $u_2^* = 5\frac{1}{3}$, $u_3^* = 4\frac{1}{3}$, $u_4^* = 7$ and $v^* = \frac{2}{3}$, which means that the requirement $T \le \alpha$ will be approximately satisfied with the certainty index $\frac{2}{3}$. □

13.5 Application of Random Variables [13,14,60]

The decision problems for the operations with random parameters are analogous to those presented in Sect. 13.2 for uncertain parameters. Let us start with the complex described by the functions

$$T_i = \varphi_i(u_i, x_i), \quad i = 1, 2, ..., k \tag{13.45}$$

where the unknown parameter x_i is assumed to be a value of a random variable $\widetilde{x}_i$ with a probability density $f_{xi}(x_i)$. The decision problem may consist in the determination of the decision u^* minimizing the expected value of the execution time T, i.e.

$$\mathrm{E}(\widetilde{T}) = \int\limits_X \max_i \varphi_i(u_i, x_i) f_x(x)\, dx \qquad (13.46)$$

where $f_x(x)$ is the joint probability distribution for $x = (x_1, x_2, \ldots, x_k) \in X$. Assume that $\widetilde{x}_i$ are independent for different i. Then

$$f_x(x) = f_{x1}(x_1) \cdot f_{x2}(x_2) \cdot \ldots \cdot f_{xk}(x_k).$$

In this case it is better to use a *distribution function*

$$F_{xi}(x_i) = P(\widetilde{x}_i \le x_i) = \int\limits_0^{x_i} f_{xi}(x_i)\, dx_i\,. \qquad (13.47)$$

Knowing (13.47) and (13.45), one can determine the distribution function $F_{Ti}(\lambda; u_i)$ for the variable $\widetilde{T}_i$, i.e.

$$F_{Ti}(\lambda; u_i) = P(\widetilde{T}_i \le \lambda).$$

Then (13.46) becomes

$$\mathrm{E}(\widetilde{T}) = \mathrm{E}(\max_i \widetilde{T}_i) = \int\limits_0^{\infty} \lambda \; d\Big(\prod_{i=1}^{k} F_{Ti}(\lambda; u_i)\Big). \qquad (13.48)$$

Example 13.4.
Let $k = 1$ and

$$T_1 = \frac{x_1}{u_1}, \qquad T_2 = \frac{2x_2}{u_2},$$

and the random variables $\widetilde{x}_1$, $\widetilde{x}_2$ have the same exponential probability density: $f_x(x) = ae^{-ax}$ for $x > 0$ and $f_x(x) = 0$ for $x \le 0$. According to (13.48) for $u_2 = U - u_1$ we shall obtain

$$\mathrm{E}(\widetilde{T}) = \frac{U^2 + 3u_1^2}{au_1(U - u_1)(U + u_1)} \stackrel{\Delta}{=} Q(u_1).$$

The condition

$$\frac{dQ(u_1)}{du_1} = 0$$

leads to the equation

$$3u_1^4 + 6U^2 u_1^2 - U^4 = 0.$$

The root of this equation satisfying the constraint $0 \le u_1 \le U$ is as follows:

$$u_1^* = U\sqrt{\frac{2}{\sqrt{3}} - 1}\,.$$

Then

$$u_2^* = U - u_1^* = U(1 - \sqrt{\frac{2}{\sqrt{3}} - 1})\,.$$ □

For the operations described by the inequalities (13.11) where x_i is a value of $\widetilde{x}_i$. According to the general formulation of the decision problem presented in Sect. 3.5, the allocation problem may be formulated as an optimization problem consisting in finding the optimal allocation u^* which maximizes the probability that u belongs to $D_u(\widetilde{x})$, i.e. the probability that the set of possible values T belongs to $[0,\alpha]$.

Optimal allocation (decision) problem: For the given φ_i, f_{xi} $(i \in \overline{1,k})$, U and find

$$u^* = \arg\max_{u \in U} p(u)$$

where

$$p(u) = v\,\{D_T(u;\widetilde{x}) \subseteq [0,\alpha]\} \tag{13.49}$$

and $D_T(u;x)$ is determined by the inequality

$$T \le \max_i \varphi_i(u_i, x_i)\,.$$

According to (13.49)

$$p(u) = P\{[\varphi_1(u_1,\widetilde{x}_1) \le \alpha] \wedge [\varphi_2(u_2,\widetilde{x}_2) \le \alpha] \wedge \ldots \wedge [\varphi_k(u_k,\widetilde{x}_k) \le \alpha]\}\,.$$

Then

$$u^* = \arg\max_{u \in \overline{U}} \prod_{i=1}^{k} p_i(u_i) \tag{13.50}$$

where

$$p_i(u_i) = P\,[\varphi_i(u_i,\widetilde{x}_i) \le \alpha] = P\,[\widetilde{x}_i \in D_{xi}(u_i)]\,,$$

$$D_{xi}(u_i) = \{x_i \in R^1 : \varphi_i(u_i,x_i) \le \alpha\}\,.$$

Finally

$$p_i(u_i) = \int_{D_{xi}(u_i)} f_{xi}(x_i)\,dx_i\,. \tag{13.51}$$

If $\varphi_i(u_i, x_i)$ is an increasing function of x_i and $\hat{x}_i(u_i)$ is the solution of Equation (13.16) then

$$p_i(u_i) = \int_0^{x_i(u_i)} f_{xi}(x_i)\,dx_i \; .$$

Introducing the distribution functions $F_{Ti}(\lambda; u_i)$ for the random variables $\varphi_i(u_i, \widetilde{x}_i) \stackrel{\Delta}{=} \widetilde{\tau}_i$ (i.e. the distribution functions used in (13.48)), we have

$$p(u) = \prod_{i=1}^{k} F_{Ti}(\lambda; u_i)$$

and

$$u^* = \arg\max_u \prod_{i=1}^{k} F_{Ti}(\lambda; u_i) .$$

The formulas (13.49), (13.50) and (13.51) are analogous to (13.12), (13.13) and (13.14), respectively. The idea of the decomposition and two-level control is analogous to that for uncertain variables, presented in Sect. 13.4.

13.6 Application to Task Allocation in a Multiprocessor System [33, 57]

It is well known that computational task allocation or scheduling in multiprocessor systems may be connected with great computational difficulties even in the case when the scheduling may be reduced to a simple assignment problem (e.g. [8]). On the other hand, the execution times may not be exactly known. The purpose of this section is to show how the uncertain variables may be applied to the allocation of computational tasks (i.e. to load distribution) in a group of parallel processors with uncertain execution times. We shall consider a rather simple allocation problem consisting in the distribution of a great number of elementary tasks (elementary programs or parts of programs).

Assume that the global computational task to be distributed may be decomposed into N separate parts (programs or parts of programs), which may be executed simultaneously by the separate processors. Each partial task is characterized by an upper bound of the execution time τ_i for the *i*-th processor ($i = 1, 2, ..., k$), and τ_i is assumed to be the same for each partial task. The decision problem consists in the determination of the numbers of the partial tasks $n_1, n_2, ..., n_k$ assigned to the processors taking into account the execution time $T = \max\{T_1, T_2, ..., T_k\}$ where T_i is the execution time for the *i*-th processor; $n_1 + n_2 + ... + n_k = N$. If N is sufficiently large, we can determine the decisions $u_i \in R^+$ (any positive numbers) satisfying the constraint

$$n_1 + n_2 + \ldots + n_k = N$$

and then obtain n_i by rounding off u_i to the nearest integer. The relational knowledge representation of the group of the processors is determined by the inequalities

$$T_i \le x_i u_i$$

and a function $T = \max\{T_1, T_2, \ldots, T_k\}$. The unknown parameters x_i are assumed to be values of uncertain variables $\bar{x}_i$ described by certainty distributions $h_{xi}(x_i)$ given by an expert estimating the execution times for the partial tasks. Then the relational knowledge representation must be completed by the distributions h_{xi}. Then we can formulate the decision problem and solve the optimal allocation problem presented in Sect. 13.2.

Consequently, the procedure of determination of u^* is the following:

1. To determine $v_i(u_i)$ according to (13.14).
2. To determine $u_1^*, u_2^*, \ldots, u_k^*$ by solving the maximization problem

$$\max_{u_1,\ldots,u_k} \min\{v_1(u_1), \ldots, v_k(u_k)\} \tag{13.52}$$

subject to the constraints $u_1 + u_2 + \ldots + u_k = N$, $u_i \ge 0$, $i \in \overline{1,k}$. The structure of the knowledge-based decision system is presented in Fig. 13.5. If α is sufficiently large then it is possible to obtain the allocation $u_1^*, u_2^*, \ldots, u_k^*$ for which the requirement $T \le \alpha$ is approximately satisfied with the certainty index $v = 1$.

Let us note that in an analogous way one can formulate and solve the allocation problem for the microprocessors with random parameters $\tilde{x}_i$ (see Sect. 13.5) and one can apply the decomposition and two-level control presented in Sect. 13.4.

Example 13.5.

Assume that $h_{xi}(x_i)$ has a parabolic form presented in Fig. 13.6. Using (13.14) we shall obtain:

$$v_i(u_i) = \begin{cases} 1 & \text{for} \quad u_i \le \dfrac{\alpha}{x_i^*} \\[2ex] -\dfrac{1}{d_i^2}\left(\dfrac{\alpha}{u_i} - x_i^*\right)^2 + 1 & \text{for} \quad \dfrac{\alpha}{x_i^*} \le u_i \le \dfrac{\alpha}{x_i^* - d_i} \\[2ex] 0 & \text{for} \quad u_i \ge \dfrac{\alpha}{x_i^* - d_i}. \end{cases}$$

For two processors the results analogous to (13.37), (13.38), (13.39) are as follows:

1. For

$$\alpha \le \frac{N(x_1^* - d_1)(x_2^* - d_2)}{x_1^* - d_1 + x_2^* - d_2}$$

$v(u) = 0$ for any u_1.

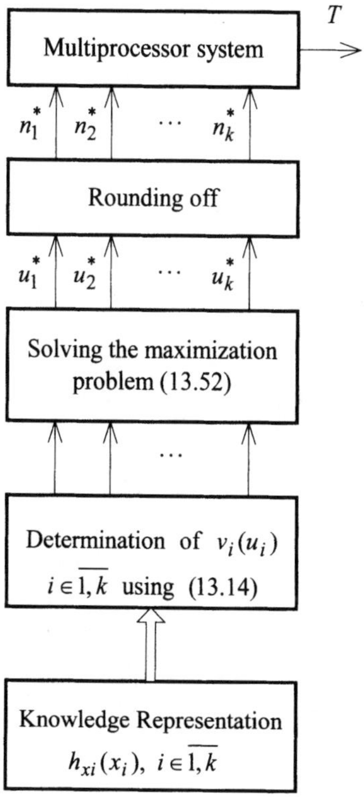

Figure 13.5. Structure of knowledge-based control system under consideration

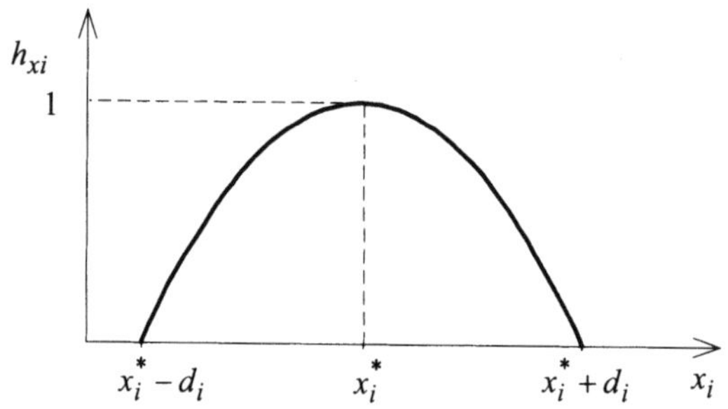

Figure 13.6. Certainty distribution in Example 13.5

2. For

$$\frac{N(x_1^* - d_1)(x_2^* - d_2)}{x_1^* - d_1 + x_2^* - d_2} \le \alpha \le \frac{N x_1^* x_2^*}{x_1^* + x_2^*}$$

u_1^* is a root of equation

$$\frac{1}{d_1}\left(\frac{\alpha}{u_1} - x_1^*\right) = -\frac{1}{d_2}\left(\frac{\alpha}{N - u_1} - x_2^*\right) \tag{13.53}$$

satisfying the condition the same as in the case (13.38), and $v(u^*) = v_1(u_1^*)$.

3. For

$$\alpha \ge \frac{N x_1^* x_2^*}{x_1^* + x_2^*}$$

$v(u^*) = 1$ for any u_1 satisfying condition the same as in the case (13.39).

Under the assumption (13.40) Equation (13.53) becomes

$$\frac{\alpha}{u_1 x_1^*} + \frac{\alpha}{(N - u_1) x_2^*} = 2 \tag{13.54}$$

and

$$v(u^*) \stackrel{\Delta}{=} v^* = 1 - \gamma^2 \left(\frac{\alpha}{x_1^* u_1^*} - 1\right)^2 . \tag{13.55}$$

For example, if $N = 40$, $\alpha = 30$, $x_1^* = 1.5$, $x_2^* = 2$, $\gamma = 1.5$ then using (13.54) and (13.55) we obtain $n_1^* = 28$, $n_2^* = 12$ and $v^* = 0.8$, which means that 28 partial tasks should be assigned to the first processor and 12 should be assigned to the second processor, and the requirement concerning the execution time will be approximately satisfied with the certainty index 0.8. The relationship between $v(u^*)$ and the parameters x_1^*, γ is illustrated in Fig. 13.7 where $x_{\min} = \alpha x_2^* (N x_2^* - \alpha)^{-1}$. □

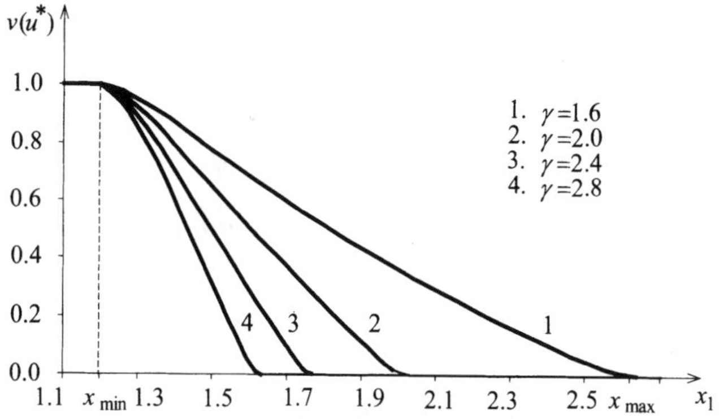

Figure 13.7. Certainty index $v(u^*)$ as a function of x_1^* for fixed γ

13.7 Learning Algorithms

Let us assume that there is no a priori knowledge of the unknown parameters x_i in the form of h_{xi} or f_{xi}, but it is possible to apply learning process using the results of current observations. Denote the unknown value of x_i by $\bar{x}_i$. According to the general approach described in Chapter 11, the learning process will consist in a *step by step* estimation of $\bar{x} = (\bar{x}_1, \ldots, \bar{x}_k)$ as a result of knowledge validation and updating, and using the successive estimates to the current determination of the allocation. Let us denote by $u^{(n)}$ and $T^{(n)}$ the allocation $u = (u_1, \ldots, u_k)$ and the execution time T in the n-th step, respectively. The considerations will be presented for the knowledge representation in the form of inequalities (13.11) and for the learning process based on the knowledge of the decisions (see Sect. 11.2). The knowledge of the decisions is determined by the greatest set $D_u(x)$ of the allocations $u = (u_1, u_2, \ldots, u_k)$ such that the implication $u \in D_u(x) \to y \le \alpha$ is satisfied and $D_u(x) \subseteq \overline{\Delta}_u$ where

$$\overline{\Delta}_u = \{u \in U : u_1 + \ldots + u_k = N \; ; \quad \bigwedge_{i \in \overline{1,k}} u_i \ge 0\} .$$

It is easy to note that $D_u(x) = \Delta_u(x) \cap \overline{\Delta}_u$ where $\Delta_u(x)$ is a set of possible decisions without the constraints, i.e.

$$\Delta_u(x) = \{u \in U : \bigwedge_{i \in \overline{1,k}} \varphi_i(u_i, x_i) \le \alpha\} .$$

Denote by $S_n = (u^{(1)}, u^{(2)}, \ldots, u^{(n)})$ the sequence of the decisions, by $\bar{u}^{(j)}$ the allocation $u^{(j)}$ for which $T^{(j)} \le \alpha$, by $\hat{u}^{(j)}$ the allocation $u^{(j)}$ for which $T^{(j)} > \alpha$, and introduce the following sets:

$$\overline{D}_x(n) = \{x \in X : \overline{u}^{(j)} \in \Delta_u(x) \text{ for each } \overline{u}^{(j)} \text{ in } S_n\},$$

$$\hat{D}_x(n) = \{x \in X : \hat{u}^{(j)} \in U - \Delta_u(x) \text{ for each } \hat{u}^{(j)} \text{ in } S_n\}.$$

As the estimate of $\overline{x}$ one may propose the set $\Delta_x(n) \overset{\Delta}{=} \overline{D}_x(n) \cap \hat{D}_x(n)$ and under some general assumptions one can prove the convergence of $\Delta_x(n)$ to $\{\overline{x}\}$ (a singleton). The estimate $\Delta_x(n)$ and the proof of the convergence are based on the results for the general formulation of the relational knowledge representation and the decision problem presented in Sect. 11.2. In our case the knowledge representations and the results have a specific form. It is worth noting that the estimates of x_i may be determined independently for separate processors:

$$\Delta_{x,i}(n) = \overline{D}_{x,i}(n) \cap \hat{D}_{x,i}(n)$$

where

$$\overline{D}_{x,i}(n) = \{x_i : H_i(\overline{u}_i^{(j)}, x_i) \le \alpha \text{ for each } \overline{u}_i^{(j)} \text{ in } S_n\},$$

$$\hat{D}_{x,i}(n) = \{x_i : H_i(\hat{u}_i^{(j)}, x_i) > \alpha \text{ for each } \hat{u}_i^{(j)} \text{ in } S_n\}.$$

The value $x_i^{(n)}$ may be chosen randomly from $\Delta_{x,i}(n)$. It is possible to present the estimation algorithm in a recursive form and consequently the *allocation algorithm with learning* has the following form (for $n > 1$):

1. Apply the allocation $u^{(n)}$ and measure the execution times $T_i^{(n)}$ for each processor.
2. If $T_i^{(n)} \le \alpha$

Knowledge validation for $u_i^{(n)} = \overline{u}_i^{(n)}$

Prove whether

$$\bigwedge_{x_i \in \overline{D}_{x,i}(n-1)} [H_i(u_i^{(n)}, x_i) \le \alpha].$$

If yes then $\overline{D}_{x,i}(n) = \overline{D}_{x,i}(n-1)$. If not, determine the new set $\overline{D}_{x,i}(n)$.

Knowledge updating for $u_i^{(n)} = \overline{u}_i^{(n)}$

$$\overline{D}_{x,i}(n) = \{x_i \in \overline{D}_{x,i}(n-1) : H_i(u_i^{(n)}, x_i) \le \alpha\},$$

$$\hat{D}_{x,i}(n) = \hat{D}_{x,i}(n-1).$$

3. If $T_i^{(n)} > \alpha$

Knowledge validation for $u_i^{(n)} = \hat{u}_i^{(n)}$

Prove whether

$$\bigwedge_{x_i \in \hat{D}_{x,i}(n-1)} [H_i(u_i^{(n)}, x_i) > \alpha].$$

If yes then $\hat{D}_{x,i}(n) = \hat{D}_{x,i}(n-1)$. If not, determine the new set $\hat{D}_{x,i}(n)$.

Knowledge updating for $u_i^{(n)} = \hat{u}_i^{(n)}$

$$\hat{D}_{x,i}(n) = \{x_i \in \hat{D}_{x,i}(n-1) : H_i(u_i^{(n)}, x_i) > \alpha\},$$

$$\overline{D}_{x,i}(n) = \overline{D}_{x,i}(n-1).$$

4. Determine $\Delta_{x,i}(n) = \overline{D}_{x,i}(n) \cap \hat{D}_{x,i}(n)$.

5. Choose randomly $x_i^{(n)}$ from $\Delta_{x,i}(n)$.

The steps 2, 3, 4, 5 should be performed for $i = 1, 2, \ldots, k$.

6. Put $x^{(n)} = (x_1^{(n)}, \ldots, x_k^{(n)})$ into the set $\Delta_u(x^{(n-1)})$ in place of x_{n-1}.

7. Choose randomly the decision $u^{(n+1)}$ from the set $D_u(x^{(n)}) = \Delta_u(x^{(n)}) \cap \overline{\Delta}_u$.

The block scheme of the control system for the control of the task allocation with the learning process is presented in Fig. 13.8. where the relation $R(x)$ denotes the set of inequalities (13.11). The plant in Fig. 13.8 is the multiprocessor system described in Sect. 13.6, but the considerations presented here are concerned with any complex of parallel operation described in Sect. 13.1. The controlling part consists of the basic program generating the solution $D_u(x^{(n)})$, the knowledge base, the learning program (the validation and updating of the knowledge) and the generators of random numbers G_1 and G_2 for the random choosing of $x^{(n)}$ from $\Delta_x(n)$ and of $u^{(n)}$ from $D_u(x^{(n-1)})$, respectively.

For (13.11) in the form $T_i \le x_i u_i$ we obtain

$$\overline{D}_{x,i} = [0, x_{\max,i}^{(n)}], \qquad \hat{D}_{x,i} = [x_{\min,i}^{(n)}, \infty),$$

$$\Delta_{x,i}(n) = (x_{\min,i}^{(n)}, x_{\max,i}^{(n)})$$

where

$$x_{\max,i}^{(n)} = \frac{\alpha}{\max_j \overline{u}_i^{(j)}}, \qquad x_{\min,i}^{(n)} = \frac{\alpha}{\min_j \hat{u}_i^{(j)}}.$$

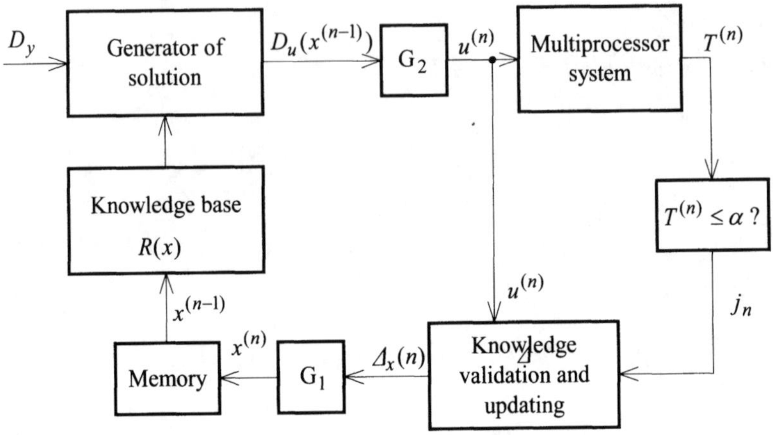

Figure 13.8. Control system with learning

Steps 2 and 3 of the learning algorithm are now as follows:
2. If $T_i^{(n)} \le \alpha$ (i.e. $u_i^{(n)} = \overline{u}_i^{(n)}$), prove whether

$$u_i^{(n)} \le \frac{\alpha}{x_{\max,i}^{(n-1)}}.$$

If yes then $x_{\max,i}^{(n)} = x_{\max,i}^{(n-1)}$. If not then

$$x_{\max,i}^{(n)} = \frac{\alpha}{u_i^{(n)}}.$$

Put $x_{\min,i}^{(n)} = x_{\min,i}^{(n-1)}$.

3. If $T_i^{(n)} > \alpha$ (i.e. $u_i^{(n)} = \hat{u}_i^{(n)}$), prove whether

$$u_i^{(n)} > \frac{\alpha}{x_{\min,i}^{(n-1)}}.$$

If yes then $x_{\min,i}^{(n)} = x_{\min,i}^{(n-1)}$. If not then

$$x_{\min,i}^{(n)} = \frac{\alpha}{u_i^{(n)}}.$$

Put $x_{\max,i}^{(n)} = x_{\max,i}^{(n-1)}$.

14 Pattern Recognition

Knowledge-based recognition (or classification) under uncertainty may be considered as a good example of the general problems and methods presented in the previous chapters. In this chapter recognition problems based on the relational and logical knowledge representation are described. Classical methods based on probabilistic description and new approaches (application of uncertain variables are the learning process consisting in the knowledge validation and updating) are presented and discussed in a uniform way, as specific analysis and decision problems with the descriptions of the uncertainty considered in Chapters 3, 5, 6, 7, 11.

14.1 Pattern Recognition Based on Relational Knowledge Representation

Let an object to be recognized or classified be characterized by a vector of features $u \in U$ which may be observed, and the index of a class j to which the object belongs; $j \in \{1, 2, ..., M\} \stackrel{\Delta}{=} J$, M is a number of the classes. The set of the objects may be described by a relational knowledge representation $R(u, j) \in U \times J$, which is reduced to the sequence of sets

$$D_u(j) \subset U, \quad j = 1, 2, ..., M, \tag{14.1}$$

i.e.

$$D_u(j) = \{u \in U: (u, j) \in R(u, j)\}.$$

Assume that as a result of the observation it is known that $u \in D_u \subset U$. The recognition problem may consist in finding the set of all possible indexes j, i.e. the set of all possible classes to which the object may belong [41, 54, 95].

Recognition problem: For the given sequence $D_u(j),\ j \in \overline{1, M}$ and the result of the observation D_u find the smallest set $D_j \subset J$ for which the implication

$$u \in D_u \rightarrow j \in D_j$$

is satisfied. Then

$$D_j = \{j \in J:\ D_u \cap D_u(j) \neq \varnothing\} \tag{14.2}$$

where $\varnothing$ denotes an empty set. In particular, if $D_u = \{u\}$, i.e. we obtain the exact result of the measurement, then

$$D_j = \{j \in J:\ u \in D_u(j)\}\,.$$

It is worth noting that our recognition problem may be considered as a specific analysis problem for a relational plant (see Sect. 2.2). In our case the set of the objects is understood as a "plant" with input u and output j, and the particular object is represented by a pair $(u, j) \in R(u, j)$. For a sequence of the objects appearing in the successive moments n, $(u_n, j_n) \in R(u, j)$ for every n. The relation $R(u, j)$ is the knowledge of the recognition or the knowledge of the "plant". The basic idea of the knowledge-based recognition is illustrated in Fig. 14.1. If the relation R is reduced to the function $j = \Phi(u)$ then for $D_u\{u\}$ the set (14.2) is reduced to one element: $D_j = \{\Phi(u)\}$, i.e. $j = \Phi(u)$ is a *deterministic recognition algorithm*, given directly by an expert.

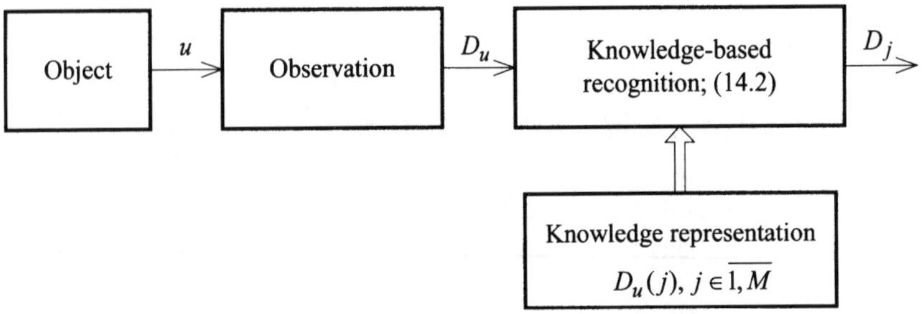

Figure 14.1. Knowledge-based recognition

An expert may formulate the knowledge of the recognition in another way, as a set of implications

$$u \in D_{up} \rightarrow j \in D_{jp}, \quad p = 1, 2, ..., N, \quad D_{up} \subset U, \quad D_{jp} \subset J \tag{14.3}$$

where $D_{up} \cap D_{us} = \varnothing$ for $p \neq s$. In this case

$$D_j = \{j \in J: \bigvee_{p \in \overline{1, N}} (D_u \cap D_{up} \neq \varnothing) \wedge (j \in D_{jp})\}, \tag{14.4}$$

what means that D_j is a union of the sets D_{jp} assigned to the sets D_{up} having common points with the set D_u. For $D_u = \{u^*\}$ and $u^* \in D_{up}$ we have $D_j = D_{jp}$ where D_{jp} is assigned to D_{up} by the implication (14.3). It is easy to see that, in general, the description of the knowledge in the form (14.3) is less precise than the description (14.1). For the function $j = \Phi(u)$ the both descriptions are equivalent when

$$D_{uj} = D_u(j) = \{u \in U : \Phi(u) = j\}.$$

14.2 Application of the Logic-algebraic Method

The relations describing the set of objects may have a form of logical formulas concerning u, j and $w \in W$ where w is a vector of additional variables. The purpose of this section is to show how the general approach based on the logical knowledge representation presented in Chapter 7 may be applied to the knowledge-based recognition. In our considerations the knowledge representation of a set of objects is expressed as a set of statements which are called formulas. We shall use the following notation (see Sect. 7.1):

1. $\alpha_{up}(u)$ – simple formula (i.e. simple property) concerning u; $p = 1, 2, ..., n_1$, e.g. "$u \in \Delta_u$" where Δ_u is a subset of U.
2. $\alpha_{wr}(u, w)$ – simple formula concerning u and w; $r = 1, 2, ..., n_2$.
3. $\alpha_{js}(j)$ – simple formula "the object belongs to class s" or shortly "$j = s$"; $s = 1, 2, ..., M$.
4. $\alpha_u = (\alpha_{u1}, \alpha_{u2}, ..., u_{un_1})$ – subsequence of simple formulas concerning u.
5. $\alpha_w = (\alpha_{w1}, \alpha_{w2}, ..., \alpha_{wn_2})$ – subsequence of simple formulas concerning u and w.
6. $\alpha_j = (\alpha_{j1}, \alpha_{j2}, ..., \alpha_{jM})$ – subsequence of simple formulas concerning j.
7. $\alpha(u, w, j) \stackrel{\Delta}{=} (\alpha_1, \alpha_2, ..., \alpha_n) = (\alpha_u, \alpha_w, \alpha_j)$ – sequence of all simple formulas in the knowledge representation, $n = n_1 + n_2 + M$.
8. $F_i(\alpha)$ – the i-th fact given by an expert. It is a logic formula composed of a subsequence of α and the logic operations: $\vee$ – *or*, $\wedge$ – *and*, $\neg$ – *not*, $\rightarrow$ – *if...then*; $i = 1, 2, ..., k$.
9. $F(\alpha) = F_1(\alpha) \wedge F_2(\alpha) \wedge ... \wedge F_k(\alpha)$.
10. $F_u(\alpha_u)$ – property concerning the features u, i.e. the logic formula using α_u.
11. $F_j(\alpha_j)$ – property concerning j.
12. $a_m \in \{0,1\}$ – logic value of the simple property α_m; $m = 1, 2, ..., n$.
13. $a = (a_1, a_2, ..., a_n)$ – zero-one sequence of the logic values.
14. $a_u(u)$, $a_w(u, w)$, $a_j(j)$ – zero-one subsequences of the logic values corresponding to $\alpha_u(u)$, $\alpha_w(u, w)$, $\alpha_j(j)$.

15. $F(a)$ – the logic value of $F(\alpha)$.

The implications (14.3) are examples of the facts. All facts given by an expert are assumed to be true, i.e. $F(a)=1$. The description

$$<(\alpha_u,\alpha_w,\alpha_j),\ F(\alpha_u,\alpha_w,\alpha_j)> \stackrel{\Delta}{=} \text{KP}$$

is a *logical knowledge representation* of the set of objects to be recognized. This is a specific form of a general formulation presented in Sect. 7.1. In our case $y=j$ is an integer, $Y=J=\{1,2,...,M\}$, the simple formulas concerning $y=j$ have specific form "$j=s$", simple formulas α_{wr} does not depend on j, and

$$w[(j=1)\vee(j=2)\vee...\vee(j=M)]=1, \tag{14.5}$$

$$w[(j=s_1)\wedge(j=s_2)]=0 \quad \text{for} \quad s_1\neq s_2, \tag{14.6}$$

according to the assumption that an object belongs to one and only one class from *J*. For the given forms of α each fact determines a relation

$$R_i(u,w,j)=\{(u,w,j)\in U\times W\times J:\ F_i(a_u,a_w,a_j)=1\},$$

and

$$R(u,j)=\{(u,j):\ \bigvee_{w\in W}[(u,w,j)\in\bigcap_{i=1}^{k}R_i]\}.$$

Assume that the result of observation is expressed as a true formula $F_u(\alpha_u)$, i.e. $F_u(a_u)=1$ or $u\in D_u$ where

$$D_u=\{u\in U:\ F_u(a_u)=1\}.$$

The solution $D_j=\{j_1,j_2,...,j_n\}\subset J$ now has the form

$$F_j(\alpha_j)=\bigvee_{m=1}^{n}[(\alpha_{js}\ \text{for}\ s=j_m)\wedge\bigwedge_{s\neq jm}^{M}\neg\alpha_{js}] \tag{14.7}$$

according to the notation $\alpha_{js}=$ "$(j=s)$" and the assumptions (14.5), (14.6). The recognition problem is now formulated as follows: given *F* and F_u, find F_j (14.7) with the smallest number n such that the implication $F(\alpha_u,\alpha_w,\alpha_j)\wedge F_u(\alpha_u)\rightarrow F_j(\alpha_j)$ is true. To solve the problem, one can apply the logic-algebraic method presented in Sect. 7.3. Denote by S_u the set of all a_u, by S_w – the set of all sequences a_w and by $\hat{S}_j$ the set of all sequences a_j containing one and only one element equal to 1. Now, our problem may be formulated as follows: given F and F_u, find the set $\check{S}_j \subset \hat{S}_j$ of all a_j for which there exist $a_u \in S_u$ and $a_w \in S_w$ satisfying the equations

$$F_u(a_u)=1\,, \quad F(a_u,a_w,a_j)=1\,. \tag{14.8}$$

The solution D_j is then determined by $\check{S}_j$, i.e. $s = j_m$ if the j_m-th element in $a_j \in \check{S}_j$ is equal to 1. Note that F_u and F are now the algebraic expressions in two-value logical algebra and our problem is reduced to the finding of all solutions a_j of the algebraic equations (14.8). In a typical situation the result of observation is a 0-1 sequence $c=(c_1,c_2,...,c_{n_1})$ where $c_p \in \{0,1\}$ is the value of the variable a_{up}, $p \in \overline{1,n_1}$, i.e. the logic value of the statement

$$\alpha_{up}(u), \quad p \in \overline{1,n_1} \quad \text{or} \quad u \in \overline{D}_{up}$$

where $\overline{D}_{up} = \{u \in U:\ a_{up}=1\}$. In other words, they are the answers "yes" or "no" for the question "has the symptom α_{up} been observed?", $p \in \overline{1,n_1}$. Equations (14.8) are now reduced to

$$F(c,a_w,a_j)=1 \tag{14.9}$$

for the given c and

$$\check{S}_j = \{a_j \in \hat{S}_j : \bigvee_{a_w \in S_w} [F(c,a_w,a_j)=1]\}\,. \tag{14.10}$$

The direct approach consists in testing (14.9) for all sequences $a_w \in S_w$, $a_j \in \hat{S}_j$. To decrease the computational difficulties, the decomposition and the recursive procedure presented in Sect. 7.3 may be applied.

Example 14.1.

The simple formulas are the following:

$\alpha_{u1}(u) =$ "$u \in D_{u1}$" or "the symptom 1 is observed",

$\alpha_{u2}(u) =$ "$u \in D_{u2}$" or "the symptom 2 is observed",

$\alpha_{wi} =$ "$w \in D_{wi}$" or "the state i occurs" $i \in \overline{1,7}$,

$\alpha_{w8} =$ "$(u,w) \in D$", $\alpha_{j1} =$ "$j=1$", $\alpha_{j2} =$ "$j=2$".

The facts are the following:

$F_1 = [(\alpha_{u1} \wedge \alpha_{w3}) \vee \neg\alpha_{w1}] \rightarrow \alpha_{w4}$,

$F_2 = (\alpha_{w1} \wedge \alpha_{w7}) \vee \neg\alpha_{w3}$,

$F_3 = (\alpha_{j1} \wedge \alpha_{w1}) \rightarrow \alpha_{w2}$,

$F_4 = (\alpha_{w4} \wedge \neg\alpha_{w7}) \vee \alpha_{w5}$,

$$F_5 = (\alpha_{u2} \wedge \alpha_{w6}) \rightarrow (\alpha_{w4} \wedge \alpha_{w8}),$$
$$F_6 = (\alpha_{u1} \wedge \alpha_{w2}) \rightarrow (\alpha_{w4} \wedge \alpha_{w6}),$$
$$F_7 = (\alpha_{w3} \wedge \alpha_{w2}) \vee \alpha_{j2}.$$

For example, F_1 means: “if (symptom 1 is observed and state 3 occurs) or state 1 does not occur, then state 4 will occur”. Assume that both symptoms have been observed, i.e. $a_{u1} = a_{u2} = 1$. Then $F_1(a_{w1}, a_{w3}, a_{w4}) = (a_{w3} \vee \neg a_{w1}) \rightarrow a_{w4}$, $F_5(a_{w4}, a_{w6}, a_{w8}) = a_{w6} \rightarrow (a_{w4} \wedge a_{w8})$, $F_6(a_{w2}, a_{w4}, a_{w6}) = a_{w2} \rightarrow (\neg a_{w4} \wedge a_{w6})$, F_2, F_3, F_4, F_7 are determined by the formulas described above. It is easy to note that the result of the decomposition used is as follows:

$$\overline{F}_1(\overline{a}_j, \overline{a}_1) = F_3(a_{w1}, a_{w2}, a_{j1}) \wedge F_7(a_{w2}, a_{w3}, a_{j2}),$$
$$\overline{a}_1 = (a_{w1}, a_{w2}, a_{w3}),$$
$$\overline{F}_2(\overline{a}_1, \overline{a}_2) = F_1(a_{w1}, a_{w3}, a_{w4}) \wedge F_2(a_{w1}, a_{w3}, a_{w7}) \wedge F_6(a_{w2}, a_{w4}, a_{w6}),$$
$$\overline{a}_2 = (a_{w4}, a_{w6}, a_{w7}),$$
$$\overline{F}_3(\overline{a}_2, \overline{a}_3) = F_4(a_{w4}, a_{w5}, a_{w7}) \wedge F_5(a_{w4}, a_{w6}, a_{w8}),$$
$$\overline{a}_3 = (a_{w5}, a_{w8}),\ S_3 = \{(1,1), (1,0), (0,1), (0,0)\}.$$

Using the recursive procedure described in Sect. 7.3 with $N = 3$, it is easy to obtain: $\overline{S}_2 = \{(1,1,0), (1,0,0), (0,0,0), (1,1,1), (1,0,1), (0,0,1)\}$, $\overline{S}_3 = \{(0,0,0), (1,0,0), (1,0,1)\}$, $\overline{S}_0 = \check{S}_j = \{(0,1)\}$. The set $\check{S}_j$ has one element only and the solution is $D_j = \{2\}$, i.e. $j = 2$. □

14.3 Application of Uncertain Variables [54]

Now let us assume that the knowledge representation contains a vector of unknown parameters $x \in X$ and x is assumed to be a value of an uncertain variable $\overline{x}$ described by a certainty distribution $h_x(x)$ given by an expert. The recognition problem is now formulated as a specific analysis problem (version I) considered in Sect. 5.4.

Recognition problem for uncertain parameters: For the given sequence $D_u(j;x)$, $h_x(x)$, D_u and the set $\hat{D}_j \subset J$ given by a user one should find the certainty index that the set $\hat{D}_j$ approximately belongs to the set of all possible classes

$$D_j(x) = \{j \in J : D_u \cap D_u(j;x) \neq \varnothing\}. \tag{14.11}$$

It is easy to see that

$$v[\hat{D}_j \,\tilde{\subseteq}\, D_j(\bar{x})] = v[\bar{x} \,\tilde{\in}\, D_x(\hat{D}_j)] \qquad (14.12)$$

where

$$D_x(\hat{D}_j) = \{x \in X : \; \hat{D}_j \subseteq D_j(x)\}\,. \qquad (14.13)$$

Then

$$v[\hat{D}_j \,\tilde{\subseteq}\, D_j(\bar{x})] = \max_{x \in D_x(\hat{D}_j)} h_x(x)\,. \qquad (14.14)$$

In particular, for $\hat{D}_j = \{j\}$ one can formulate the optimization problem consisting in the determination of a class j maximizing the certainty index that j belongs to the set of all possible classes.

Optimal recognition problem: For the given sequence $D_u(j\,;x)$, $h_x(x)$ and D_u one should find j^* maximizing

$$v[j \,\tilde{\in}\, D_j(\bar{x})] \stackrel{\Delta}{=} v(j)\,.$$

Using (14.12), (14.13) and (14.14) for $\hat{D}_j = \{j\}$, we obtain

$$v(j) = v[\bar{x} \,\tilde{\in}\, D_x(j)] = \max_{x \in D_x(j)} h_x(x) \qquad (14.15)$$

where

$$D_x(j) = \{x \in X : \; j \in D_j(x)\} \qquad (14.16)$$

and $D_j(x)$ is determined by (14.11). Then

$$j^* = \arg\max_j v(j) = \arg\max_j \max_{x \in D_x(j)} h_x(x)\,. \qquad (14.17)$$

Assume that the different unknown parameters are separated in the different sets $D_u(j)$, i.e. the knowledge representation is described by the sets $D_u(j\,;x_j)$ where $x_j \in X_j$ are subvectors of x, different for the different j. Assume also that $\bar{x}_j$ and $\bar{x}_i$ are independent uncertain variables for $i \neq j$ and $\bar{x}_j$ is described by the certainty distribution $h_{xj}(x_j)$. In this case, according to (14.11)

$$j \in D_j(x) \Leftrightarrow D_u \cap D_u(j\,;x_j) \neq \varnothing\,.$$

Then

$$v(j) = v[\bigvee_{u \in D_u} u \,\tilde{\in}\, D_u(j\,;\bar{x}_j)] = v[\bar{x}_j \,\tilde{\in}\, D_{xj}(j)] \qquad (14.18)$$

where

$$D_{xj}(j) = \{x_j \in X_j : \bigvee_{u \in D_u} u \in D_u(j; x_j)\}. \quad (14.19)$$

Finally

$$j^* = \arg\max_j \max_{x_j \in D_{xj}(j)} h_{xj}(x_j). \quad (14.20)$$

In particular, for $D_u = \{u\}$ (14.11), (14.18) and (14.19) become

$$D_j(x) = \{j \in J : u \in D_u(j; x)\},$$

$$v(j) = v[u \mathrel{\tilde{\in}} D_u(j; \bar{x}_j)] = v[\bar{x}_j \mathrel{\tilde{\in}} D_{xj}(j)] = \max_{x_j \in D_{xj}(j)} h_{xj}(x_j), \quad (14.21)$$

$$D_{xj}(j) = \{x_j \in X_j : u \in D_u(j; x_j)\}. \quad (14.22)$$

The procedure of finding j^* based on the knowledge representation $< D_u(j; x),\ j \in \overline{1, M};\ h_x(x) >$ or the block scheme of the corresponding recognition system is illustrated in Fig. 14.2. The solution may be not unique, i.e. $v(j)$ may take the maximum value for the different j^*. The result $v(j) = 0$ for each $j \in J$ means that the result of the observation $u \in D_u$ is not possible or there is a contradiction between the result of the observation and the knowledge representation given by an expert.

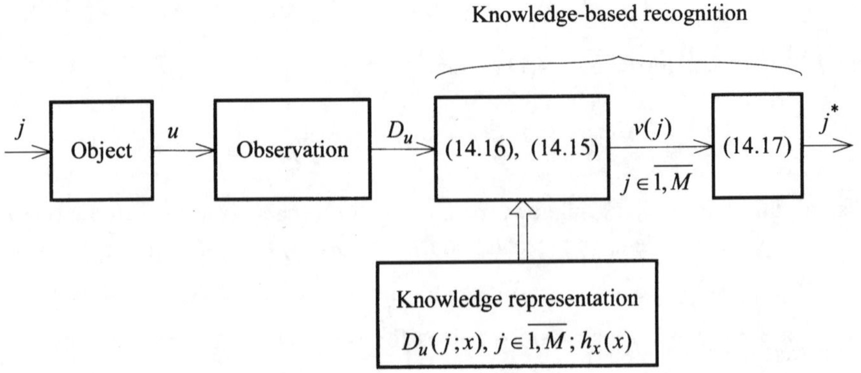

Figure 14.2. Block scheme of a recognition system

If $\bar{x}$ is considered as a C-uncertain variable then

$$j_c^* = \arg\max_j v_c(j)$$

where

$$v_c(j) = \frac{1}{2}\{v[\bar{x} \,\tilde{\in}\, D_x(j)] + 1 - v[\bar{x} \,\tilde{\in}\, \overline{D}_x(j)]\},$$

$\overline{D}_x(j) = X - D_x(j)$. Finally

$$v_c(j) = \frac{1}{2}[\max_{x \in D_x(j)} h_x(x) + 1 - \max_{x \in \overline{D}_x(j)} h_x(x)]. \qquad (14.23)$$

The certainty indexes $v_c(j)$ corresponding to (14.18) and (14.21) have the analogous form.

Example 14.2.

Let $u, x_j \in R^1$, the sets $D_u(j; x_j)$ be described by the inequalities

$$x_j \le u \le 2x_j, \quad j = 1, 2, ..., M$$

and the certainty distributions $h_{xj}(x_j)$ have a parabolic form for each j (Fig. 14.3):

$$h_{xj}(x_j) = \begin{cases} -(x_j - d_j)^2 + 1 & \text{for } d_j - 1 \le x_j \le d_j + 1 \\ 0 & \text{otherwise} \end{cases}$$

where $d_j > 1$.

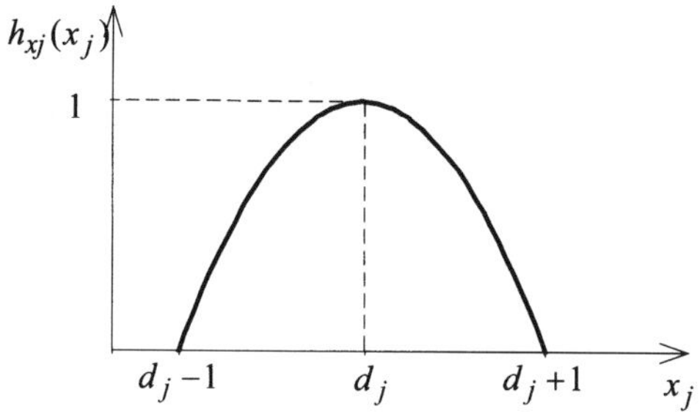

Figure 14.3. Parabolic certainty distribution

In this case the sets (14.22) for the given u are described by the inequality

$$\frac{u}{2} \le x_j \le u.$$

Applying (14.21), one obtains $v(j)$ as a function of d_j illustrated in Fig. 14.4:

$$v(j) = \begin{cases} 0 & \text{for} \quad d_j \le \frac{u}{2} - 1 \\ -(\frac{u}{2} - d_j)^2 + 1 & \text{for} \quad \frac{u}{2} - 1 \le d_j \le \frac{u}{2} \\ 1 & \text{for} \quad \frac{u}{2} \le d_j \le u \\ -(u - d_j)^2 + 1 & \text{for} \quad u \le d_j \le u + 1 \\ 0 & \text{for} \quad d_j \ge u + 1 . \end{cases}$$

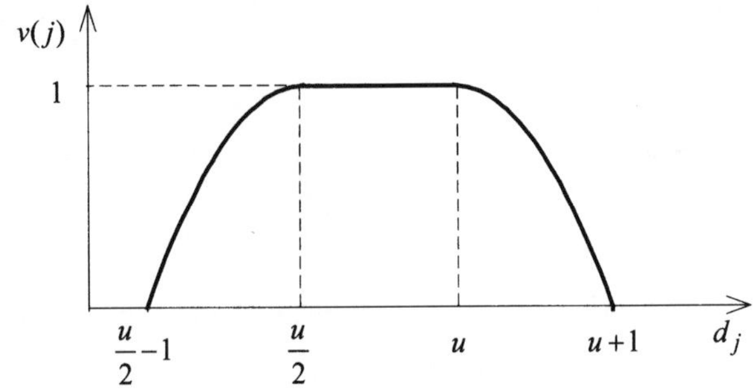

Figure 14.4. Relationship between v and the parameter of certainty distribution

For example, for $M = 3$, $u = 5$, $d_1 = 2$, $d_2 = 5.2$, $d_3 = 6$ we obtain $v(1) = 0.75$, $v(2) = 0.96$ and $v(3) = 0$. Then $j^* = 2$, which means that for $u = 5$ the certainty index that $j = 2$ belongs to the set of the possible classes has the maximum value equal to 0.96. For $d_1, d_2, d_3 \in [\frac{u}{2}, u]$ one obtains $j^* = 1$ or 2 or 3 and $v(j^*) = 1$.

Let us consider $\bar{x}$ as a *C*-uncertain variable for the same numerical data. To obtain $v_c(j)$ according to (14.23) it is necessary to determine

$$v[\bar{x}_j \,\tilde{\in}\, \overline{D}_{xj}(j)] = \max_{x_j \in \overline{D}_{xj}(j)} h_{xj}(x_j) \stackrel{\Delta}{=} v_n(j) . \tag{14.24}$$

In our case the set $\overline{D}_{xj}(j) = X_j - D_{xj}(j)$ is determined by the inequalities

$$x_j < \frac{u}{2} \qquad \text{or} \qquad x_j > u .$$

Using (14.24), we obtain $v_n(1) = v_n(2) = v_n(3) = 1$. Then

$$v_c(j) = \frac{1}{2}[v(j) + 1 - v_n(j)] = \frac{1}{2} v(j) , \tag{14.25}$$

i.e. $v_c(1) = 0.375$, $v_c(2) = 0.48$, $v_c(3) = 0$ and $j_c^* = 2$ with the certainty index

$v_c(j^*) = 0.48$.

For $d_1 = 3$, $d_2 = 3.2$, $d_3 = 4$ we obtain $v(1) = v(2) = v(3) = 1$ and

$$v_n(1) = -(2.5-3)^2 + 1 = 0.75,$$

$$v_n(2) = -(2.5-3.2)^2 + 1 = 0.51,$$

$$v_n(3) = 0.$$

Then

$$v_c(1) = \frac{1}{2}(1+1-0.75) = 0.625,$$

$$v_c(2) = \frac{1}{2}(1+1-0.51) = 0.745,$$

$$v_c(3) = 1$$

and $j_c^* = 3$ with the certainty index $v_c(j_c^*) = 1$. □

Example 14.3.

Assume that in Example 14.2 the certainty distributions have an exponential form:

$$h_{xj}(x_j) = e^{-(x_j-d_j)^2}.$$

Applying (14.21), one obtains $v(j)$ as a function of d_j:

$$v(j) = \begin{cases} e^{-(\frac{u}{2}-d_j)^2} & \text{for} \quad d_j \le \frac{u}{2} \\ 1 & \text{for} \quad \frac{u}{2} \le d_j \le u \\ e^{-(u-d_j)^2} & \text{for} \quad d_j \ge u. \end{cases}$$

For $M = 3$, $u = 5$, $d_1 = 2$, $d_2 = 5.2$, $d_3 = 6$ we obtain

$$v(1) = e^{-0.25}, \quad v(2) = e^{-0.4}, \quad v(3) = e^{-1}.$$

Then $j^* = 2$ with the certainty index $v(j^*) = e^{-0.4} = 0.67$. For $d_1, d_2, d_3 \in [\frac{u}{2}, u]$ one obtains $j^* = 1$ or 2 or 3 and $v(j^*) = 1$.

Now let us consider $\bar{x}$ as a C-uncertain variable. Using (14.24), we obtain

$$v_n(j) = \begin{cases} 1 & \text{for} \quad d_j \le \frac{u}{2} \\ e^{-(\frac{u}{2}-d_j)^2} & \text{for} \quad \frac{u}{2} \le d_j \le \frac{3}{4}u \\ e^{-(u-d_j)^2} & \text{for} \quad \frac{3}{4}u \le d_j \le u \\ 1 & \text{for} \quad d_j \ge u\,. \end{cases}$$

Then the formula

$$v_c(j) = \frac{1}{2}[v(j) + 1 - v_n(j)]$$

gives the following results:

$$v_c(j) = \begin{cases} \frac{1}{2}e^{-(\frac{u}{2}-d_j)^2} & \text{for} \quad d_j \le \frac{u}{2} \\ 1-\frac{1}{2}e^{-(\frac{u}{2}-d_j)^2} & \text{for} \quad \frac{u}{2} \le d_j \le \frac{3}{4}u \\ 1-\frac{1}{2}e^{-(u-d_j)^2} & \text{for} \quad \frac{3}{4}u \le d_j \le u \\ \frac{1}{2}e^{-(u-d_j)^2} & \text{for} \quad d_j \ge u\,. \end{cases}$$

Substituting the numerical data $u = 5$, $d_1 = 2$, $d_2 = 5.2$, $d_3 = 6$, one obtains

$$v_c(1) = \frac{1}{2}e^{-0.25}, \qquad v_c(2) = \frac{1}{2}e^{-0.04}, \qquad v_c(3) = \frac{1}{2}e^{-1}.$$

Then $j_c^* = j^* = 2$ with the certainty index $v_c(j_c^*) = \frac{1}{2}e^{-0.4} = 0.335$. The results for $d_1 = 3$, $d_2 = 3.2$ and $d_3 = 4$ are as follows:

$$v_c(1) = 1 - \frac{1}{2}e^{-0.25}, \qquad v_c(2) = 1 - \frac{1}{2}e^{-0.49}, \qquad v_c(3) = 1 - \frac{1}{2}e^{-1}$$

and $j_c^* = 3$ with the certainty index $v_c(j_c^*) = 1 - \frac{1}{2}e^{-1} = 0.816$.

In this particular case the results j^* and j_c^* are the same for the different forms of the certainty distribution (see Example 14.2). □

14.4 Application of Random Variables

The considerations for random parameters in the relational knowledge representation are analogous to those for uncertain variables presented in the previous section. Assume that the sets $D_u(j;x)$ depend on a vector of unknown parameters $x \in X$ which is a value of a random variable $\widetilde{x}$ with a probability density $f_x(x)$. The recognition problem is now formulated as a specific analysis problem considered in Sect. 3.5.

Recognition problem for random parameters: For the given sequence $D_u(j;x)$, $f_x(x)$, D_u and the set $\hat{D}_j \subset J$ given by a user one should find the probability that the set $\hat{D}_j$ belongs to the set of all possible classes (14.11).

Then

$$P[\hat{D}_j \subseteq D_j(\widetilde{x})] = P[\widetilde{x} \in D_x(\hat{D}_j)] = \int\limits_{D_x(\hat{D}_j)} f_x(x)dx$$

where $D_x(\hat{D}_j)$ and $D_j(x)$ are determined by (14.13) and (14.11), respectively. In particular, for $\hat{D}_j = \{j\}$ one can formulate the optimization problem consisting in the determination of a class j maximizing the probability that j belongs to the set of all possible classes, or the probability that the set $D_j(\widetilde{x})$ contains the set $\hat{D}_j$.

Optimal recognition problem: For the given sequence $D_u(j;x)$, $f_x(x)$ and D_u one should find j^* maximizing

$$P[j \in D_j(\widetilde{x})] = P[\widetilde{x} \in D_x(j)] \stackrel{\Delta}{=} p(j) \tag{14.26}$$

where $D_x(j)$ is determined by (14.16). Then

$$j^* = \arg\max_j p(j) = \arg\max_j \int\limits_{D_x(j)} f_x(x)dx\,. \tag{14.27}$$

Assume that the different unknown parameters are separated in the different sets and the knowledge representation is described by $D_u(j;x_j)$ where $x_j \in X_j$ are subvectors of x, different for the different j. Assume also that $\widetilde{x}_j$ and $\widetilde{x}_i$ are independent for $i \neq j$ and $\widetilde{x}_j$ is described by the probability density $f_{xj}(x_j)$. In this case

$$j^* = \arg\max_j p(j) = \arg\max_j \int\limits_{D_{xj}(j)} f_{xj}(x_j)dx_j$$

where $D_{xj}(j)$ is determined by (14.19). In particular, for $D_u = \{u\}$, $D_{xj}(j)$ is determined by (14.22). The knowledge-based procedure of finding j^* is illustrated in Fig. 14.5.

Remark 14.1. It is worth noting that, in general, $p(1) + p(2) + \ldots + p(M) \leq M$. It follows from the fact that

$$P(\bigwedge_{j\in\overline{1,M}} D_j(x)=\varnothing)\geq 0\,, \tag{14.28}$$

which means that for the result of the observation $u \in D_u$ and for all $j \in \overline{1,M}$, $(u,j) \notin R(u,j;x)$, or the probability of a contradiction between the result of the observation and the knowledge representation R may be greater than zero. □

Example 14.4.

For $D_u(j;x_j)$ the same as in Example 14.2 assume that $\tilde{x}_j$ have exponential probability densities

$$f_{xj}(x_j)=\begin{cases}\lambda_j e^{-\lambda_j x_j} & \text{for} \quad x_j \geq 0\\ 0 & \text{for} \quad x_j < 0.\end{cases}$$

Then

$$p(j)=\int_{\frac{u}{2}}^{u} f_{xj}(x_j)dx_j = \exp(-\frac{\lambda_j u}{2})-\exp(-\lambda_j u)\,. \tag{14.29}$$

For example, for $M=3$, $u=8$, $\lambda_1=0.2$, $\lambda_2=0.4$, $\lambda_3=0.6$ we obtain $p(1)=0.25$, $p(2)=0.16$, $p(3)=0.08$. Then $j^*=1$, i.e. for $u=8$ the probability that $j=1$ belongs to the set of the possible classes has the maximum value equal to 0.25. For $u=20$ we obtain $p(1)=0.12$, $p(2)=0.02$, $p(3)=0.0025$, which means that in this case the probability (14.28) is relatively large. Using (14.29) it is easy to show that

$$\max_{(\lambda_j u)} p(j)=0.25\,.$$

□

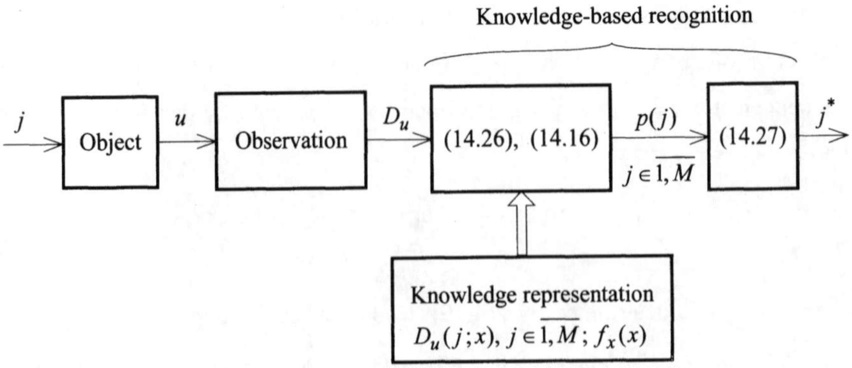

Figure 14.5. Block scheme of a recognition system

14.5 Non-parametric Problems

Let us present together basic non-parametric recognition problems (i.e. the problems based on non-parametric descriptions) corresponding to the general analysis problems presented in Sect. 6.5. The problem for the relational knowledge representation without unknown parameters x has been already presented in Sect. 14.1.

A . Description based on random variables

Assume that (u, j) are values of random variables $(\tilde{u}, \tilde{j})$, and $\tilde{u}$ is a continuous random variable. The knowledge of the set of objects, directly corresponding to the description in Sects. 3.7 and 3.8 for a plant with an output y, has a form of conditional probabilities

$$P(\tilde{j} = j \mid u) \stackrel{\Delta}{=} p_j(u), \quad j = 1,2,\ldots,M .$$

The analysis problem for the given u consists in the determination of

$$j^* = \arg\max_j p_j(u) \stackrel{\Delta}{=} \Psi(u)$$

where Ψ denotes the *deterministic recognition algorithm.*

The knowledge of the set of objects may be given in form of a joint probability distribution for $(\tilde{u}, \tilde{j})$, i.e.

$$f_u(u \mid j), \quad P(\tilde{j} = j) \stackrel{\Delta}{=} p_j, \quad j = 1,2,\ldots,M \tag{14.30}$$

where $f_u(u \mid j)$ is a conditional probability density. Using (14.30), one can determine

$$p_j(u) = \frac{p_j f_u(u \mid j)}{\sum_{j=1}^{M} p_j f_u(u \mid j)} . \tag{14.31}$$

It is easy to see that for the determination of j^* it is sufficient to use the numerator of (14.31) and

$$j^* = \arg\max_j p_j(u) = \arg\max_j p_j f_u(u \mid j) \stackrel{\Delta}{=} \Psi(u) . \tag{14.32}$$

This is a very well known algorithm of a *Bayesian classifier*, based on Bayes rule (14.31). The problem may be generalized by introducing the deterministic recognition algorithm $i = \Psi(u)$ and a *loss function* $L(i, j)$ where $i, j \in \{1,2,\ldots,M\}$ and $L(i, j) = 0$ for $i = j$.

Optimal recognition problem: For the given probability distributions (14.30) and the function $L(i,j)$ one should determine the algorithm Ψ minimizing the expected value of $L[\Psi(\tilde{u}),\tilde{j}]$:

$$\underset{u,j}{\mathrm{E}}[L[\Psi(\tilde{u}),\tilde{j}]] \stackrel{\Delta}{=} \mathcal{R} = \int\limits_U \sum_{j=1}^{M} L[\Psi(u),j] p_j f_u(u \mid j)\, du\,.$$

The expected loss $\mathcal{R}$ is called a *risk*. The minimization of the risk $\mathcal{R}$ with respect to Ψ may be reduced to the minimization of a *conditional risk*

$$r(u,i) = \sum_{j=1}^{M} L(i,j) p_j(u)$$

with respect to *i*. Then

$$i^* = \arg\min_i r(u,i) \stackrel{\Delta}{=} \Psi(u)\,. \tag{14.33}$$

If

$$L(i,j) = \begin{cases} 1 & \text{for} \quad i \neq j \\ 0 & \text{for} \quad i = j \end{cases}$$

then $r(u,i) = 1 - p_i(u)$ and the algorithm Ψ in (14.33) is reduced to the algorithm Ψ in (14.32).

The probability distributions (14.30) may be easier (than directly $p_i(u)$) to obtain by statistical investigations, in the form of empirical distributions. If p_j are not available then one may determine

$$j^* = \arg\max_j f_u(u \mid j) \stackrel{\Delta}{=} \Psi(u)\,. \tag{14.34}$$

It is worth noting that the determination of $i = \Psi(u)$ may be considered as an estimation problem in which i is an estimate of j, and $i = j^*$ determined by (14.34) is an estimate obtained by using the *maximum likelihood method.*

B. Description based on uncertain variables

One of the possible formulations of the recognition problem is the following: for the given conditional certainty distributions $h_j(j \mid u)$, $j \in \overline{1,M}$ – one should determine

$$j^* = \arg\max_j h_j(j \mid u) \stackrel{\Delta}{=} \Psi(u)\,.$$

It should be noted that $v(\bar{j} \cong j)$ (i.e. the certainty index that the class to which the object belongs "is approximately equal to j") has no evident practical interpretation. It follows from the fact that in the case under consideration not only the set of approximate values used by an expert but also the set of real exact values of the uncertain variable is a discrete set $J=\{1,2,...,M\}$. For the given sequence $h_u(u\,|\,j)$, $j\in\overline{1,M}$ – an approach analogous to (14.34) consists in the determination of

$$j^* = \arg\max_j h_u(u\,|\,j) \stackrel{\Delta}{=} \Psi(u) \tag{14.35}$$

i.e. class j maximizing the certainty index that $\bar{u}\cong u$.

C. Description based on fuzzy variables

One of the possible formulations of the recognition problem is the following: for the determined soft properties $\varphi_u(u)$, $\varphi_j(j)$ and the given membership functions $\mu_j(j\,|\,u)=w[\varphi_u(u)\rightarrow\varphi_j(j)]$, $j\in\overline{1,M}$ – one should determine

$$j^* = \arg\max_j \mu_j(j\,|\,u) \stackrel{\Delta}{=} \Psi(u).$$

It should be noted that in the case under consideration $\varphi_j(j)$ does not concern the size of j (e.g. "j is small"), but may be understood as "j is possible". An approach analogous to (14.34) and (14.35) has two versions (see Sect. 9.5):

1. For the given sequence $\mu_u(u\,|\,j)=w[\varphi_j(j)\rightarrow\varphi_u(u)]$, $j\in\overline{1,M}$

$$j^* = \arg\max_j \mu_u(u\,|\,j).$$

2. For the given sequence $\mu_{uj}(u,j)=w[\hat{j}=j\rightarrow\varphi_u(u)]$, $j\in\overline{1,M}$

$$j^* = \arg\max_j \mu_{uj}(u,j).$$

It is possible to use more complicated descriptions of the knowledge corresponding to the descriptions presented for a fuzzy controller in Sect. 9.5 with (u,j) in place of (s,u), and formulate respective recognition problems, i.e. the analysis problems for the plant with input u and output j. It is important to note that in our case the defuzzification consists in the determination of j^* maximizing the respective membership function instead of the determination of the mean value.

14.6 Learning Algorithms

For the recognition described by a relation with unknown parameters one can apply learning algorithms described in Chapter 11. The set of objects is described by a relation $R(u, j; c)$ or by the sets $D_u(j; c)$, $j \in \overline{1, M}$, where $c \in C$ is a vector of unknown parameters, and we have no a priori knowledge of c in the form of a certainty distribution or probability density. We assume that $D_u(j; c)$ are continuous and closed domains in U and the vector c has the value $c = \bar{c}$. If we have the learning sequence

$$(u_1, j_1), (u_2, j_2), ..., (u_n, j_n), \quad \bigwedge_i (u_i, j_i) \in R(u, j; \bar{c})$$

then we can propose a current *step by step* estimation of $\bar{c}$.

In each step, one should prove if the current pair in the learning sequence "belongs" to the knowledge representation determined before this step (*knowledge validation*) and, if not, one should modify the current estimation of parameters in the knowledge representation (*knowledge updating*). We shall present two versions of the estimation. In version 1 the relation R is considered as a whole and in version 2 the sets D_u are considered separately.

Version 1. Let us introduce the set

$$D_c(n) = \{c \in C : \bigwedge_{i \in \overline{1,n}} [u_i \in D_u(j_i; c)]\}. \tag{14.36}$$

The boundary $\Delta_c(n)$ of the set $D_c(n)$ is proposed as the estimation of $\bar{c}$. Assume that the pairs (u_i, j_i) in the learning sequence occur randomly with probabilities p_j for j and probability densities $f_j(u)$ for u.

Theorem 14.1. If $p_j > 0$ and $f_j(u) > 0$ for every $(u, j) \in R$ and $R(u, j; c) \neq R(u, j; \bar{c})$ for every $c \neq \bar{c}$ then $\Delta_c(n)$ converges to $\{\bar{c}\}$ with probability 1. □

The proof is analogous to that presented for Theorem 11.1 in Sect. 11.1. The determination of $\Delta_c(n)$ may be presented in the form of the following recursive algorithm for $n > 1$:

Knowledge validation

Prove if

$$\bigwedge_{c \in D_c(n-1)} [u_n \in D_u(j_n; c)]. \tag{14.37}$$

If yes then $D_c(n) = D_c(n-1)$ and $\Delta_c(n) = \Delta_c(n-1)$. If not then one should determine the new $D_c(n)$ and $\Delta_c(n)$, i.e. update the knowledge.

Knowledge updating

Determine new $D_c(n)$ and $\Delta_c(n)$:

$$D_c(n) = \{c \in D_c(n-1) : u_n \in D_u(j_n; c)\}. \tag{14.38}$$

For $n = 1$

$$D_c(1) = \{c \in C : u_1 \in D_u(j_1;c)\}.$$

The result of the knowledge validation may be used after the learning process (i.e. for $i > n$) or in the current determination of the set D_j. In the first case, for the result of the observation D_u, one should choose randomly c_n from $\Delta_c(n)$ and determine

$$D_j(c_n) = \{j \in J : D_u \cap D_u(j;c_n) \neq \varnothing\} \qquad (14.39)$$

(see (14.2)). In the second case the sets $D_j(c_n)$ are determined currently for the successive moments n and the observations D_{un}, simultaneously with the knowledge updating. To present the *recognition algorithm with learning* one should add to the knowledge updating algorithm the following steps:

1. Choose randomly c_n from $\Delta_c(n)$ and put in $D_u(j;c)$ in place of c.
2. Introduce the result of the observation $D_u = D_{un}$ and determine $D_j(c_n)$ according to (14.39).
3. Choose randomly $j = i_n$ from $D_j(c_n)$.

The learning recognition system is presented in Fig. 14.6 where G_1 and G_2 denote generators for random choosing of c_n form $\Delta_c(n)$ and i_n from $D_j(c_n)$.

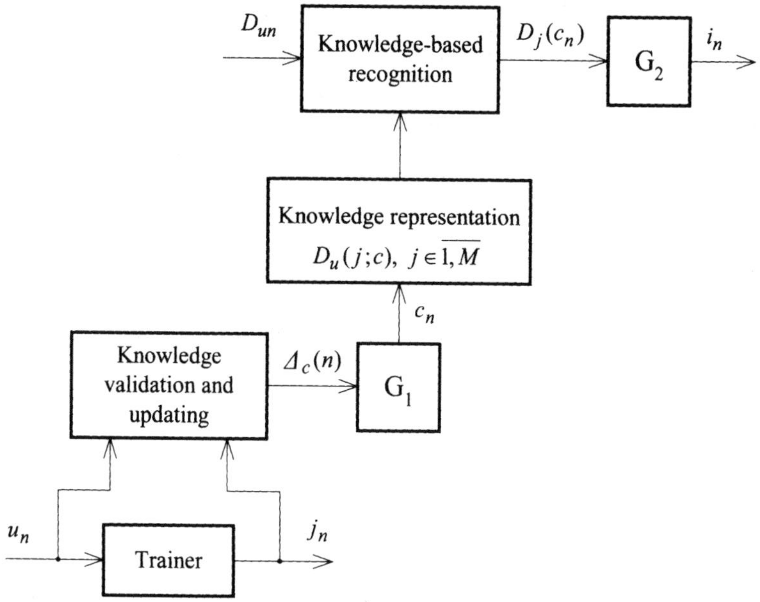

Figure 14.6. Learning recognition system

Version 2. Assume that the different unknown parameters are separated in the different sets $D_u(j)$, i.e. the knowledge representation is described by the sets $D_u(j;c_j)$ where $c_j \in C_j$ are subvectors of c, different for the different j. Now the estimations $\Delta_{cj}(n)$ of $\bar{c}_j$ may be determined separately for each class. For the class j, denote by $\bar{u}_{ji}$ the subsequence of u_i for which $j_i = j$, by $\hat{u}_{ji}$ the subsequence for which $j_i \neq j$, and introduce the sets corresponding to (14.36):

$$\overline{D}_{cj}(n) = \{c_j \in C_j : \text{each } \bar{u}_{ji} \in D_u(j;c_j)\}, \tag{14.40}$$

$$\hat{D}_{cj}(n) = \{c_j \in C_j : \text{each } \hat{u}_{ji} \in U - D_u(j;c_j)\}. \tag{14.41}$$

The set

$$\overline{D}_{cj}(n) \cap \hat{D}_{cj}(n) \stackrel{\Delta}{=} \Delta_{cj}(n) \tag{14.42}$$

is proposed as the estimation of $\bar{c}_j$. The convergence condition and the recursive algorithms for the determination of c_{jn} are analogous to those in version 1. It is easy to note that the presented approach may be extended for the general case considered in version 1 where the same parameters c may be used in the description of $D_u(j)$ for the different j. In this case we determine the estimations of $\bar{c}$ in the form of $\Delta_{cj}(n)$ separately for each class using (14.40), (14.41) and (14.42) with c in place of c_j. Consequently,

$$\Delta_c(n) = \bigcap_{j=1}^{M} \Delta_{cj}(n).$$

In general, version 2 gives better estimation than version 1, i.e. the set $\Delta_c(n)$ obtained in version 2 is smaller than in version 1.

Example 14.5.

Let $M = 2$, $D_u(1;c)$ be a hyperball $u^{\mathrm{T}} u \le c^2$ and $D_u(2;c) = U - D_u(1;c)$. Then using version 2, we obtain

$$\overline{D}_c(n) = [c_{\min,n}, \infty), \quad \hat{D}_c(n) = [0, c_{\max,n}), \quad \Delta_c(n) = [c_{\min,n}, c_{\max,n})$$

where

$$c_{\min,n}^2 = \max_i \bar{u}_i^{\mathrm{T}} \bar{u}_i, \quad c_{\max,n}^2 = \min_i \hat{u}_i^{\mathrm{T}} \hat{u}_i.$$

For simplicity assume $u = (u^{(1)}, u^{(2)})$. The recognition algorithm with learning is the following:

Introduce (u_n, j_n) from the learning sequence, $n > 0$.

If $j_n = 1$

Knowledge validation for $u_n = \overline{u}_n$

Prove if

$$\bigwedge_{c \in \overline{D}_c(n-1)} [u_n \in D_u(1;c)]$$

i.e.

$$(u_n^{(1)})^2 + (u_n^{(2)})^2 \le c_{\min,n-1}^2 .$$

If yes then $\overline{D}_c(n) = \overline{D}_c(n-1)$, i.e. $c_{\min,n} = c_{\min,n-1}$. If not, determine new $\overline{D}_c(n)$.

Knowledge updating for $u_n = \overline{u}_n$

$$\overline{D}_c(n) = \{c \in \overline{D}_c(n-1) : u_n \in D_u(1;c)\}$$

i.e

$$c_{\min,n}^2 = (u_n^{(1)})^2 + (u_n^{(2)})^2 .$$

Put $\hat{D}_c(n) = \hat{D}_c(n-1)$, i.e. $c_{\max,n} = c_{\max,n-1}$.

If $j_n = 2$

Knowledge validation for $u_n = \hat{u}_n$

Prove if

$$\bigwedge_{c \in \hat{D}_c(n-1)} [u_n \in D_u(2;c)]$$

i.e.

$$(u_n^{(1)})^2 + (u_n^{(2)})^2 > c_{\max,n-1}^2 .$$

If yes then $\hat{D}_c(n) = \hat{D}_c(n-1)$, i.e. $c_{\max,n} = c_{\max,n-1}$. If not, determine new $\hat{D}_c(n)$.

Knowledge updating for $u_n = \hat{u}_n$

$$\hat{D}_c(n) = \{c \in \hat{D}_c(n-1) : u_n \in D_u(2;c)\}$$

i.e.

$$c_{\max,n}^2 = (u_n^{(1)})^2 + (u_n^{(2)})^2 .$$

Put $\overline{D}_c(n) = \overline{D}_c(n-1)$, i.e. $c_{\min,n} = c_{\min,n-1}$.

Choose randomly c_n from

$$\Delta_c(n) = \overline{D}_c(n) \cap \hat{D}_c(n) = [c_{\min,n}, c_{\max,n}).$$

Introduce the result of observations D_{un}.

Determine $D_j(c_n)$:

$$D_j(c_n) = \begin{cases} \{1\} & \text{if } D_{un} \subseteq D_u(1; c_n), \\ \{2\} & \text{if } D_{un} \subseteq D_u(2; c_n), \\ \{1,2\} & \text{otherwise}. \end{cases}$$

Conclusions

This book has been concerned with a wide class of uncertain systems described by traditional mathematical models and by relational knowledge representations. For the systems with different formal models of uncertainty, a uniform description of analysis and decision making problems has been presented and discussed. A special emphasis has been put on new approaches:

1. Uncertain variables and their applications.
2. Learning concepts consisting in *step by step* knowledge validation and updating.
3. Soft variables as a tool for unification and generalization of non-parametric problems.

The uncertain variables have been proved to be a convenient tool for handling the analysis and decision problems based on an uncertain knowledge given by an expert in the form of certainty distributions. It has been shown how the uncertain variables, learning concepts and other approaches presented in the book may be applied to special cases such as dynamical closed-loop control systems, pattern recognition under uncertainty and control of the complex of operations with uncertain execution times.

It should be noted that the considerations based on the descriptions of random, uncertain and fuzzy variables presented in the book (probability distributions, certainty distributions and membership functions) may be very complicated from the computational point of view. For the cases more complicated than those described in the examples, in order to obtain effective solutions it may be necessary to apply special numerical methods. On the other hand, the solutions may be very sensitive to the values of parameters in the descriptions listed above. This is specially important in the case of certainty distributions and membership functions, i.e. more or less arbitrary descriptions given by an expert and expressing his / her individual, subjective opinion. That is why the simulations have a specially important role in the evaluation of the decision quality, and the adaptation of the decision algorithms to real decision plants or their simulators are recommended.

The purpose of the book was to give a uniform framework for analysis and decision problems concerning different forms of knowledge representation (including traditional functional models) and various descriptions of uncertainty. More advanced cases described in the literature (based on probabilistic and fuzzy formalisms) and not presented in the book, as well as new problems concerning the class of uncertain systems under consideration, may be put into this framework.

It seems to be important to indicate the following new problems and trends:

1. A combination of the learning process with a priori information in the form of certainty distributions, i.e. *step by step* updating of a priori information, using the

results of current observations. Such an idea is very well known in the classical probabilistic decision theory.

2. An extension of the considerations presented in the book to more complicated cases of a complex and distributed knowledge, and to complex systems with hybrid uncertainties, i.e. with different forms of the uncertainty in different parts of the system e.g. [50].

3. An interconnection between different concepts of the complex knowledge decomposition and respective formulations of the knowledge-based decision problems.

References

1. Amato F, Mattei M, Pironti A. (1998) A note on quadratic stability of uncertain linear discrete-time systems, IEEE Trans. on AC, 43:227–229
2. Åström KJ (1970) Introduction to Stochastic Control Theory. Academic Press, N. York
3. Åström KJ, Wittenmork B (1995) Adaptive Control, 2nd edition. Addison-Wesley, Reading, Massachusetts
4. Ayyub BM, Gupta MM (eds) (1994) Uncertainty Modeling and Analysis: Theory and Applications. North Holland, Amsterdam
5. Bargiela A, Pedrycz W (2003) Granular Computing. An Introduction. Kluwer Academic Publishers, Boston, Dordrecht, London
6. Bauer PH, Premaratne K, Duran J (1993) A necessary and sufficient condition for robust asymptotic stability of time-invariant discrete systems. IEEE Trans. on AC, 38:1427–1430
7. Beferhat S, Dubois D, Prade H (1999) Possibilistic and standard probabilistic semantics of conditional knowledge bases. J. Logic Comput., 9:873–895
8. Blazewicz J, Ecker KH, Pesch E, Schmidt G, Węglarz J (1996) Scheduling Computer and Manufacturing Processes. Springer-Verlag, Heildelberg, Berlin
9. Boutilier C, Goldszmidt M (eds) (2000) Uncertainty in Artificial Intelligence. Morgan Kaufmann, San Mateo, CA
10. Bubnicki Z (1964) On the stability condition of nonlinear sampled-data systems. IEEE Trans. on AC, 9:280–281
11. Bubnicki Z (1967) On the convergence condition in discrete optimisation systems. Automat. Remote Control (Automatika i Telemekhanika), 10:115–123
12. Bubnicki Z (1968) On the linear conjecture in the deterministic and stochastic stability of discrete systems. IEEE Trans. on AC, 13:199–200
13. Bubnicki Z (1978) Two-level optimization and control of the complex of operations. In: Proc. of VII World IFAC Congress, Vol 2. Pergamon Press, Oxford
14. Bubnicki Z (1980) Identification of Control Plants. Elsevier, Oxford, Amsterdam, New York
15. Bubnicki Z, Lebrun A (1980) Stochastic approach to two-level optimization in the complex of operations. Lectures Notes in Control and Information Sciences, 23 Springer-Verlag, Berlin
16. Bubnicki Z, Staroswiecki M, Lebrun A (1981) Optimal planning of production under uncertain operating conditions. In: Proc. VIII World IFAC Congress, Vol. 9, Oxford: Pergamon Press, Kyoto
17. Bubnicki Z (1990) Introduction to Expert Systems. Polish Scientific Publishers, Warsaw (in Polish)

18. Bubnicki Z (1992) Decomposition of a system described by logical model. In: Trappl R (ed.) Cybernetics and Systems Research, Vol 1. World Scientific, Singapore, pp. 121–128

19. Bubnicki Z (1996) Logic-algebraic foundations of a class of knowledge based control systems. In: Jamshidi M, Yuh J and Dauchez P (eds) Intelligent Automation and Control. Recent trends in development and applications. In: Proc. of the World Automation Congress, Vol. 4, Albuquerque: TSI Press, Montpellier, pp. 89–94

20. Bubnicki Z (1997) Knowledge updating in a class of knowledge-based learning control systems. Systems Science, 23:19–36

21. Bubnicki Z (1997) Logic-algebraic approach to a class of knowledge based fuzzy control systems. In: Proc. of the European Control Conference ECC 97, Vol. 1. Brussels

22. Bubnicki Z (1997) Logic-algebraic method for a class of dynamical knowledge-based systems. In: Sydow A (ed.) Proc. of the 15th IMACS World Congress on Scientific Computation, Modelling and Applied Mathematics, Vol 4. Wissenschaft und Technik Verlag, Berlin, pp. 101–106

23. Bubnicki Z (1997) Logic-algebraic method for a class of knowledge based systems. In: Pichler F, Moreno-Diaz R (eds) Computer Aided Systems Theory. Lecture Notes in Computer Science, Vol 1333. Springer-Verlag, Berlin, pp. 420–428

24. Bubnicki Z (1998) Logic-algebraic method for knowledge-based relation systems. Systems Analysis Modelling Simulation, 33:21–35

25. Bubnicki Z (1998) Uncertain logics, variables and systems. In: Guangquan L (ed.) Proc. of the 3rd Workshop of International Institute for General Systems Studies. Tianjin People's Publishing House, Tianjin, pp. 7–14

26. Bubnicki Z (1998) Uncertain variables and logic-algebraic method in knowledge based systems. In: Hamza MH (ed.) Proc. of IASTED International Conference on Intelligent Systems and Control. Acta Press, Zurich, pp. 135–139

27. Bubnicki Z (1998) Uncertain variables and their applications in uncertain control systems. In: Hamza M H (ed.) Modelling, Identification and Control. Acta Press, Zurich, pp. 305–308

28. Bubnicki Z (1999) Learning control systems with relational plants. In: Proc. of the European Control Conference ECC 99. Karlsruhe

29. Bubnicki Z (1999) Learning processes and logic-algebraic method in knowledge-based control systems. In: Tzafestas S G, Schmidt G (eds) Progress in System and Robot Analysis and Control Design. Lecture Notes in Control and Information Sciences, Vol 243. Springer-Verlag, London, pp. 183–194

30. Bubnicki Z (1999) Uncertain variables and learning algorithms in knowledge-based control systems. Artificial Life and Robotics, 3:155–159

31. Bubnicki Z (2000) General approach to stability and stabilization for a class of uncertain discrete nonlinear systems. International Journal of Control, 73:1298–1306

32. Bubnicki Z (2000) Knowledge validation and updating in a class of uncertain distributed knowledge systems. In: Shi Z, Faltings B, Musen M (eds) Proc. of 16th IFIP World Computer Congress. Intelligent Information Processing. Publishing House of Electronics Industry, Beijing, pp. 516–523

33. Bubnicki Z (2000) Learning process in an expert system for job distribution in a set of parallel computers. In: Proc. of the 14th International Conference on Systems Engineering, Vol 1. Coventry, pp. 78–83

34. Bubnicki Z (2000) Learning processes in a class of knowledge-based systems. Kybernetes, 29:1016–1028

35. Bubnicki Z (2000) Uncertain variables in the computer aided analysis of uncertain systems. In: Pichler F, Moreno-Diaz R, Kopacek P (eds) Computer Aided Systems Theory. Lecture Notes in Computer Science, Vol 1798. Springer-Verlag, Berlin, pp. 528–542

36. Bubnicki Z (2000) Learning algorithms for a class of knowledge-based systems with dynamical knowledge representation. Systems Science, 26:15–27

37. Bubnicki Z (2000) Learning algorithms in a class of knowledge-based systems. Artificial Life and Robotics, 4:22–26

38. Bubnicki Z (2001) A unified approach to decision making and control in knowledge-based uncertain systems. In: Dubois Daniel M (ed.) Computing Anticipatory Systems: CASYS'00 – Fourth International Conference. American Institute of Physics, Melville, N. York, pp. 545–557

39. Bubnicki Z (2001) A unified approach to descriptive and prescriptive concepts in uncertain decision systems. In: Proc. of the European Control Conference ECC 01. Porto, pp. 2458–2463

40. Bubnicki Z (2001) Application of uncertain variables and logics to complex intelligent systems. In: Sugisaka M, Tanaka H (eds) 2001 Proc. of the 6th International Symposium on Artificial Life and Robotics, Vol 1. Tokyo, pp. 220–223

41. Bubnicki Z (2001) The application of learning algorithms and uncertain variables in the knowledge-based pattern recognition. Artificial Life and Robotics, 5:67–71

42. Bubnicki Z (2001) Uncertain logics, variables and systems. In: Bubnicki Z, Grzech A (eds) Proc. of the 14th International Conference on Systems Science, Vol I. Wroclaw, pp. 34–49

43. Bubnicki Z (2001) Uncertain variables – a new tool for analysis and design of knowledge-based control systems. In: Hamza MH (ed.) Modelling, Identification and Control, Vol II. Acta Press, Zurich, pp. 928–930

44. Bubnicki Z (2001) Uncertain variables and their applications for a class of uncertain systems. International Journal of Systems Science, 32:651–659

45. Bubnicki Z (2002) A probabilistic approach to stability and stabilization of uncertain discrete systems. In: Proc. of 5th IFAC Symposium "Nonlinear Control Systems", Vol. 2. Elsevier Science, Oxford, pp. 1071–76

46. Bubnicki Z (2002) Uncertain variables and their application to decision making. IEEE Trans. on SMC, Part A: Systems and Humans, 31:587–596

47. Bubnicki Z (2002) A unified approach to descriptive and prescriptive concepts in uncertain decision systems. Systems Analysis Modelling Simulation, 42:331–342

48. Bubnicki Z (2002) Application of uncertain variables for a class of intelligent knowledge-based assembly systems. In: Proc. of the 7th International Symposium on Artificial Life and Robotics, Vol.1, Oita, Japan, pp. 98–101

49. Bubnicki Z (2002) Application of uncertain variables in allocation problem for a complex of parallel operations. Foundations of Computing and Decision Sciences, 27:3–15

50. Bubnicki Z (2002) Application of uncertain variables to decision making in a class of distributed computer systems. In: Musen M, Neumann B and Studer R (eds) Proc. of 17th IFIP World Computer Congress. Intelligent Information Processing, Kluwer Academic Publishers, Montreal, pp. 261–264

51. Bubnicki Z (2002) Learning process in a class of computer integrated manufacturing systems with parametric uncertainties. Journal of Intelligent Manufacturing, 13:409–415

52. Bubnicki Z (2002) Stability and stabilization of discrete systems with random and uncertain parameters. In: Proc. of 15th Triennial World Congress of IFAC, Barcelona

53. Bubnicki Z (2002) Uncertain logics, variables and systems. International Journal of Computing Anticipatory Systems, 11:259–274

54. Bubnicki Z (2002) Uncertain Logics, Variables and Systems. Springer-Verlag, Berlin, London

55. Bubnicki Z (2002) Uncertain variables and their applications for control systems. Kybernetes, 31:1260–1273

56. Bubnicki Z (2002) Uncertain variables and their applications to stability analysis and stabilization of uncertain discrete systems. In: Proc. of 5th Portuguese Conference on Automatic Control CONTROLO 2002, Aveiro, Portugal, pp. 629–634

57. Bubnicki Z (2003) Application of uncertain variables and learning algorithm to task allocation in multiprocessor systems. In: Proc. of the 8th International Symposium on Artificial Life and Robotics, Vol. 1, Oita, Japan, pp. I–19–I–22

58. Bubnicki Z (2003) Application of uncertain variables in a class of control systems with uncertain and random parameters. In: Proc. of the European Control Conference ECC 03, Cambridge

59. Bubnicki Z (2003) Soft variables as a generalization of uncertain, random and fuzzy variables. In: Rutkowski L and Kacprzyk J (eds) Advances in Soft Computing. Neural Network and Soft Computing, Physica-Verlag, Heidelberg, pp. 4–17

60. Bubnicki Z (2003) Control Theory and Algorithms. Polish Scientific Publishers, Warsaw (in Polish)

61. Buckner GD (2002) Intelligent bounds on modeling uncertainty: applications to sliding mode control. IEEE Trans. SMC, Part C: Applications and Reviews, 32:113–124

62. Coutinho D, Trofino A, Fu Minyue (2002) Guaranteed cost control of uncertain nonlinear systems Via Polynomial Lyapunov Functions. IEEE Trans. AC, 47:1575–1580

63. De Caluwe R (ed.) (1997) Fuzzy and Uncertain Object-oriented Databases: Concepts and Models. World Scientific, Singapore

64. Dubois D, Prade H (1988) Possibility Theory – An Approach to the Computerized Processing of Uncertainty. Plenum Press, N. York

65. Dubois D, Wellman MP, D'Ambrosio B, Smets P (eds) (1992) Uncertainty in Artificial Intelligence. Morgan Kaufmann, San Mateo, CA

66. Golberg MA (1984) An Introduction to Probability Theory with Statistical Applications. Plenum Press, N. York

67. Grabisch M, Murofushi T, Sugeno M (eds) (1999) Fuzzy Measures and Integrals. Physica-Verlag, Heidelberg
68. Isermann R (1991) Digital Control Systems. Vol. 2, Stochastic Control, Multivariable Control, Adaptive Control, Applications. Springer-Verlag, Berlin
69. Kacprzyk J (1997) Multistage Fuzzy Control. John Wiley & Sons, Chichester
70. Kaszkurewicz E, Bhaya A. (1993) Robust stability and diagonal Lyapunov functions. SIAM J. Matrix Anal. Appl., 14:508–520
71. Kaufmann A, Gupta MM (1985) Introduction to Fuzzy Arithmetic: Theory and Applications. Van Nostrand Reinhold, N. York
72. Kearfott RB, Kreinovich V (eds) (1996) Applications of Interval Computations. Kluwer Academic Publishers, Dordrecht
73. Klimov GP (1986) Probability Theory and Mathematical Statistics. MIR, Moscow
74. Klir GJ, Folger TA (1988) Fuzzy Sets, Uncertainty, and Information. Prentice-Hall, Englewood Cliffs, NJ
75. Klir GJ, Yuan B (1995) Fuzzy Sets and Fuzzy Logic: Theory and Applications. Prentice-Hall, Englewood Cliffs, NJ
76. Kouvaritakis B, Latchman H (1985) Singular-value and eigenvalue techniques in the analysis of systems with structured perturbations. International Journal of Control, 41:1381–1412
77. Krsti M, Hua Deng (1998) Stabilization of Nonlinear Uncertain Systems. Springer-Verlag, London
78. Kruse R, Schwecke E, Heinsohn J (1991) Uncertainty and Vagueness in Knowledge Based Systems: Numerical Methods. Springer-Verlag, Berlin
79. Kumar PR, Varaiya P (1986) Stochastic Systems: Estimation, Identification, and Adaptive Control. Prentice Hall, Englewood Cliffs, NJ
80. Landau ID, Lozano R, M'Saad M (1998) Adaptive Control. Springer-Verlag, London
81. Morgan MG, Henrion M (1990) Uncertainty: A Guide to Dealing with Uncertainty in Quantitative Risk and Policy Analysis. Cambridge Univ. Press, Cambridge
82. Orski D (1998) Bubnicki method for decision making in a system with hybrid knowledge representation. In: Bubnicki Z, Grzech A (eds) Proc. of the 13th International Conference on Systems Science, Vol 2. Wroclaw, pp. 230–237
83. Pawlak Z (1991) Rough Sets. Theoretical Aspects of Reasoning about Data. Kluwer Academic Publishers, Dordrecht
84. Pedrycz W, Gomide F (1998) Fuzzy Sets. MIT Press, Cambridge MA
85. Pozniak I (1995) Application of Bubnicki method to knowledge-based computer load sharing. In: Bubnicki Z, Grzech A (eds) Proc. of the 12th International Conference on Systems Science, Vol 3. Wroclaw, pp. 290–297
86. Pozniak I (1996) Knowledge-based algorithm by using Bubnicki method to improve efficiency of parallel performing the complex computational jobs. In: Proc. of the 11th International Conference on Systems Engineering, Las Vegas, pp. 817–822
87. Qu Zhibua, Xu Jianxin (2002) Asymptotic learning control for a class of cascaded nonlinear uncertain systems. IEEE Trans. AC, 47:1369–1376
88. Qu Zhihua (1998) Robust Control of Nonlinear Uncertain Systems. John Wiley & Sons, N. York

89. Ranze KC, Stuckenschmidt H (1998) Modelling uncertainty in expertise. In: Cuena J (ed.) Proc. of the XV. IFIP World Computer Congress, Information Technologies and Knowledge Systems. Österreichische Computer Gesellschaft, Vienna, pp. 105–116

90. Rapior P (1998) The Bubnicki method in knowledge based admission control for ATM networks. In: Bubnicki Z, Grzech A (eds) 1998 Proc. of the 13th International Conference on Systems Science, Vol 2. Wroclaw, pp. 238–243

91. Slowinski R, Hapke M (eds) (2000) Scheduling Under Fuzziness (Studies in Fuzziness and Soft Computing). Springer-Verlag, Berlin, London

92. Slowinski R, Teghem J (eds) (1990) Stochastic Versus Fuzzy Approaches to Multiobjective Mathematical Programming Under Uncertainty. Kluwer Academic Publishers, Dordrecht

93. Sobel KM, Banda SS, Yeh H (1989) Robust control for linear systems with structured state space uncertainty. International Journal of Control, 50:1997–2004

94. Stroock DW (1994) Probability Theory, an Analytic View. Cambridge Univ. Press, Cambridge

95. Szala M (2002) Two-level pattern recognition in a class of knowledge-based systems. Knowledge-Based Systems, 15:95–101

96. Yager RR (2002) Uncertainty representation using fuzzy measures. IEEE Trans. SMC, Part B: Cybernetics, 32:13–20

97. Yager RR, Kacprzyk J, Fedrizzi M (1994) Advances in the Dempster-Shafer Theory of Evidence. John Wiley & Sons, N. York

98. Yager RR (2000) Fuzzy modeling for intelligent decision making under uncertainty. IEEE Trans. on SMC, Part B: Cybernetics, 30:60–70

99. Zadeh LA (1978) Fuzzy sets as a basis for a theory of possibility. Fuzzy Sets and Systems, 1:3–28

100. Zadeh L, Kacprzyk J (eds) (1992) Fuzzy Logic for the Management of Uncertainty. John Wiley & Sons, N. York

101. Zadeh L, Kacprzyk J (1999) Computing with Words in Information / Intelligent Systems, Vol. 1–2, Physica-Verlag, Heidelberg

102. Zhang S, Mizukami K (1999) Robust stabilization conditions of discrete-time large-scale uncertain systems. International Journal of Systems Science, 30:11–18

103. Zimmermann HJ (1987) Fuzzy Sets, Decision Making, and Expert Systems. Kluwer Academic Publishers, Boston

104. Zimmermann HJ (2001) Fuzzy Set Theory and Its Applications, 4th edition. Kluwer Academic Publishers, Boston

Index